Guido Kanschat · Erik Meinköhn · Rolf Rannacher ·
Rainer Wehrse

Editors

Numerical Methods in Multidimensional Radiative Transfer

Guido Kanschat
Dept. of Mathematics
Texas A&M University
College Station TX 7843-3368
USA
kanschat@tamu.edu

Rolf Rannacher
Interdisziplinäres Zentrum
f. Wissenschaftliches Rechnen (IWR)
Universität Heidelberg
Im Neuenheimer Feld 368
69121 Heidelberg
Germany
rannacher@iwr.uni-heidelberg.de

Erik Meinköhn
Rainer Wehrse
Zentrum f. Astronomie
Institut f. Theoretische Astrophysik
der Universität Heidelberg
Albert-Ueberle-Str. 2
69120 Heidelberg
Germany
wehrse@ita.uni-heidelberg.de

ISBN 978-3-540-85368-8   e-ISBN 978-3-540-85369-5

DOI 10.1007/978-3-540-85369-5

Library of Congress Control Number: 2008937506

Mathematics Subject Classification (2000): 35M10

© 2009 Springer-Verlag Berlin Heidelberg

This work is subject to copyright. All rights are reserved, whether the whole or part of the material is concerned, specifically the rights of translation, reprinting, reuse of illustrations, recitation, broadcasting, reproduction on microfilm or in any other way, and storage in data banks. Duplication of this publication or parts thereof is permitted only under the provisions of the German Copyright Law of September 9, 1965, in its current version, and permission for use must always be obtained from Springer. Violations are liable for prosecution under the German Copyright Law.

The use of general descriptive names, registered names, trademarks, etc. in this publication does not imply, even in the absence of a specific statement, that such names are exempt from the relevant protective laws and regulations and therefore free for general use.

Cover design: VTEX, Vilnius

Printed on acid-free paper

9 8 7 6 5 4 3 2 1

springer.com

Preface

Traditionally, radiative transfer has been the domain of astrophysicists and climatologists. In nuclear technology one has been dealing with the analogous equations of neutron transport. In recent years, applications of radiative transfer in combustion machine design and in medicine became more and more important.

In all these disciplines one uses the radiative transfer equation to model the formation of the radiation field and its propagation. For slabs and spheres effective algorithms for the solution of the transfer equation have been available for quite some time. In addition, the analysis of the equation is quite well developed. Unfortunately, in many modern applications the approximation of a 1D geometry is no longer adequate and one has to consider the full 3D dependencies. This makes the modeling immensely more intricate. The main reasons for the difficulties result from the fact that not only the dimension of the geometric space has to be increased but one also has to employ two angle variables (instead of one) and very often one has to consider frequency coupling (due to motion or redistribution in spectral lines). In actual calculations this leads to extremely large matrices which, in addition, are usually badly conditioned and therefore require special care. Analytical solutions are not available except for very special cases.

Although radiative transfer problems are interesting also from a mathematical point of view, mathematicians have largely neglected the transfer equation for a long time. As a consequence, in each discipline various codes have been developed that—although they mostly served their purpose—were not optimal with respect to computing time and memory requirement as well as accuracy. Furthermore, the appropriate analytical basis had hardly been worked out.

It was therefore fortunate that the Deutsche Forschungsgemeinschaft for many years supported a graduate school and a special research project in Heidelberg in which mathematicians and astrophysicists collaborated on algorithms for the solution of the multidimensional radiative transfer equation. In this framework in 1994 and 2003 two interdisciplinary workshops took place in Heidelberg. During the latter, the idea evolved to publish a book containing a collection of papers that illuminate multifaceted aspects of radiative transfer and in particular the progress in the numerical solution of the multidimensional radiative transfer equation. It resulted in the present volume.

The editors thank the Deutsche Forschungsgemeinschaft (SFB 359) for its longterm support for developing modern algorithms for the solution of the 3D radiative transfer equation.

Heidelberg, August 2008
G. Kanschat
E. Meinköhn
R. Rannacher
R. Wehrse

Contents

Introduction: The Radiation Field and its Transfer Equation
Guido Kanschat, Erik Meinköhn, Rolf Rannacher, and Rainer Wehrse .. 1

Stochastic Properties of the Radiative Transfer Equation
Wilhelm von Waldenfels 19

An Approach to Neutrino Radiative Transfer in Supernova Simulations
Christian Y. Cardall 27

A Finite Element Method for the Even-Parity Radiative Transfer Equation Using the P_N Approximation
Stephen Wright, Simon Arridge, and Martin Schweiger 39

Solution of Radiative Transfer Problems with Finite Elements
Guido Kanschat 49

A General-Purpose Finite Element Method for 3D Radiative Transfer Problems
Erik Meinköhn 99

Radiative Transfer in 4D: The Inclusion of Kinematical Information
Maarten Baes 175

A Problem-Orientable Numerical Algorithm for Modeling Multi-Dimensional Radiative MHD Flows in Astrophysics – the Hierarchical Solution Scenario
A. Hujeirat 185

Rapidly-Converging Methods for Solving Multilevel Transfer Problems
Eugene H. Avrett 217

Radiative Transfer in NLTE Model Atmospheres
Jiří Kubát 227

The Solution of the Radiative Transfer Equation in Axial Symmetry
Daniela Korčáková and Jiří Kubát 237

Multidimensional Radiation Hydrodynamics
Wolfgang Kalkofen 247

Probing the Initial Conditions for Star Formation with Monte Carlo Radiative Transfer Simulations
Dimitris Stamatellos and Anthony P. Whitworth 259

Radiative Transfer Through the Intergalactic Medium
Avery Meiksin 271

Radiative Transfer with Finite Elements: Application to the Lyα Emission of High-Redshift Galaxies
Sabine Richling 279

Radiative Transfer Problem in Dusty Galaxies: Ray-Tracing Approach
Dmitrij Semionov and Vladas Vansevičius 289

Shape Reconstruction for an Inverse Radiative Transfer Problem Arising in Medical Imaging
Oliver Dorn 299

Introduction:
The Radiation Field and its Transfer Equation

Guido Kanschat[1], Erik Meinköhn[2,3], Rolf Rannacher[2], and Rainer Wehrse[3]

[1] Department of Mathematics, Texas A&M University, College Station, TX 77843, USA
[2] Institut f. Angewandte Mathematik, Im Neuenheimer Feld 294, D-69120 Heidelberg, Germany
[3] Institut f. Theoretische Astrophysik, Albert-Ueberle-Str. 4, D-69120 Heidelberg, Germany

1 Introduction

Light is important in many fields of science and technology (in addition to everyone's daily life) because it can be used as an important tool for information and may provide significant amounts of energy and momentum. Although the complete description of light is extremely complicated (cf. Mandel and Wolf, 1995), it is often sufficient to characterize it just by the specific intensity, which is a function of position, direction and frequency (or wavelength), for details see Section 2. The main advantages of using specific intensities are (i) that they are comparatively easy to measure (observe) and (ii) that there exists a transport equation that allows the modeling from local quantities (as e.g. extinction coefficients, emissivities).

2 Description of the Radiation Field by Specific Intensities

In astrophysics one deals usually with *natural* or *chaotic light*, i.e with light that is derived from thermal sources and that is unpolarized. The radiation field is then described by the *specific intensity* $I(\mathbf{x}, \mathbf{n}, \nu, t)$, which is a function of the position in space $\mathbf{x}$, the direction $\omega = (\cos\phi\sin\theta, \sin\phi\sin\theta, \cos\theta)^t$. The subscript refers to an interval in the frequency. It is usually defined by (cf. Unsöld, 1958) in terms of the energy of radiation that flows through a plane $d\sigma$ within the time interval dt. The radiation is assumed to have frequencies $\nu - d\nu/2 \ldots \nu + d\nu/2$ and filling a cone $d\omega = d\sin\theta d\theta d\phi$ around the direction ω. The plane is assumed to be in the x-y plane (other choices are here also possible, obviously).

$$dE = I d\nu \cos\theta d\sigma d\omega dt \qquad (1)$$

If we employ the particle picture, i.e. we deal with *photons*, the specific intensity is related to the photon distribution function $\Phi(\mathbf{x}, \mathbf{p}, t)$ by

$$I(\mathbf{x}, \mathbf{n}, \nu, t) = \frac{h^4 \nu^3}{c^2} \Phi(\mathbf{x}, \mathbf{p}, t) \tag{2}$$

where $\mathbf{p} = \frac{h\nu}{c}\mathbf{n}$ is the photon momentum, h Planck's constant, and c is the speed of light.

If we employ the wave picture, the specific intensity is given in terms of the electric and magnetic fields, $\mathbf{E}(\mathbf{x}, \tilde{t})$ and $\mathbf{H}(\mathbf{x}, \tilde{t})$ or their Fourier transforms $\mathcal{E}(\mathbf{x}, 2\pi\nu, t)$ and $\mathcal{H}(\mathbf{x}, 2\pi\nu, t)$.

$$I(\mathbf{x}, \mathbf{n}, \nu, t) = I(z, \phi, \theta, \nu, t) = \frac{c}{\Delta t \Delta \nu \Delta \phi \sin\theta \Delta\theta \Delta x \Delta y}$$

$$\cdot \int_{x-\Delta x/2}^{x+\Delta x/2} \int_{y-\Delta y/2}^{y+\Delta y/2} \int_{\theta-\Delta\theta/2}^{\theta+\Delta\theta/2} \int_{\phi-\Delta\phi/2}^{\phi+\Delta\phi/2} \int_{\nu-\Delta\nu/2}^{\nu+\Delta\nu/2} \int_{t-\Delta t/2}^{t+\Delta t/2}$$

$$\mathcal{E}(x', y', z, 2\pi\nu', t') \times \mathcal{H}(x', y', z, 2\pi\nu', t') \cdot \mathbf{n} dt' d\nu' \sin\theta' d\theta' dy' dx'$$

$$\propto \frac{1}{\Delta t} \int_{t-\Delta t/2}^{t+\Delta t/2} \mathbf{E}^2(\mathbf{x}, t) dt = <\mathbf{E}^2(\mathbf{x}, t)> \tag{3}$$

Here we have assumed that the referenced plane is the xy-plane. Since implicitly the validity of geometrical optics is assumed Δx and Δy have to be much larger than the wavelength $\lambda = c/\nu$. The time intervals $\Delta \tilde{t}$ and Δt have to be chosen so that many maxima and minima are comprised. Since observation times are always longer than a few milliseconds this is not a significant restriction. Figures 1 to 2 show an example for the time variation of the electric field vector and the resulting specific intensity.

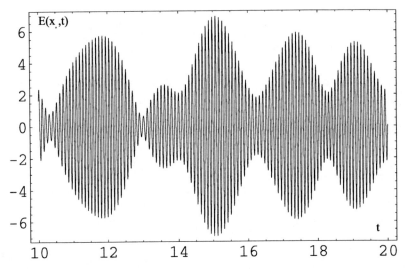

Fig. 1. Example for the variation of electric field strength at position $\mathbf{x}$ with time t (not to scale)

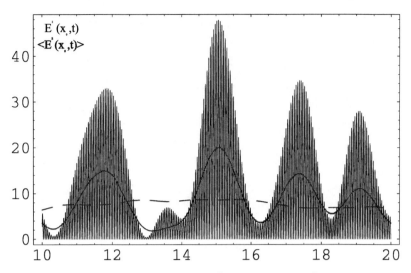

Fig. 2. Example for the time variation of $E^2(\mathbf{x},t)$ and of $<E^2(\mathbf{x},t)>$ at a position $\mathbf{x}$ for the electric field displayed in Fig. 1

Note that the specific intensity is only the lowest order term in the description of the statistical properties of light, for a full description of the radiation the coherence matrices g (cf.) would be needed. Of importance are also the first three moments: the mean intensity

$$J(\mathbf{x},\nu,t) = \frac{1}{4\pi}\int_{S^2} I(\mathbf{x},\mathbf{n},\nu,t)d\omega, \tag{4a}$$

the flux

$$\mathbf{F}(\mathbf{x},\nu,t) = \int_{S^2} \mathbf{n} I(\mathbf{x},\mathbf{n},\nu,t)d\omega \tag{4b}$$

and the tensor of radiation pressure $\mathbf{K}$ with

$$K_{ij} = \int_{S^2} n_i n_j I(\mathbf{x},\mathbf{n},\nu,t)d\omega. \tag{4c}$$

3 The Transfer Equation

3.1 Basic Physical Properties

The equation of radiative transfer for the monochromatic, specific intensity can be derived in several different ways: phenomenologically, the monochromatic intensity varies along a ray, subject to the process of absorption, described by the opacity κ_ν, scattering, described by the cross section σ_ν, and emission, described by the function η_ν. The absorption and scattering coefficients may be combined into the extinction coefficient $\chi_\nu = \kappa_\nu + \sigma_\nu$. The inverse of the extinction coefficient is the monochromatic mean free path length $\tilde{\lambda}_\nu = 1/\chi_\nu$.

The intensity along the ray direction $\mathbf{n}$ and in an element ds ($\ll \lambda$) is reduced by extinction,

$$(dI/ds)_- = -\chi I, \tag{5}$$

and increased by emission,

$$(dI/ds)_+ = \eta. \tag{6}$$

For a stationary radiation field the combined description of theses effects results in the monochromatic radiative transfer equation

$$\frac{dI(\mathbf{x},\mathbf{n})}{ds} = -\chi I(\mathbf{x},\mathbf{n}) + \eta(\mathbf{x},\mathbf{n}). \tag{7}$$

A measure of the decay of the intensity traveling from point $\mathbf{x}_1$ to $\mathbf{x}_2$ is the optical thickness τ defined by

$$\tau(\mathbf{x}_1,\mathbf{x}_2) = \int_0^d \chi_\nu(\mathbf{x}_1 + s\mathbf{n})\, ds, \tag{8}$$

where $d = d(\mathbf{x}_1,\mathbf{x}_2)$ is the distance between the two points and $\mathbf{n}$ is the unit vector pointing from $\mathbf{x}_1$ towards $\mathbf{x}_2$. For its relevance see the formal solutions (13) later in this section.

The derivative along the ray d/ds is independent of the coordinate system and, therefore, may be expressed in terms of Cartesian coordinates as

$$\frac{dI}{ds} = \left(\frac{\partial x}{\partial s}\right)\left(\frac{\partial I}{\partial x}\right) + \left(\frac{\partial y}{\partial s}\right)\left(\frac{\partial I}{\partial y}\right) + \left(\frac{\partial z}{\partial s}\right)\left(\frac{\partial I}{\partial z}\right) = \mathbf{n}\cdot\nabla I. \tag{9}$$

This introduction of the radiative transfer equation is based on heuristic principles of classical radiometry, i.e., on intuitively appealing considerations of energy balance and the simple phenomenological notions of light rays and ray pencils.

Another (phenomenological) way to introduce the radiative transfer equation is to employ Einstein's concept of photons and describe the radiation field in terms of a "photon gas" which satisfies the linear Boltzmann equation (cf. [18]). In this derivation the photons are considered as (non-interacting) particles that can be localized. The Boltzmann equation is one of the central aspects of transport theory, which studies the transport of generic particles without regard for their physical meaning [8]. In terms of the photon distribution function Φ the kinetic equation (or Boltzmann equation) for the photon gas can be written

$$\frac{\partial \Phi}{\partial t} + c\mathbf{n} \cdot \nabla \Phi = \left(\frac{\delta \Phi}{\delta t}\right)_+ - \left(\frac{\delta \Phi}{\delta t}\right)_- \tag{10}$$

where the terms on the right-hand side describe the creation and destruction of photons. Note that there is no force term since photons have zero rest mass. By means of Eq. (2) Eq. (10) can now be written

$$\frac{\partial I}{\partial t} + \mathbf{n} \cdot \nabla I = KI, \tag{11}$$

where K is the collision operator describing the interaction between photons and surrounding matter (see subsection 4.2).

A recent approach introduces the equation for radiative transfer by means of a stochastic model (see [26] in this volume). This approach sheds new light on the nature of the transfer equation by showing that it can be regarded as the differential equation for the potential of a Markov process.

The first two derivations are short and intuitively appealing, and are therefore frequently presented in the literature. However, this heuristic way to introduce the radiative transfer equation, based on the concept of photons as particles or photon (particle) rays, has been often criticized for its phenomenological character, lack of solid physical background, and unknown range of applicability (see [14]). It becomes even more delicate when polarization effects and the effects of nonsphericity and orientation of the scatterers are taken into account. Furthermore, it has led to the widespread ignorance of the fact that the real derivation of the radiative transfer equation and the clarification of the physical meaning of the quantities that enter this equation must be based on the classical, i.e., non-quantum, limit of the quantum theory of radiation or the quantum electrodynamics. This classical limit is described by Maxwell's equations for the electromagnetic fields, which do not involve particles. Under certain approximations transfer equations of the type (7) can be derived directly from Maxwell's equation.

One of the fundamental physical difficulties inherent in the solution of radiative transfer problems is the existence of scattering terms, which decouple the radiation field from local sources and sinks, and involve global transport of light over large distances. The diffuse regime of light transport in scattering media is formulated in three inequalities (see [25] and references therein)

$$\lambda \ll l \ll L \ll L_{\text{abs}}, \tag{12}$$

where λ is the wavelength, l the mean free path, L the sample or configuration size, and L_{abs} the absorption length. The first inequality ensures that localization effects are small (for details see [25]); the second inequality implies that many scattering events occur if the wave traverses the system/configuration; the third inequality ensures that not all radiation is absorbed.

The description of radiation transport can occur on roughly three length scales (see [25]):

- *Macroscopic:* On scales much larger than the mean free path the diffuse mean intensity $J(\mathbf{x})$ satisfies a diffusion equation. This approximation turns out to be very accurate in (almost) homogeneous media where the extinction coefficient χ is so high that over a mean free path $l = 1/\chi$ the extinction coefficient does not vary much. The diffusion coefficient D enters as a system parameter that has to be calculated on mesoscopic length scales.
- *Mesoscopic:* On length scales of the mean free path l, the problem is described by the radiative transfer equation (7) or Schwarzschild-Milne equation (cf. Eq. (14)). At this level one needs as input the mean free path l and the speed of transport v, which should be derived from microphysics. In the diffusive regime, where the radiation field is isotropic, this approach leads to the diffusion coefficient $D = vl/3$.
- *Microscopic:* The appropriate wave equation, i.e. Maxwell's equations, is used on this length scale. The precise location and shapes of scatterers are assumed to be known. Together with the wave nature they determine the interference effects of scattered waves. In light scattering systems the scatterers often have a size in the micron regime, comparable to the wavelength λ, which could lead to important resonance effects. The drawback of the microscopic approach is that it is too detailed. In practice the precise shapes, positions and orientations of the scatterers often are not known and a mesoscopic or macroscopic description is necessary.

The approximations inherent in the classical radiative transfer equation defined by Eq. (7) fall into two classes. First, the approximations that are incorporated to achieve a simplification of the radiative transfer problem. These approximations can be relaxed and the corresponding physical effects, e.g., due to polarization, refraction and dispersion, can be incorporated into the transfer equation (7). The second type of approximations are more fundamental in nature and are indispensable for the derivation of the radiative transfer equation from Maxwell's equations. It is obvious that the latter assumptions restrict the range of validity and therefore the applicability of this equation.

During the past three decades, there has been significant progress in studies of the wave content of the radiative transfer theory, which has resulted in a much better understanding of the basic assumptions leading to the radiative transfer equation (see [25, 16]). The derivations of the radiative transfer equation (7) are based on geometric optics expansions [23] or Wigner transforms (see [4, 22] and references therein), and on explicit microphysical considerations from statistical electromagnetics [16, 25]. For the interesting case of heterogeneous media, geometric optics methods are not very well-adapted to analysis of wave propagation, since multiple scattering of waves quickly creates a superposition of an infinite number of fronts propagating in essentially every direction. Phase space based methods, such as developed in [22, 4], are often preferable. An essential step in that approach to deriving radiative transfer equations from wave equations is the introduction of the Wigner distribution [28] which is a good candidate for analyzing the evolution of wave energy in phase space. However, these phase space based methods presented in [22, 4] have been limited to isotropic disordered media and one has to look closely at the details of the analysis to see the breakdown of the transport approximation and the onset of the wave localization (for details see [22]). The explicit microphysical derivation of the radiative transfer equation, including polarization effects, from statistical electromagnetics in the case of arbitrarily shaped and arbitrarily oriented particles gives an even more detailed description of the fundamental assumption leading to the transfer equation [16, 17]. As a consequence, the radiative transfer equation cannot be expected to perform well for densely packed media and does not describe important interference effects such as coherent backscattering (for details see [16] and references therein). The scattering particles have to be randomly positioned and separated widely enough that each of them is located in the far-field zones of all other particles.

3.2 Difficulties in the Solution

The determination of the specific intensity from the radiative transfer equation may be difficult on account of the following complications:

- Depending on the situation to be modeled, the transfer equation can take several different forms, representing different types of equations and requiring different algorithms for the solution.
- It is often necessary to solve radiative transfer problems in composite domains with mixed transport regimes, so that in some parts of the domain diffusion is prevalent, whereas transport dominates elsewhere. In case of such a mixed regime, the application of standard solvers and of standard preconditioning methods for the discrete problem is usually marred by an extremely slow rate of convergence.
- All intensities incident on static configurations must usually be given as boundary values (except for media of infinite optical depth). For moving configurations the distribution of the boundary values depends on the velocity field and may be quite complicated.
- Except for pure absorption cases, radiative transfer problems are not initial value but boundary value problems. An inaccurate formulation of the boundary may lead to spurious solutions that let computer codes fail.
- The particular coupling of the time, space and frequency variables in the transport operator and of angle and frequency in the scattering term may prevent the use of standard methods and standard program libraries of numerical mathematics. Further complications arise from variations of the coefficients in the transfer equation over many orders of magnitude, with strong gradients and rapid fluctuations with frequency.
- Radiative transfer problems may have high dimension (the Stokes vector may depend on 3 spatial, 2 angle, 1 time and 1 frequency variable) requiring very large memory capacities for the numerical solution.

3.3 Simplifications in the Solution

If the extinction coefficient and the emissivity in Eq. (7) are given functions – as e.g. in an LTE situation – we have a simple initial value problem along a ray starting at the boundary point $\mathbf{x}_0$ of the domain. It has the well known general solution at $\mathbf{x} = \mathbf{x}_0 + d\mathbf{n}$

$$I(\mathbf{x}) = I(\mathbf{x}_0)e^{-\tau(\mathbf{x}_0,\mathbf{x})} + \int_0^d e^{-\tau(\mathbf{x}_s,\mathbf{x})}\eta(\mathbf{x}_s)ds, \tag{13}$$

with $\mathbf{x}_s = \mathbf{x}_0 + s\mathbf{n}$. If η is a linear functional of I it is an integral equation for the specific intensity usually called formal solution. If we have coherent and isotropic scattering the transfer equation is equivalent to the Schwarzschild–Milne equation for the mean intensity, i.e. the formal solution integrated over the sphere. If we assume zero boundary conditions and change from polar to Cartesian coordinates, the Schwarzschild–Milne equation reads

$$J(\mathbf{x}) = \int_\Omega \frac{e^{-\tau(\mathbf{x},\mathbf{y})}}{|x-y|^{D-1}} \eta(\mathbf{y})\, d\mathbf{y}, \tag{14}$$

with D specifying the dimension of the geometry. This equation has central importance in 1D radiative transfer but limited relevance in multidimensional cases. The Schwarzschild–Milne equation has only been solved exactly in a limited number of cases. Analytical results have been obtained for isotropic scattering of scalar waves (Ref. aus Amic et al, S. 4916 and [19, 20]), and in the case where the scattering cross section depends linearly on the cosine of the scattering angle (Ref. aus Amic et al, S. 4916).

If we assume homogeneous media and an emissivity η that does not vary with location the formal solution in Eq. (13) simplifies to

$$I(\mathbf{x}) = I(\mathbf{x}_0)e^{-\tau(\mathbf{x}_0,\mathbf{x})} + S\left(1 - e^{-\tau(\mathbf{x}_0,\mathbf{x})}\right), \tag{15}$$

with the source function $S = \eta/\chi$. Thus, for optically thick objects with $\tau(\mathbf{x}_0,\mathbf{x}) \gg 1$ the solution above reduces to

$$I(\mathbf{x}) \approx S, \tag{16}$$

whereas in the optically thin case with $\tau(\mathbf{x}_0,\mathbf{x}) \ll 1$ we have

$$I(\mathbf{x}) \approx I(\mathbf{x}_0) + (S - I(\mathbf{x}_0))\,\tau(\mathbf{x}_0,\mathbf{x}). \tag{17}$$

3.4 Diffusion Case

On scales much larger than the mean free path the radiation field can be described by the diffuse mean intensity $J(\mathbf{x})$ that satisfies a diffusion equation. This approximation turns out to be very accurate in (almost) homogeneous media where the extinction coefficient χ is so high that over a mean free path $l = 1/\chi$ the extinction coefficient does not vary much. The diffusion coefficient D enters as a system parameter that has to be calculated on mesoscopic length scales. If the radiation propagation velocity, i.e. the speed of light c, is scaled to unity, the calculation of the diffusion coefficient yields $D = l/3$. For a scattering medium, the emissivity in Eq. (7) reads

$$\eta(\mathbf{x}) = \chi(\mathbf{x})S(\mathbf{x}) = \chi(\mathbf{x})\left((1-\varepsilon)J(\mathbf{x}) + \varepsilon B(\mathbf{x})\right), \tag{18}$$

where $S(\mathbf{x})$ is the source function, $B(\mathbf{x})$ a source term called the Planck function, and $\varepsilon = \kappa/\chi$ the "thermalization factor". The asymptotic diffusion limit of Eq. (14) for the situation described by inequality (12) then results in a second-order diffusion equation

$$-\nabla_x \left(\frac{1}{3\sigma(\mathbf{x})}\nabla_x J(\mathbf{x})\right) + \kappa(\mathbf{x})J(\mathbf{x}) = \kappa(\mathbf{x})B(\mathbf{x}). \tag{19}$$

This diffusion equation is a classical equation that fully neglects interference effects inherent in wave propagation. Within astrophysical literature one generally (often) refers to the diffusion limit as the case where simple expressions for the radiation quantities are derived at positions far away from the surface (i.e. the boundary of the domain) in a medium

- that is optically very thick ($\chi \gg 1$),
- in which the effects of the boundaries are negligible,
- in which the extinction coefficient hardly varies over a photon mean free path, and
- in which the variation of the source function can be approximated by a linear function in the neighborhood of the point where the radiative quantities are to be calculated.

If the last inequality in (12) is violated, the solution of Eq. (14) does not approach the diffusion limit, but a *local limit*. For moderately and even for highly scattering media, this local limit causes a localization of the transfer problem, where the radiation field is strongly coupled to the local material conditions, and no global light transport occurs. Thus, the mean intensities are very close to the Planck function $B(T(\mathbf{x}))$ of the local temperature $T(\mathbf{x})$, which implies that LTE conditions prevail and the source function is identical to the Planck function,

$$S(\mathbf{x}) = B(T(\mathbf{x})). \tag{20}$$

In terms of a small parameter $\delta \to 0$, the local limit is characterized by the two coefficients

$$\kappa \sim \mathcal{O}(1/\delta) \quad \text{and} \quad \varepsilon \sim \mathcal{O}(1). \tag{21}$$

The rigorous asymptotic derivation of the diffusion equation Eq. (19) from transfer theory is performed in the limit $\varepsilon \to 0$ as required by (12) (for details

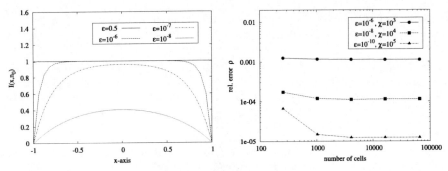

Fig. 3. *Left:* the specific intensity $I(\mathbf{x}, \mathbf{n}_0)$ is plotted along the x-axis ($y = 0$) of the 2D-domain $\Omega = [-1, 1]^2$. The light propagation direction $\mathbf{n}_0$ points from $x = -1$ to $x = +1$, which is parallel to the x-axis. The density structure is homogeneous, so that the coefficients have constant values, with $\chi = 10^4$ and values of ε as specified in the inserts. *Right:* the plots of the relative error $\rho = (J(\mathbf{x}_0) - I(\mathbf{x}_0, \mathbf{n}_0))/J(\mathbf{x}_0)$, which gives a measure of the relative deviation from the assumption $I(\mathbf{x}, \mathbf{n}) = J(\mathbf{x})$ at a particular point $\mathbf{x}_0 = (0, 0)$ far from the surface boundary and in a particular direction $\mathbf{n}_0$, against the number of cells for various structured spatial grids for three different assignments for the values of the coefficients ϵ and χ

see [7]). In terms of the small parameter δ, this (global) limit, that is generally referred to as *diffusion limit*, is characterized by the two coefficients

$$\kappa \sim \mathcal{O}(\delta) \quad \text{and} \quad \varepsilon \sim \mathcal{O}(\delta^2), \qquad (22)$$

indicating a transfer problem in an almost purely scattering, optically thick medium. The random scattering of light is accurately described by a diffusion process, although the radiation field decouples from local sources and sinks, and global light transport over large distances is involved. It is through these terms that the presence of an open boundary influences the radiation field deep inside the medium at large optical depths, and causes large departures of the mean intensity from local values of the Planck function (see Fig. 3),

$$J(\mathbf{x}) \neq B(T(\mathbf{x})). \qquad (23)$$

For a constant Planck function $B = 1$, the left-hand part of Fig. 3 displays the specific intensity $I(\mathbf{x}, \mathbf{n}_0)$, along the x-axis ($y = 0$) of the 2D-domain $\Omega = [-1, 1]^2$. The light propagation direction $\mathbf{n}_0$ points from $x = -1$ to $x = +1$, which is parallel to the x-axis. The density structure is homogeneous, so that the coefficients have constant values, with $\chi = 10^4$ and values of ε as specified in the inserts. In the local limit (solid line with $\varepsilon = 0.5$) we have $I = J = B = 1$. While ε approaches the diffusion limit, the presence of the open boundary becomes exceedingly "visible" even for points deep inside the medium. Finally, the diffusion limit (dotted line with $\varepsilon = 10^{-8}$) is reached and the mean intensity differs from the local values of the Planck function throughout the whole medium, i.e. $I = J \neq B$. The right-hand part of Fig. 3

Table 1. Form of the transport operator, depending on geometry and differential velocity of the medium

Geometry	static	slow	relativistic
1D/2-stream	$\pm \frac{d}{ds}$	$\pm \frac{\partial}{\partial s} + \frac{\partial \beta}{\partial s}\frac{\partial}{\partial \xi}$	$\frac{\partial}{\partial s} \pm \gamma^2 \frac{\partial \beta}{\partial s}\frac{\partial}{\partial \xi}$
slab	$\mu \frac{d}{dz}$	$\mu \frac{\partial}{\partial z} + \mu^2 \frac{\partial \beta}{\partial z}\frac{\partial}{\partial \xi}$	$\frac{\partial}{\partial z} - (1-\mu^2)\gamma^2 \frac{\partial \beta}{\partial z}\frac{\partial}{\partial \mu} + \gamma^2 \mu \frac{\partial \beta}{\partial z}\frac{\partial}{\partial \xi}$
spherical	$\mu \frac{\partial}{\partial r} + \frac{1-\mu^2}{r}\frac{\partial}{\partial \mu}$	static $+ \mu^2 \frac{\partial \beta}{\partial r}\frac{\partial}{\partial \xi}$	
2D/3D	$\mathbf{n}\cdot\nabla$	$\mathbf{n}\cdot\nabla - \mathbf{n}\cdot\nabla(\mathbf{n}\cdot\beta)\frac{\partial}{\partial \xi}$	

plots the relative error $\rho = (J(\mathbf{x}_0) - I(\mathbf{x}_0, \mathbf{n}_0))/J(\mathbf{x}_0)$, which gives a measure of the relative deviation from the assumption $I(\mathbf{x}_0, \mathbf{n}_0) = J(\mathbf{x}_0)$ at a particular point $\mathbf{x}_0 = (0,0)$ far from the surface boundary and in a particular direction $\mathbf{n}_0$, against the number of cells for various structured spatial grids for three different assignments for the values of the coefficients ϵ and χ. When a family of solutions to the radiative transfer equation (7) with parameters $\chi \sim \mathcal{O}(1/\delta)$ and $\epsilon \sim \mathcal{O}(\delta^2)$ for $\delta \to 0$ is considered, the difference to the solution to the diffusion equation (19) admits the estimate $I(\mathbf{x}, \mathbf{n}) - J(\mathbf{x}) = \mathcal{O}(\delta)$ in the interior of the domain Ω (see e.g. [7]). The relative error ρ plotted in Fig. 3 confirms this estimate for (almost) every spatial grid and approaches the correct value of $\delta = 1/\chi$ for the three different values of χ, respectively.

4 Classification of Models

Since the simulation of a general radiative transfer model is exorbitantly expensive and indeed still way beyond the capabilities of existing computing power, the choice of the model required for a certain situation is most important. Depending on the physical regime, we can distinguish several hierarchies of models, affecting the transport operator and the collision operator, respectively.

4.1 The Transport Operator

The form of the transport operator depends on the geometry of the medium and its differential motion. An overview is given in Table 1. Here, μ is the projection of the solid angle $\mathbf{n}$ onto the interval $[0, 1]$, $\beta = v/c$ is the relativistic velocity and $\gamma = 1/\sqrt{1-\beta^2}$.

The different geometries are characterized as follows:

2-stream A true one-dimensional model, where the directions are only upward and downward with unit velocity. While this model is a mathematical

slab A medium varying in z-direction, but stretching infinitely and homogeneously in the xy-plane.
spherical A medium with spherical symmetry, like a star.
cylindrical Reduction of the 3D model to a geometry varying in the xy-plane, but stretching infinitely with constant parameters in z-direction.
2D The two-dimensional equivalent of the full model. While this has the same mathematical properties as the full 3D model, it is not actually an approximation of a 3D geometry by symmetry.
3D The model for an arbitrary geometrical setting if no general relativistic effects are considered.

equivalent to the higher dimensional models with similar properties, it is *not* an approximation to a higher dimensional model obtained through symmetry.

From the analytical point of view, the lower dimensional models are split in two groups. On one side are the 2-stream and the 2D models: like the 3D model, all photon velocities considered have absolute value unity (in units of light speeds), which allows for proofs of existence and uniqueness of solutions[7]. On the other hand, the slab, spherical and cylindrical models are actual reductions of the full three-dimensional problem by exploiting symmetry. For them, mathematical analysis is only available through the 3D problem.

As soon as the media exhibits motions of its parts relative to each other, Doppler shift has to be taken into account as in column "slow" of Table 1. These shifts depend on the differential of the velocity and can be expressed by a derivative with respect to the frequency of the radiation; in order to obtain the simple terms in our table, the frequency ν is replaced by the logarithmic frequency $\xi = \log \nu$. For relativistic velocities above roughly 10% of the speed of light, aberration effects must be taken into account as in the last column of Table 1. Note that the empty boxes to the lower right can be filled accordingly.

4.2 The Collision Operator

The collision operator K in Eq. (11) models the physical processes of absorption, emission and scattering of photons. It is usually a combination of the following terms

- By thermal absorption, photons are removed from the photon gas and their energy is converted to thermal energy of the surrounding matter. With the absorption cross section κ_ν, this term has the form

$$KI_\nu = -\kappa_\nu I_\nu.$$

- Thermal emission is modeled by the Planck function $B_\nu(T)$ and depends on the temperature of the surrounding matter:

$$KI_\nu = \kappa_\nu B_\nu(T).$$

- Coherent scattering with a scattering phase function p is modeled by the integral operator

$$KI(\mathbf{n},\nu) = -\sigma_\nu I(\mathbf{n},\nu) + \frac{\sigma_\nu}{4\pi}\int_{S^2} p(\mathbf{n},\mathbf{n})I(\mathbf{n}')d\omega. \qquad (24)$$

- Complete redistribution with a profile function Φ

$$KI(\mathbf{n},\nu) = -\Phi(\nu)I(\mathbf{n},\nu) + \frac{\Phi(\nu)}{4\pi}\int_{S^2}\int_0^\infty \Phi(\nu')p(\mathbf{n},\mathbf{n})I(\nu',\mathbf{n}')d\nu'd\omega \qquad (25)$$

- Partial redistribution with redistribution function R:

$$KI(\mathbf{n},\nu) = -\sigma_\nu I(\mathbf{n},\nu) + \frac{1}{4\pi}\int_{S^2}\int_0^\infty R(\nu,\nu',\mathbf{n},\mathbf{n}')I(\nu',\mathbf{n}')d\nu'd\omega. \qquad (26)$$

5 Numerical Solution of the Transfer Equation: A Brief Overview

Numerical schemes for radiative transfer problems can be split in two major groups. The first consists of particle schemes, following photons on their paths through scattering matter. Here, elementary physical processes are reproduced during the simulation and no further modeling is necessary. Nevertheless, additional models may be used inside the numerical scheme in order to speed up the method considerably.

The second group of methods aims at discretizing a partial differential equation or an integro-differential equation as a model for radiative transfer. First, a suitable model equation must be chosen; this may be the linear Boltzmann equation for radiative transfer, its integral form, which eliminates the angular dependence of the simulated solution or a set of moments obtained from the Boltzmann equation. For any of these equations, a suitable discretization scheme must be chosen.

5.1 Monte–Carlo Methods

In a general Monte-Carlo scheme, random photons are shot into the domain of computation, either from internal sources or from the boundary. After a random distance depending on the mean free path length, these photons are randomly scattered or absorbed. This process is iterated until the photon leaves the domain or is absorbed.

Naturally, this process becomes very slow when the mean free path lengths are short; then, a photon may be scattered a several thousand times before leaving the domain again. Therefore, several improvements of this scheme have been developed. For instance, the photons, which are binary objects, may be replaced by intensity packets that are not completely absorbed at a random stage, but get there value decreased according to the absorption coefficient. Other models exist to even increase the average travel distance between two events far beyond the mean free path length.

With these methods, physical phenomena very hard to handle by discretization schemes can be implemented quite easily. One example is resonant scattering, when a photon at a high frequency is absorbed and reemitted as two photons of lower frequencies.

5.2 Discretization Schemes

Milne Equation

In the case of isotropic scattering, the Boltzmann equation for radiative transfer can be transformed into the integral equation (14) operating in space only. This integral equation can then be discretized using finite elements or a collocation method, the integral equation equivalent of finite differences (see e.g. [9, 21]).

While the integral equation model reduces the dimension of the discretized down to two or three, the resulting matrix, as usual with integral equations, is dense, such that the memory requirements may be even higher than with a direct discretization of the Boltzmann equation. The kernel of the integral is concentrated around the origin; it will be the more localized the shorter the mean free path length is. Therefore, the mesh width must not exceed this length. On the other hand, long distance effects due to low scattering may be modeled quite well by this scheme.

Boltzmann Equation

Discretization methods can be grouped roughly into three categories, namely finite difference, finite volume and finite element schemes. While it is possible to write all discretizations as a finite difference method, the derivation and mathematical tools for analysis of these methods are quite different.

The angular variable is most often discretized by a collocation method, replacing the integral over the unit sphere by a quadrature rule. These schemes are usually referred to as discrete ordinate methods [3]. In particular for large scattering cross sections, these quadrature rules must ensure energy conservation in order to produce even qualitatively correct solutions. An alternative to discrete ordinate methods are so called P_N schemes involving spherical harmonics.

Usually, finite difference schemes are those which are implemented easiest, at least on regular meshes. They only use point values of functions in mesh points and difference quotients between those. On the other hand, their mathematical structure is very weak, such that optimal error estimates are hard to develop; a theory for a posteriori error estimates which has led to highly efficient adaptive finite element schemes does not exist for finite differences.

After angular discretization, a system of transport equations in space must be solved. A scheme which has received much attention in the literature is the method of short characteristics [27]. There, the value at a mesh point is determined by following the characteristic in opposite ray direction backwards until the it leaves the cell. The value at this point is interpolated from neighboring points and then the transport equation is solved exactly along the characteristic. Accordingly, the quality of this method depends crucially on the interpolation chosen. By truncating characteristics at cell boundaries, these schemes avoid the global coupling generated by standard characteristics methods; therefore, a marching algorithm can be used to solve the transport problems successively.

Backward finite differencing schemes like the upwind scheme used in [24] are closely related to short characteristics, but they replace the exact solution along the characteristic by a finite difference approximation.

Standard (symmetric) finite difference schemes cannot be used for transport problems as well as standard finite elements. This is due to the properties of transport as a first order differential operator. Stable finite difference methods can be obtained by using upwind schemes and similar modifications must be applied to the finite element method.

Finite volume methods are based on the conservation properties of the differential equation and construct solutions as functions with constant values on small control volumes; reconstruction techniques have been developed in order to achieve higher order methods.

Finite element methods are based on a weak formulation of the equation and the method itself is constructed by projection of this formulation on a piecewise polynomial function space. As a consequence, the construction of the discrete form of the differential operator is quite complex and involves the integration over mesh cells. On the other hand, the error of these methods exhibits a structure called Galerkin orthogonality, allowing for efficient a priori and a posteriori error estimates. It is also very simple to devise higher order finite element schemes. When applied to the spacial differential operator of the Boltzmann equation, standard finite element methods fail; they are replaced by streamline diffusion schemes [6, 9] or discontinuous Galerkin methods [1, 13, 10].

- The size of the discretization is enormous. While in other areas of scientific computation, three is considered higher dimensional, here the simplest model in two space dimensions yields a three dimensional domain of computation, while the full model with dependence on time and light frequency is seven dimensional. Therefore, devising a discretization scheme is only a first step in acquiring an approximate solution. In order to compute this approximation, high performance computers are required together with the associated programming techniques.
- For very high values of the scattering coefficient (σ_ν in (24), the problem becomes singularly perturbed, that is, if we consider limits of the solution

for σ_ν going to infinity, this limit is not the solution of the equation without the transport term [12, 5]. A numerical scheme must be able to reproduce functions close to this limit. While the conservation properties of finite element discretizations seem to be sufficient for this property, it must be verified explicitly for finite difference schemes.
- Apart from the size of the resulting discrete systems, these systems become extremely ill-conditioned close to the singularly perturbed limit. Special preconditioners must be designed to counter this effect (see [2] and [11, 15] in this volume).

Because of these peculiarities, special schemes exploiting the structure of the equation have been developed. These usually use different discretizations with respect to the space, angular, frequency and time domains, where standard timestepping methods are used for the last.

6 Open Problems and Outlook

One-dimensional radiative transfer problems for media in non-relativistic media with given extinction coefficients and well behaved redistribution functions pose essentially no problems any more. Problems still arise, however, at least for grid methods, for plasmas with velocities $\gamma \gg 1$, since then aberration and advection terms have to be taken into account also, and in the high energy regime where in a scattering event the energies of the incoming and outgoing photon may differ considerably.

Another range of problems refers to the inclusion of a large number of spectral lines. Although recently progress has been made by the invocation of the Poisson Point Process for non-scattering and (differentially) slowly moving media in LTE the NLTE case involving many a atomic and/or molecular levels has not found its final solution. Therefore, a number of astronomically interesting cases as e.g. winds from hot stars is awaiting a satisfying solution. In addition to the problems mentioned in the previous sections which refer to multidimensional optically thick, scattering dominated cases, the proper and balanced coupling of radiation transfer to hydrodynamics has still not been achieved. Another problem that often limits the accuracy of radiative modeling is the lack of good laboratory data for transition probabilities and line broadening coefficients as well as for collision cross-sections (in the case of NLTE). It is often caused by the restraint of funding agencies worldwide to support corresponding projects. Fortunately, for the solution of the other problems mentioned one may be optimistic due to the progress in numerical mathematics and due to the steeply increasing availability of fast computers with very large memory.

References

1. M.L. Adams. Discontinuous finite element transport solutions in thick diffusive problems. *Nuclear Sci. Engng.*, 137(3):298–333, 2001.
2. M.L. Adams and E.W. Larsen. Fast iterative methods for discrete-ordinates particle transport calculations. *Progress in Nuclear Energy*, 40(1):3–159, 2002.
3. M. Asadzadeh, P. Kumlin, and S. Larsson. The discrete ordinates method for the neutron transport equation in an infinite cylindrical domain. *Math. Models Methods Appl. Sci.*, 2(3):317–338, 1992.
4. G. Bal. Radiative transfer equations with varying refractive index: a mathematical perspective. *J. Opt. Soc. Amer. A.* to appear.
5. M. Borysiewicz, J. Mika, and G. Spiga. Asymptotic analysis of the linear boltzmann equation. *Math. Meth. Appl. Sci.*, 3:405–423, 1981.
6. A. Brooks and T.J.R. Hughes. Streamline upwind/Petrov-Galerkin formulation for convection dominated flows with particular emphasis on the incompressible Navier-Stokes equations. *Comput. Methods Appl. Mech. Engrg.*, 32:199–259, 1982.
7. R. Dautray and J.-L. Lions. *Mathematical Analysis and Numerical Methods for Science and Technology*, volume 6. Springer, 2000.
8. J.J. Duderstadt and W.R. Martin. *Transport Theory*. Wiley, Chichester, 1979.
9. C. Johnson and J. Pitkäranta. Convergence of a fully discrete scheme for two-dimensional neutron transport. *SIAM J. Numer. Anal.*, 20(5):951–966, October 1983.
10. G. Kanschat. *Disctontinuous Galerkin Methods for Viscous Flow*. Deutscher Universitätsverlag, Wiesbaden, 2007.
11. G. Kanschat. Solution of radiative transfer problems with finite elements. In *Numerical Methods in Multidimensional Radiative Transfer*, page 49 Springer, 2008.
12. E.W. Larsen, J.E. Morel, and W.F. Miller Jr. Asymptotic solutions of numerical transport problems in optically thick, diffusive regions. *Comp. Phys.*, 69:283–324, 1987.
13. P. LeSaint and P.-A. Raviart. On a finite element method for solving the neutron transport equation. In C. de Boor, editor, *Mathematical aspects of finite elements in partial differential equations*, pages 89–123, New York, 1974. Academic Press.
14. L. Mandel and E. Wolf. *Optical Coherence and Quantum Optics*. Cambridge University Press, 1995.
15. E. Meinköhn. A general-purpose finite element method for 3d radiative transfer problems. In *Numerical Methods in Multidimensional Radiative Transfer*, page 99 Springer, 2008.
16. M.I. Mishchenko. Vector radiative transfer equation for arbitrarily shaped and arbitrarily oriented particles: a microphysical derivation from statistical electromagnetics. *Appl. Opt.*, 41:7114–7135, 2002.
17. M.I. Mishchenko. Microphysical approach to polarized radiative transfer: extension to the case of an external observation point. *Appl. Opt.*, 42:4963–4967, 2003.
18. J. Oxenius. *Kinetic Theory of Particles and Photons Theoretical Foundations of non-LTE Plasma Spectroscopy*. Springer, 1986.
19. J. Pitkäranta. On the differential properties of solutions to fredholm equations with weakly singular kernels. *J. Inst. Math. Appl.*, 24:109–119, 1979.

20. J. Pitkäranta. Estimates for the derivatives of solutions to weakly singular fredholm integral equations. *SIAM J. Math. Anal.*, 11(6):952–968, November 1980.
21. J. Pitkäranta and L.R. Scott. Error estimates for the combined spatial and angular approximation of the transport equation for slab geometry. *SIAM J. Numer. Anal.*, 20(5):922–950, 1983.
22. L. Ryzhik, G. Papanicolaou, and J.B. Keller. Transport equations for elastic and other waves in random media. *Wave Motion*, 24:327–270, 1996.
23. J. Tualle and E. Tinet. Derivation of the radiative transfer equation for scattering media with a spatially varying refractive index. *Optics Comm.*, 228:33–38, 2003.
24. S. Turek. An efficient solution technique for the radiative transfer equation. *Imp. Comput. Sci. Engrg.*, 5:201–214, 1993.
25. M.C.W. van Rossum and T.M. Nieuwenhuizen. Multiple scattering of classical waves: Microscopy, mesoscopy, and diffusion. *Rev. Mod. Phys.*, 71(1):313–371, 1999.
26. W. von Waldenfels. Stochastic properties of the radiative transfer equation. In *Numerical Methods in Multidimensional Radiative Transfer*, page 19 Springer, 2008.
27. R. Wehrse and W. Kalkofen. Advances in radiative transfer. *Astron. Astrophys. Rev.*, 13(1/2):3–29, 2006.
28. E. Wigner. On the quantum correction for thermodynamic equilibrium. *Phys. Rev.*, 40:749–759, 1932.

Stochastic Properties of the Radiative Transfer Equation

Wilhelm von Waldenfels

Institut für Angewandte Mathematik der Universität Heidelberg,
Im Neuenheimer Feld 294 69120 Heidelberg Germany

Summary. Radiative transfer is connected with Markov processes in two ways. The usual transfer equation corresponds to a process with five dimensional state space in continuous time, the Milne's equation corresponds to process in three dimensional state space in discrete time.

1 Introduction

The stochastic nature of radiative transfer is implicitly contained in the classical deductions of the radiative transfer equation [1]. It is the basis of many Monte Carlo simulations. Thereabout exists a huge literature. We establish here the connection with the theory of Markov processes in an explicit way.

There are two Markov processes connected with radiative transfer. The first one is in continuous time and lists the probabilities, that the photon either flies unhindered by an atom or hits an atom and is by the atom either scattered or absorbed. The usual Monte Carlo methods consist in following some sample paths of this process and making a statistic about them. The potential equation of this process is the usual radiative transfer equation. The state space is the cartesian product of a domain in three dimensional space, describing the positions of the photons, with the unit sphere of the three dimensional space, describing the directions. So the state space is five dimensional.

The other Markov process is in discrete time. It is connected with the Milne's equation. It describes the jumping of the photon from atom to atom, where it is either scattered or absorbed. The state space is three dimensional. To our knowledge this process has not yet been used in Monte Carlo simulations. It might be advantageous, to do so. It presupposes, however, isotropic scattering.

Radiative transfer is an important stochastic process. The solution of the radiative transfer equation for an isotropic planparallel semi-infinite equation by Hopf and Wiener [4][5] gave rise to a huge literature in probability theory. The Wiener-Hopf factorization is a favorite tool in the theory of random walks [2], the connection with theoretical astrophysics, however, is widely forgotten among probabilists.

2 The Notion of the Potential of a Transient Markov Process

In order to avoid analytical complications we treat only Markov processes with a finite state space $S = \{1, \cdots, d\}$ [6]. A *probability* on S is sequence $(p_i; i = 1, \cdots, d)$ such that $p_i \geq 0$ and $\sum p_i = 1$. A *transition probability* is a matrix $P = (P_{i,j}; i, j = 1, \cdots, d)$, such that $P_{i,j} \geq 0$ and $\sum_i P_{i,j} = 1$ for all $j = 1, \cdots d$.

We discuss at first stationary Markov processes $X_n; n = 0, 1, \cdots$ in discrete time. It describes e.g. the zig-zagging of a particle between d different states. It is given by a probability π on S, called initial probability, and by a transition probability P. The process is determined by the probabilities

$$\mathbb{P}\{X_0 = i_0, X_1 = i_1, \cdots, X_n = i_n\} = P_{i_n, i_{n-1}} P_{i_{n-1}, i_{n-2}} \cdots P_{i_1, i_0} \pi_{i_0}$$

The probability of coming in one step from j to i is P_{ij}, the probability of coming in n steps from j to i is $(P^n)_{ij}$.

We assume now, that the Markov process is *transient*. In the simplest case there exists an element in S, called graveyard or absorbing element with the property, that the process stays in the graveyard, if it enters there once and that the process will end up in the graveyard with probability 1, starting from any point in S. Assume that the graveyard is the point d. Then P has the shape

$$P = \begin{pmatrix} Q & 0 \\ q & 1 \end{pmatrix} \tag{1}$$

That the process ends up in d with probability 1, implies

$$Q^n \to 0 \text{ for } n \to 0.$$

Hence

$$(1 - Q)(1 + Q^2 \cdots Q^n) = 1 - Q^n \to 1,$$

so $1 - Q$ is invertible and

$$\Pi = 1 + Q + Q^2 + \cdots = (1 - Q)^{-1}. \tag{2}$$

The matrix Π is the *potential* of the transient Markov process. The matrix element Π_{ij} is the mean number of times, the process stays in i starting from j.

We switch to another interpretation. Assume $d-1$ containers, each containing the amount of liquid $I_{n,i}$ at time n with $i = 1, \cdots, d-1; n = 0, 1, 2, \cdots$. Assume

$$I_{n+1,i} = \sum_{j=1}^{d-1} P_{ij} I_{j,n} + S_i.$$

So in any time unit an amount of liquid is transferred to the other containers and a source is furnishing additional liquid. In vector notation

$$I_{n+1} = QI_n + S. \qquad (3)$$

So

$$I_n = S + QS + \cdots + Q^{n-1}S + Q^n I_0 \to \Pi S.$$

The stationary solution is $I = \Pi S$. It satisfies the *potential equation*

$$I = PI + S. \qquad (4)$$

For continuous time $t \geq 0$ a Markov process $X_t, t \geq 0$ is given by the initial probability π and by the *rate matrix* $A = A_{ij}; i, j = 1, \cdots, d$ with the properties $A_{ij} \geq 0$ for $i \neq j$ and $\sum_i A_{ij} = 0$ for all j. Hence

$$A_j = -A_{jj} = \sum_{i: i \neq j} A_{ij} \geq 0.$$

The exponentials $P_t = \exp At$ are transition probabilities. The process X_t is described by

$$\mathbb{P}\{X_{t_0} = i_0, X_{t_1} = i_1, \cdots, X_{t_n} = i_n\}$$
$$= (P_{t_n - t_{n-1}})_{i_n, i_{n-1}} (P_{t_{n-1} - t_{n-2}})_{i_{n-1}, i_{n-2}} \cdots (P_{t_1 - t_0})_{i_1, i_0} \pi_{i_0}.$$

The matrix element $(P_t)_{ij}$ is the probability, that the process is in i after a time t starting in j. The process jumps back and forth between the points $1, \cdots, d$. It can be conveniently described in a differential form using conditional probability

$$\mathbb{P}\{X_{t+dt} = i | X_t = j\} = \begin{cases} 1 - A_j dt & \text{for } i = j \\ A_{ij} dt & \text{for } i \neq j. \end{cases} \qquad (5)$$

We assume again, that the process is transient in the sense discussed above. Then A has the shape

$$A = \begin{pmatrix} B & 0 \\ b & 0 \end{pmatrix}, \qquad (6)$$

with

$$e^{Bt} \to 0 \text{ for } t \to \infty.$$

As
$$B \int_0^\infty e^{Bt} dt = \int_0^\infty \frac{d}{dt} e^{Bt} dt = -1$$
the matrix B is invertible and
$$\Pi = \int_0^\infty e^{Bt} dt = -B^{-1}. \tag{7}$$

The matrix Π is the *potential* of the transient Markov process and Π_{ij} is mean time, the process spends in i starting from j.

We switch again to the problem with $d-1$ containers. Equation (3) is replaced by
$$\frac{dI(t)}{dt} = BI + S \tag{8}$$
and
$$I(t) \to I = \Pi S \text{ for } t \to \infty,$$
and
$$BI + S = 0. \tag{9}$$

This is the *potential equation* for the Markov process.

Of course the classical potential equation
$$\Delta u + \varrho = 0,$$
where Δ is the Laplacian is a potential equation of a Markov process in the sense discussed above. Here the state space is a domain in some $\mathbb{R}^d$ and the Markov process is Brownian motion.

3 The Radiative Transfer Equation is the Potential Equation of a Markov Process in Continuous Time

The state of a photon is given by its location $z \in G$, where G is ome domain in $\mathbb{R}^3$ and by its direction $n \in \mathbb{S}^2$, where $\mathbb{S}^2$ is the unit sphere in $\mathbb{R}^3$. We consider only photons with one fixed frequency and assume distances long against wave length. So a photon can be assumed to have a fixed location and a fixed direction. If the photon is absorbed or goes across the boundary of G, it goes to the grave yard Δ.

We define the state space for the Markov process X_t by
$$\Xi = (G \times \mathbb{S}^2) \cup \Delta. \tag{10}$$

So X_t is either in $G \times \mathbb{S}^2$ or in Δ. If it is once in Δ, it stays there for all times. In the other case, $X_t = (Z_t, N_t)$, where Z_t takes its values in G and N_t takes its values in the unit sphere. Assume $X_t = (z, n)$ with z in the interior of G and any direction n, then

$$X_{t+dt} = \begin{cases} (z+ndt, n) & \text{with probability } 1-(\sigma+\chi)dt \\ (z, n') & \text{with probability } \sigma dt d\omega(n')/4\pi \\ \Delta & \text{with probability } \chi dt. \end{cases} \quad (11)$$

The first line corresponds to the case, that the photon does not hit an atom during dt, the second line corresponds to the case, that the photon hits an atom and is scattered into the direction n', the third line corresponds to the case, that the photon hits an atom and is absorbed. We assume isotropic scattering. $d\omega(n)$ is the surface element on the unit sphere.

Assume a smooth test function φ on Ξ, such that $\varphi(\Delta) = 0$ and $\varphi(z,n)$ vanishes, if z is not contained in a compact subset of the interior of G. Denote by $p_t(z,n)$ the probability density of (Z_t, N_t) on $G \times \mathbb{S}^2$ with respect to the usual volume element $d\tau(z)$ and the surface element $d\omega(z)$, then

$$\mathbb{E}\varphi(X_{t+dt}) = \int\int p_{t+dt}(z,n)\varphi(z,n)d\tau(z)d\omega(n)$$
$$= \int\int \mathbb{E}\{\varphi(X_{t+dt})|X_t = (z,n)\}p_t(z,n)d\tau(z)d\omega(n)$$
$$= \int\int p_t(z,n)d\tau(z)d\omega(n)$$
$$(\varphi(z+ndt,n)(1-(\sigma+\chi)dt) + (\sigma dt)/(4\pi)\int d\omega(n')\varphi(z,n'))$$
$$= \int\int p_t(z,n)d\tau(z)d\omega(n)$$
$$(\varphi(z,n) + dt\,(n\cdot\nabla\varphi(z,n) - (\sigma+\chi)\varphi(z,n) + (\sigma dt)/(4\pi)\int d\omega(n')\varphi(z,n')))$$

By partial integration

$$\tfrac{1}{dt}\int\int d\tau(z)d\omega(n)(p_{t+dt}(z,n) - p_t(z,n))\varphi(z,n)$$
$$= \int\int d\tau(z)d\omega(n)\mathfrak{D}p_t(z,n))\varphi(z,n) ,$$

where $\mathfrak{D}$ is the linear integro-differential operator given by

$$\mathfrak{D}f(z,n) = -n\cdot\nabla f(z,n) - (\sigma+\chi)f(z,n) + \frac{\sigma}{4\pi}\int d\omega(n')f(z,n'). \quad (12)$$

So

$$\frac{\partial p_t(z,n)}{\partial t} = \mathfrak{D}p_t(z,n). \quad (13)$$

Obviously $\mathfrak{D}$ plays the role of the matrix B in equation (6). Switching to intensities and adding sources we obtain the corresponding potential equation

$$\mathfrak{D}I(z,n) + S(z,n) = 0. \quad (14)$$

Finally putting the nabla term to the left hand side

$$n\cdot\nabla I(z,n) = -(\sigma+\chi)I(z,n) + \sigma J(z) + S(z,n), \quad (15)$$

with

$$J(z) = 1/(4\pi)\int d\omega(n)I(z,n) \quad (16)$$

which is nothing other than the usual radiative transfer equation. The sources at the boundary are taken care by the boundary conditions

$$I(z,n) = I_0(z,n) \quad (17)$$

for *ingoing* directions n.

4 Milne's Equation is the Potential Equation of a Markov Process in Discrete Time

The radiative transfer equation (15) yields

$$d/(ds)I(z+sn,n) = n \cdot \nabla I(z+sn,n)$$
$$= -(\sigma+\chi)I(z+sn,n) + \sigma J(z+sn) + S(z+sn,n)$$

Integrate the differential equation

$$I(z+sn,n) = e^{-(\sigma+\chi)(s+s_0)}I(z-s_0 n, n)$$
$$+ \int_{-s_0}^{s} ds' e^{-(\sigma+\chi)(s-s')}(\sigma J(z+s'n) + S(z+s'n,n))$$

Assume that G is convex and bounded. If $(z,n) \in G \times \mathbb{S}^2$ then there exists a unique real number s_0, call it $s_0(z,n)$, such that $z - s_0 n$ is on the boundary. Replace in the last equation s by 0 and s_0 by $s_0(z,n)$ and obtain

$$I(z,n) = e^{-(\sigma+\chi)s_0}I(z - s_0(z,n)n, n)$$
$$+ \int_{-s_0(z,n)}^{0} ds' e^{-(\sigma+\chi)(s-s')}(\sigma J(z+s'n) + S(z+s'n,n)).$$

Integrating over the unit sphere one gets

$$J(z) = \int_G d\tau(z') \frac{1}{4\pi|z-z'|^2} e^{-(\sigma+\chi)|z-z'|} \sigma J(z') + \tilde{S}(z) \tag{18}$$

with

$$\tilde{S}(z) = \frac{1}{4\pi} \int d\omega(n) \left(e^{-(\sigma+\chi)s_0} I_0(z-s_0(z,n),n) \right.$$
$$\left. + \int_{-s_0(z,n)}^{0} ds' e^{(\sigma+\chi)s'} S(z+s'n,n) \right) \tag{19}$$

This is the Milne's equation [3].

Define a Markov process $Y_0, Y_1, \cdots$ with state space $G \cup \Delta$ in discrete time. The initial probability is arbitrary. The transition probability is given by the following statement: i) If $Y_n = \Delta$, then $Y_{n+1} = \Delta$. ii) If $Y_n = z_0 \in G$, then

$$Y_{n+1} \begin{cases} \in d\tau(z) & \text{with probability } \frac{\sigma}{4\pi|z-z_0|^2} e^{-(\sigma+\chi)|z-z_0|} d\tau(z) \text{ for } z \in G \\ = \Delta & \text{with probability } \int_G \frac{\chi}{4\pi|z-z_0|^2} e^{-(\sigma+\chi)|z-z_0|} d\tau(z) \\ & + \int_{G^c} \frac{\sigma+\chi}{4\pi|z-z_0|^2} e^{-(\sigma+\chi)|z-z_0|} d\tau(z). \end{cases} \tag{20}$$

Here G^c is the complement of G. The Milne's equation is the potential equation of this Markov process. The analogue of the matrix Q of equation (1) is the kernel $q(z, z_0)$ with $z, z_0 \in G$

$$q(z, z_0) = \frac{\sigma}{4\pi|z-z_0|^2} e^{-(\sigma+\chi)|z-z_0|}. \tag{21}$$

If G were the whole space and $\chi = 0$, then we had a random walk before us

$$Y_n = Y_0 + U_1 + \cdots U_n,$$

where the U_i are independent and identically distributed. Their distribution has the density

$$p_U(u) = \frac{\sigma}{4\pi u^2} \exp\left(-\sigma |u|\right). \tag{22}$$

References

1. S. Chandrasekhar, *Radiative Transfer*, Dover, 1960
2. W. Feller, *An Introduction to Probability Theory and its Applications II*, John Wiley, 1971
3. Fuehrer?
4. N. Wiener, E. Hopf, Sitzungsberichte der Preussischen, Akad. Phys.-Math. Klasse, p. 696 (1931)
5. E. Hopf, *Mathematical Problems of Radiative Equilibrium* (Cambridge, Math. Tract, No. 31) Cambridge, 1934
6. J.G. Kemenyi, J.L. Snell, *Finite Markov Chains*, Springer, 1976

An Approach to Neutrino Radiative Transfer in Supernova Simulations

Christian Y. Cardall[1,2]

[1] Physics Division, Oak Ridge National Laboratory, Oak Ridge, TN 37831-6354
   cardallcy@ornl.gov
[2] Department of Physics and Astronomy, University of Tennessee, Knoxville, TN 37996-1200

1 Core-collapse Supernovae: A Numerical Challenge

The approach to radiative transfer described in this contribution is being developed for use in simulations of core-collapse supernovae. These events are the deaths of stars more than about eight times as massive as the Sun, caused by the catastrophic collapse of the star's core. This collapse is triggered in part by electron capture on heavy nuclei, resulting in the emission of weakly-interacting particles called "neutrinos." Early in the collapse process these neutrinos escape freely, but eventually densities are sufficiently high that even neutrinos are trapped. Collapse is finally halted when central densities reach a few times the matter density of atomic nuclei. The newly-born neutron star—the compact object resulting from this process—is a hot thermal bath of dense nuclear matter, electrons, positrons, neutrinos, and antineutrinos. Neutrinos and antineutrinos, having the weakest interactions among the species present, are the most efficient means of cooling; their emission accounts for virtually all of the gravitational potential energy released during collapse.

Because neutrinos dominate the energetics of the supernova process, neutrino radiative transfer is a central feature of the core collapse phenomenon. In particular, energy transfer from neutrino radiation to infalling stellar matter may be crucial to the supernova explosion mechanism [1, 2]. In addition to energy, neutrinos exchange a quantity called "lepton number" with the fluid; this composition variable affects the fluid's equation of state, and also has a strong influence on the relative abundances of nuclear species synthesized in the supernova environment.

The radiative transfer of energy and lepton number is of particular interest in the semi-transparent region near the neutron star surface. This is a region of transition from the optically thick interior—where the diffusive neutrino field is nearly isotropic—to the optically thin exterior, where the neutrino radiation becomes strongly forward-peaked. Hence energy- and angle-dependent neutrino transport is key to accurate modeling of core-collapse supernovae.

At least three groups have published reports of spherically symmetric core-collapse supernova simulations with energy- and angle-dependent neutrino transport. Two of these groups—centered at the Max Planck Institute for Astrophysics [3] and the University of Arizona [4]—employed a method in which angle-integrated and angle-dependent neutrino transport equations are iterated to simultaneous convergence.[3] In contrast, the group centered at Oak Ridge National Laboratory [5] performed a direct solution of the angle-dependent transport equation, fully discretized in all variables in all terms. All these efforts were grid-based (with the first two groups employing "tangent rays" to discretize the momentum angle).

The iteration of angle-integrated and angle-dependent transport equations—which might be called an "iterated moment method"—works as follows. Zeroth and first angular moments of the neutrino transport equations are formed, with energy dependence retained. The (energy-dependent) zeroth and first angular moments of the neutrino distribution thus become variables to be evolved. The second and third angular moments also appear in these equations; to close the system at the first moment, the higher angular moments are expressed as numerical factors (the so-called "Eddington factors") multiplying the zeroth moment. The Eddington factors can be computed from the solution of the angle-dependent transport equation; this is solved with a simplified collision integral, which is expressed in terms of the zeroth and first angular moments of the neutrino distribution. In summary: The moment equations need Eddington factors for closure, which are obtained from the solution of the (simplified) angle-dependent transport equation; while the angle-dependent transport equation requires the zeroth and first moments for its simplified collision integral. This system is iterated to convergence.

The "direct method" of solving the transport equation is not subject to a structural limitation of the iterated moment method. In the direct method, all terms are discretized in all variables—time, space, energy, and angles. In particular the angle dependencies of neutrino scattering and pair production terms are fully represented, while in the iterated moment method only the $l = 0, 1$ terms in a Legendre expansion of these collision kernels are employed; these are the only terms in the expansion that can be constructed from the zeroth and first moments of the neutrino distribution. (In principle, the Eddington factors for the second and third moments could be used to get two additional terms in the Legendre expansions, but this has not been implemented to date.) In the supernova environment, these truncated angular expansions are might not be accurate representations of neutrino pair production and neutrino scattering by electrons and positrons—important processes in determining the spectra of neutrinos emerging from the nascent neutron star. (Another approximation of undemonstrated safety—simplification or neglect of angular aberration in the angle-dependent transport equation [3, 4]—seems less fundamental to the iterated moment method itself.)

---

[3] See also H.T. Janka's contribution to this volume.

While energy- and angle-dependent neutrino transport is an important advance even in spherical symmetry, the physics of stellar core collapse demands a spatially multidimensional treatment. There is ample theoretical and observational evidence for this conclusion (see for example [6]).

Inclusion of energy- and angle-dependent neutrino transport in spatially multidimensional simulations represents a significant computational challenge. Consider the "direct method" mentioned above, in which the neutrino distribution functions and all terms in the transport equation are discretized in all variables. Assume azimuthal symmetry; let the numbers of spatial zones in (r, θ) be $(256, 128)$, and the numbers of momentum bins in energy and angle variables $(\epsilon, \vartheta, \varphi)$ be $(64, 32, 16)$. An implicit time evolution algorithm requires the solution of a large linear system, in which the matrix represents the coupling between values of the neutrino distribution function at different points on this five-dimensional grid. The spatial coupling of points having the same momenta is very sparse, involving only nearest neighbors. But neutrino scattering and pair production terms in the collision integral involve dense and extended coupling in momentum space; this means that the matrix contains 256×128 dense blocks of size $(64 \times 32 \times 16)^2$. Storing all of these dense blocks in double precision would require a few 10^{14} bytes, beyond the capacity of today's terascale machines. Moreover, the elements of the dense blocks are expensive to compute, rendering unattractive an approach in which they are generated "on the fly" several times during each solution of the linear system. These considerations motivate the algorithm presented in this paper.

2 Formalism for Radiative Transfer

Some details of a radiative transfer formalism and its finite differenced representation will now be presented. Conservative formulations of radiative transfer—motivated by the importance of accurately tracking energy and lepton number transfer in the supernova environment—are discussed first. This is followed by a discussion of the finite differencing of the terms that do not depend on the velocity of the background fluid.

2.1 Conservative Formulations of Relativistic Radiative Transfer

A radiative transfer calculation involves a variable describing the distribution in phase space of the particles comprising the radiation.

The particle distributions we consider in some detail are the scalar distribution function f and the specific number density $\mathcal{N}$; we also mention the specific energy density $\mathcal{T}$. Each of these is taken to be a function of spacetime coordinates x^μ and three-momentum variables $u^{\hat{i}}$. (Greek and latin letters are spacetime and space indices respectively. Hatted indices indicate quantities measured in an orthonormal frame comoving with the fluid with which the particle species interact, and unadorned indices indicate components with respect to a global coordinate basis.) The momentum variables $u^{\hat{i}}$ arise from a change of variables (e.g. to momentum space spherical coordinates) from $p^{\hat{i}}$, the Cartesian spatial momentum components measured in a comoving orthonormal frame.

The scalar distribution function f gives the number of particles dN in an invariant spacetime 3-volume element dV and invariant momentum space volume element dP [7]:

$$dN = f(x, \mathbf{p})(-v \cdot p) \, dV \, dP. \tag{1}$$

The quantities x, v, and p are 4-vectors, and $\mathbf{p}$ is the spatial 3-vector portion of p. The unit 4-vector v is timelike, and defines the orientation of dV:

$$dV = \sqrt{-g} \epsilon_{\mu\nu\rho\sigma} v^\mu \, d_1 x^\nu \, d_2 x^\rho \, d_3 x^\sigma, \tag{2}$$

where g is the determinant of the metric tensor (taken to have signature $-+++$) and $\epsilon_{\mu\nu\rho\sigma}$ is the Levi-Civita alternating symbol ($\epsilon_{0123} = +1$). The momentum space volume element is

$$dP = \sqrt{-g} \epsilon_{ijk} \frac{d_1 p^i \, d_2 p^j \, d_3 p^k}{(-p_0)}$$
$$= \frac{1}{\epsilon} \left| \det \left(\frac{d\mathbf{p}}{d\mathbf{u}} \right) \right| du^1 \, du^2 \, du^3, \tag{3}$$

where specialization to momentum variables $u^{\hat{i}}$ in the comoving frame has been made in the second line, and

$$\epsilon \equiv p^{\hat{0}} = \sqrt{|\mathbf{p}|^2 + m^2} \tag{4}$$

is the particle energy measured in the comoving frame (m is the particle mass). The definition of f in eq. (1) makes a convenient connection to nonrelativistic definitions of the distribution function; for an equivalent but more geometric approach see Ref. [8]. The transport equation for electrically neutral particles satisfied by f is the Boltzmann equation [7, 8, 9, 10]:

$$p^{\hat{\mu}} \mathcal{L}^\mu{}_{\hat{\mu}} \frac{\partial f}{\partial x^\mu} - \Gamma^{\hat{j}}{}_{\hat{\mu}\hat{\nu}} p^{\hat{\mu}} p^{\hat{\nu}} \frac{\partial u^{\hat{i}}}{\partial p^{\hat{j}}} \frac{\partial f}{\partial u^{\hat{i}}} = \mathbb{C}[f]. \tag{5}$$

In this expression, $\mathcal{L}^\mu{}_{\hat{\mu}}$ is the transformation between the coordinate basis and the comoving orthonormal basis; it involves a transformation to an orthonormal "lab frame" basis followed by a Lorentz boost to the orthonormal comoving frame. The connection coefficients in the orthonormal comoving basis are

$$\Gamma^{\hat{\mu}}{}_{\hat{\nu}\hat{\rho}} = \mathcal{L}^{\hat{\mu}}{}_\mu \mathcal{L}^\nu{}_{\hat{\nu}} \mathcal{L}^\rho{}_{\hat{\rho}} \Gamma^\mu{}_{\nu\rho} + \mathcal{L}^{\hat{\mu}}{}_\mu \mathcal{L}^\rho{}_{\hat{\rho}} \frac{\partial \mathcal{L}^\mu{}_{\hat{\nu}}}{\partial x^\rho}. \tag{6}$$

Because the transformation $\mathcal{L}^\mu{}_{\hat{\mu}}$ contains a Lorentz boost, the term $\partial \mathcal{L}^\mu{}_{\hat{\rho}}/\partial x^\rho$ gives rise to Doppler shifts and angular aberrations associated with this transformation. The coordinate basis connection coefficients,

$$\Gamma^{\mu}{}_{\nu\rho} = \frac{1}{2}g^{\mu\sigma}\left(\frac{\partial g_{\sigma\nu}}{\partial x^{\rho}} + \frac{\partial g_{\sigma\rho}}{\partial x^{\nu}} - \frac{\partial g_{\nu\rho}}{\partial x^{\sigma}}\right), \tag{7}$$

where $g_{\mu\nu}$ are the metric components, give rise to energy shifts and angular aberrations associated with spacetime curvature and the use of curvilinear coordinates. The right-hand side of (5) is the invariant collision integral.

The Boltzmann equation can be put in a number-conservative form [10], which motivates the definition of the specific particle number density $\mathcal{N}$. Specifically, eq. (5) can be rewritten as

$$\frac{1}{\sqrt{-g}}\frac{\partial}{\partial x^{\mu}}\left(\sqrt{-g}\, p^{\hat{\mu}}\mathcal{L}^{\mu}{}_{\hat{\mu}} f\right) +$$
$$\epsilon\left|\det\left(\frac{d\mathbf{p}}{d\mathbf{u}}\right)\right|^{-1}\frac{\partial}{\partial u^{\hat{\imath}}}\left(-\Gamma^{\hat{\jmath}}{}_{\hat{\mu}\hat{\nu}}\frac{p^{\hat{\mu}}p^{\hat{\nu}}}{\epsilon}\left|\det\left(\frac{d\mathbf{p}}{d\mathbf{u}}\right)\right|\frac{\partial u^{\hat{\imath}}}{\partial p^{\hat{\jmath}}}f\right) = \mathbb{C}[f]. \tag{8}$$

This form is called "conservative" because the left hand side is expressed as divergences in spacetime and momentum space, so that volume integrals of these terms are transparently related to surface terms. In particular, it is obvious that the momentum space divergence vanishes upon integration over dP (given by eq. (3)), yielding the equation for particle number balance:

$$\frac{1}{\sqrt{-g}}\frac{\partial}{\partial x^{\mu}}\left(\sqrt{-g}\, N^{\mu}\right) = \int \mathbb{C}[f]\, dP, \tag{9}$$

where

$$N^{\mu} = \int \mathcal{L}^{\mu}{}_{\hat{\mu}} p^{\hat{\mu}} f\, dP \tag{10}$$

are the coordinate basis components of the particle number flux vector. This motivates the definition of the specific particle number density $\mathcal{N}$, given by

$$\mathcal{N} \equiv \frac{p^{\hat{\mu}}\mathcal{L}^{0}{}_{\hat{\mu}}}{\epsilon} f. \tag{11}$$

From eqs. (10) and (11), we see that the specific number density is the contribution of each *comoving frame* momentum bin to the *lab frame* number density:

$$N^{0} = \int \mathcal{N}\, \epsilon\, dP. \tag{12}$$

In terms of $\mathcal{N}$, eq. (8) can be rewritten as

$$\frac{1}{\sqrt{-g}}\frac{\partial}{\partial x^{\mu}}\left(\sqrt{-g}\, n^{\mu}\mathcal{N}\right) +$$
$$\left|\det\left(\frac{d\mathbf{p}}{d\mathbf{u}}\right)\right|^{-1}\frac{\partial}{\partial u^{\hat{\imath}}}\left(-\Gamma^{\hat{\jmath}}{}_{\hat{\mu}\hat{\nu}}\mathcal{L}^{\hat{\mu}}{}_{\mu}n^{\mu}p^{\hat{\nu}}\left|\det\left(\frac{d\mathbf{p}}{d\mathbf{u}}\right)\right|\frac{\partial u^{\hat{\imath}}}{\partial p^{\hat{\jmath}}}\mathcal{N}\right) = \frac{1}{\epsilon}\mathbb{C}, \tag{13}$$

where we have defined

$$n^\mu \equiv \left(1, \frac{p^{\hat{\mu}} \mathcal{L}^\mu{}_{\hat{\mu}}}{p^{\hat{\nu}} \mathcal{L}^0{}_{\hat{\nu}}}\right). \tag{14}$$

Note that n^μ is not a 4-vector.

Another reformulation of the Boltzmann equation is an energy-conservative form [10], which motivates the definition of the specific particle energy density $\mathcal{T}$. This reformulation is not detailed here; suffice it to say that its momentum integral gives a transparent connection to the divergence of the particle stress-energy tensor.

The number- and energy-conservative formulations—which respectively facilitate computation of lepton number and energy transfer to essentially machine accuracy—might be used in a couple of different ways. Both formulations could be solved separately, in order to nail down the transfer of both energy and lepton number. The values of the scalar distribution function implied by these two different distributions would then serve as a consistency check. Alternatively, only one of the formulations might be solved, with the analytic relationship between the two conservative formulations [10] being used to design finite difference expressions that provide consistency with the other formulation. This latter philosophy has been employed in the work of the group centered at Oak Ridge [5].

2.2 Finite Differencing of Selected Terms

Consider (13) in two spatial dimensions in spherical coordinates, with the assumption of a static (zero velocity) background and flat spacetime:

$$\frac{\partial \mathcal{N}}{\partial t} + \frac{\cos \vartheta}{r^2} \frac{\partial}{\partial r}\left(r^2 \mathcal{N}\right) + \frac{\sin \vartheta \cos \varphi}{r \sin \theta} \frac{\partial}{\partial \theta}(\sin \theta \mathcal{N}) - \frac{1}{r \sin \vartheta} \frac{\partial}{\partial \vartheta}\left(\sin^2 \vartheta \mathcal{N}\right) - \frac{\cot \theta}{r} \frac{\partial}{\partial \varphi}(\sin \vartheta \sin \varphi \mathcal{N}) = \frac{1}{\epsilon} \mathbb{C}. \tag{15}$$

In this equation, (r, θ) are the spatial radius and polar angle, (ϑ, φ) are momentum space angles, and ϵ is the particle energy.

There are two things to keep in mind in constructing a finite-differenced representation of (15). First, one can take advantage of the conservative form to make numerical "volume integrals" transparently related to numerical "surface integrals." Second, notice that for $\mathcal{N}$ spatially homogeneous, the second and fourth terms of (15) cancel, as do the third and fifth terms. The finite difference representation should respect this cancellation.

A conservative differencing of the spatial divergence in (15) is

$$\frac{1}{V_{i',j'}} (\cos \vartheta)_{\beta'} \left[(A_r \mathcal{N})_{i+1,j'} - (A_r \mathcal{N})_{i,j'}\right] + \frac{1}{V_{i',j'}} (\sin \vartheta)_{\beta'} (\cos \varphi)_{\gamma'} \left[(A_\theta \mathcal{N})_{i',j+1} - (A_\theta \mathcal{N})_{i',j}\right], \tag{16}$$

where the geometric factors are

An Approach to Neutrino Radiative Transfer in Supernova Simulations   33

$$V_{i',j'} = 2\pi (r_{i'})^2 \sin(\theta_{j'})(\Delta r)_{i'}(\Delta\theta)_{j'}, \tag{17}$$

$$(A_r)_{i,j'} = 2\pi (r_i)^2 \sin(\theta_{j'})(\Delta\theta)_{j'}, \tag{18}$$

$$(A_\theta)_{i',j} = 2\pi r_{i'} \sin(\theta_j)(\Delta r)_{i'}. \tag{19}$$

In the finite-difference expressions in this subsection, the subscripts (i, j) index spatial zones in the (r, θ) spatial coordinate directions, and subscripts (α, β, γ) index momentum bins in the $(\epsilon, \vartheta, \varphi)$ momentum space coordinate directions. Unprimed indices denote values evaluated on the surfaces ("edges") of spatial zones or momentum bins. The values of the coordinates $(r, \theta, \vartheta, \varphi)$ on the zone and bin edges are prescribed by the user. Primed indices denote values evaluated at zone or bin centers. Given the "edge values," the coordinates of the zone centers are taken to be

$$r_{i'} = \frac{1}{2}\left(r_i^3 + r_{i+1}^3\right)^{1/3}, \tag{20}$$

$$\theta_{j'} = \arccos\left[\frac{1}{2}(\cos(\theta_j) + \cos(\theta_{j+1}))\right], \tag{21}$$

and the zone widths are $(\Delta r)_{i'} = r_{i+1} - r_i$ and $(\Delta\theta)_{j'} = \theta_{j+1} - \theta_j$. The definitions of $(\cos\vartheta)_{\beta'}$, $(\sin\vartheta)_{\beta'}$, and $(\cos\varphi)_{\gamma'}$ in (16) will be given below. Finally, the values of $\mathcal{N}$ on zone surfaces in (16) are given by a particular linear interpolation of zone center values. This linear interpolation—which depends on the neutrino mean free paths—has the effect of shifting from "diamond" differencing in diffusive regimes to "upwind" (or "donor-cell") differencing in free-streaming regions; see [5] for details.

A conservative differencing of the momentum space divergence in (15) is

$$-\frac{1}{\mathcal{V}_{\alpha',\beta',\gamma'}}\left(\frac{1}{r}\right)_{i'}[(\mathcal{A}_\vartheta \mathcal{N})_{\beta+1,\gamma'} - (\mathcal{A}_\vartheta \mathcal{N})_{\beta,\gamma'}] -$$
$$\frac{1}{\mathcal{V}_{\alpha',\beta',\gamma'}}\frac{(\cot\theta)_{j'}}{r_{i'}}[(\mathcal{A}_\varphi \mathcal{N})_{\beta',\gamma+1} - (\mathcal{A}_\varphi \mathcal{N})_{\beta',\gamma}], \tag{22}$$

where the momentum space "geometric" factors are

$$\mathcal{V}_{\alpha',\beta',\gamma'} = (\epsilon_{\alpha'})^2 \sin(\vartheta_{\beta'})(\Delta\epsilon)_{\alpha'}(\Delta\vartheta)_{\beta'}(\Delta\varphi)_{\gamma'}, \tag{23}$$

$$(\mathcal{A}_\vartheta)_{\beta,\gamma'} = (\epsilon_{\alpha'})^2[\sin(\vartheta_\beta)]^2(\Delta\epsilon)_{\alpha'}(\Delta\varphi)_{\gamma'}, \tag{24}$$

$$(\mathcal{A}_\varphi)_{\beta',\gamma} = (\epsilon_{\alpha'})^2 \sin(\vartheta_{\beta'})\sin(\varphi_\gamma)(\Delta\epsilon)_{\alpha'}(\Delta\vartheta)_{\beta'}. \tag{25}$$

The bin center values $\epsilon_{\alpha'}$ and $\vartheta_{\beta'}$ are given just as $r_{i'}$ and $\theta_{j'}$ in (20) and (21). Also, the momentum bin widths are given by the difference of the bounding edge values, just as in the case of the spatial zones. The definitions of $(1/r)_{i'}$ and $(\cot\theta)_{j'}$ will be given below. The values of $\mathcal{N}$ on momentum bin edges are given by "upwind" or "donor-cell" differencing (see [5]).

Finally, we show the remaining definitions needed to ensure that the second and fourth terms in (15), as well as the third and fifth terms, cancel for $\mathcal{N}$ spatially homogeneous. Given the differencings defined above, this is accomplished with the following definitions:

$$(\sin\vartheta)_{\beta'} = \sin(\vartheta_{\beta'}), \tag{26}$$

$$(\cos\vartheta)_{\beta'} = \frac{\sin(\vartheta_{\beta+1})^2 - \sin(\vartheta_\beta)^2}{2\sin(\vartheta_{\beta'})(\Delta\vartheta)_{\beta'}}, \tag{27}$$

$$(\cos\varphi)_{\gamma'} = \frac{\sin(\varphi_{\gamma+1}) - \sin(\varphi_\gamma)}{(\Delta\varphi)_{\gamma'}}, \tag{28}$$

$$\left(\frac{1}{r}\right)_{i'} = \frac{(r_{i+1})^2 - (r_i)^2}{2(r_{i'})^2(\Delta r)_{i'}}, \tag{29}$$

$$(\cot\theta)_{j'} = \frac{\sin(\theta_{j+1}) - \sin(\theta_j)}{\sin(\theta_{j'})(\Delta\theta)_{j'}}. \tag{30}$$

3 Radiative Transfer Algorithm and Distributed-Memory Implementation

In this section classes of operators involved in radiation transport are identified, and a strategy for implementing them on distributed-memory computer architectures is described.

The equations of neutrino radiative transfer are the transport equation for whatever radiation particle distribution function is used, together with equations that describe lepton number and energy transfer (the latter are given by appropriate momentum integrals of the transport equation). The terms in these equations correspond to operators acting on the discretized distribution function and transfer quantities. This can be expressed as

$$F[y] = 0, \tag{31}$$

where y denotes the set of unknowns at a given time step: $N_{\text{species}} \times N_{\text{space}} \times N_{\text{momentum}}$ unknown values of the distribution functions of N_{species} neutrino species in N_{space} spatial zones and N_{momentum} momentum bins, and $2 \times N_{\text{space}}$ unknown values of energy and lepton number transferred to the fluid in each spatial zone. The total operator F has various pieces:

$$F = T + S + M + C. \tag{32}$$

The time derivative operator T relates unknowns at fixed position $\mathbf{x}$ and momentum $\mathbf{u}$ at different times t. The space derivative operator S is linear, and connects nearest neighbors in $\mathbf{x}$ at fixed $\mathbf{u}$ and t; similarly, the momentum derivative operator M is linear and connects nearest neighbors in $\mathbf{u}$ at fixed $\mathbf{x}$ and t. (The operators S and M are divergences in the conservative formulations.) In the case of astrophysical neutrino transport, the collision operator C is nonlinear due to neutrino-neutrino interactions and phase space blocking associated with the Pauli exclusion principle; and because of scattering and pair production and annihilation processes, it exhibits extensive, nonlocal coupling in **u** at fixed **x** and t.

The large disparity between hydrodynamic ($\sim 10^{-3}$ s) and neutrino interaction ($\sim 10^{-10}$ s) time scales in the collapsed core of the supernova environment calls for implicit evolution of the transfer equations. This means that in a time step in which the system is evolved from time t^n to t^{n+1}, the operators S, M, and C are evaluated at t^{n+1}. The discretized transfer equations are then

$$T(y^{n+1}, y^n) + S(y^{n+1}) + M(y^{n+1}) + C(y^{n+1}) = 0, \tag{33}$$

where we have used the notation $y^n = y(t^n)$ and suppressed the dependence on space and momentum variables. The dependence of T on the values of y at only two different times indicates that a method that is first order in time is being used (specifically, backward Euler).

Because of the nonlinearity of the collision operator C, (33) is a set of nonlinear algebraic equations for the discretized values of the distribution functions and transfer variables; this system is solved with a Newton-Raphson iteration procedure. Specifically, (31) is linearized:

$$J \cdot \Delta y = -F, \tag{34}$$

where $J \equiv \partial F / \partial y$ is the Jacobian matrix. In this linearized equation, J and $-F$ are evaluated at a guess $(y^{n+1})_{\text{guess}}$ for the value of the distribution function at the new time t^{n+1}. The solution Δy of this linear system provides a new guess $(y^{n+1})_{\text{new guess}} = (y^{n+1})_{\text{guess}} + \Delta y$. This procedure is iterated to convergence of y^{n+1} to the solution of (33).

To solve the linear problem at the heart of each Newton-Raphson iteration, a simple fixed-point method is employed. The basic idea is as follows. Start with a guess $(\Delta y)_{\text{guess}}$ for the solution of (34), and compute the residual r,

$$r = (-F) - J \cdot (\Delta y)_{\text{guess}}. \tag{35}$$

Given $(J^{-1})_{\text{approx}}$, an approximate inverse of J, compute the correction c,

$$c = (J^{-1})_{\text{approx}} \cdot r. \tag{36}$$

Why the name "correction"? Note that if one computes c_{exact} with the exact inverse J^{-1}, (36), (35), and (34) give

$$\begin{aligned} c_{\text{exact}} &= J^{-1} \cdot (-F) - J^{-1} \cdot J \cdot (\Delta y)_{\text{guess}} \\ &= \Delta y - (\Delta y)_{\text{guess}}, \end{aligned} \tag{37}$$

the difference between the exact solution Δy to eq. (34) and $(\Delta y)_{\text{guess}}$. When c is computed with an approximate inverse it does not provide the exact difference between Δy and $(\Delta y)_{\text{guess}}$, but it does provide a new (and hopefully improved) guess $(\Delta y)_{\text{new guess}} = (\Delta y)_{\text{guess}} + c$. This procedure is iterated to convergence of Δy to the solution of (34).

Details of the implementation of this iterative fixed-point method for the solution of the large linear system will now be given. In practice, the inverses of two different approximate Jacobian matrices are applied in succession in each iteration (this also requires the computation of two residuals in each iteration). The first approximate Jacobian is

$$J_{\text{momentum}} = J_T + J_M + J_C, \tag{38}$$

which consists of the contributions to the Jacobian from the operators T, M, and C in (33). As previously described, this combination of operators densely couples different momenta $\mathbf{u}$ at fixed spatial position $\mathbf{x}$; hence J_{momentum} consists of N_{space} independent dense blocks—one for each spatial zone—and $(J_{\text{momentum}})^{-1}$ consists of individual inverses of these dense blocks. With the spatial grid partitioned among the many processors of a distributed-memory computer, the inversion of these separate blocks is trivially parallelized. The second approximate Jacobian is

$$J_{\text{space}} = J_T + J_S, \tag{39}$$

arising from the operators T, S in (33). By reasoning similar to above, J_{space} can be conceptualized as N_{momentum} independent matrices, but this time with sparse coupling, because the derivatives in S only require nearest neighbors in space. Having chosen to partition the spatial grid, parallel solution of these independent "spatial matrices" requires an "all-to-all" communication to give each processor all the spatial data for its share of momentum bins; but the fact the matrices are sparse makes this communication manageable.

In addition to its simple structure, this fixed-point method has an important practical advantage over other iterative linear solver algorithms. As mentioned at the end of the first section, in spatially multidimensional problems simultaneous storage of all the dense blocks is impractical. The fixed-point algorithm outlined above can be structured so that each processor can construct a few dense blocks at a time, use them in all steps required in a given iteration, and discard them. In contrast, other linear solver algorithms seem to require dense blocks to be discarded and rebuilt multiple times in each iteration.

A code that implements the algorithm described above—written in Fortran 90, and using the MPI library for message passing—is being developed and tested at Oak Ridge National Laboratory, for eventual use in core-collapse supernova simulations. The implementation has been tested in both one and two spatial dimensions on a "homogeneous sphere" problem which has an analytic solution, with good results. The test has also been generalized to an "inhomogeneous sphere" problem, with emissivity and opacity varying in spatial polar angle. With regard to performance, it is found that inversion of the dense blocks dominates the computation; communication costs are not excessive. Because dense matrix solvers (e.g. the LAPACK library) are typically highly optimized, the dominance of the computation by dense blocks ensures that computational resources are used efficiently.

References

1. Colgate, S.A., White, R.H.: The Hydrodynamic Behavior of Supernovae Explosions. Astrophys. J., **143**, 626–681 (1966)
2. Bethe, H.A., Wilson, J.R.: Revival of a stalled supernova shock by neutrino heating. Astrophys. J., **295**, 14–23 (1985)
3. Rampp, M., Janka, H.T.: Radiation hydrodynamics with neutrinos. Variable Eddington factor method for core-collapse supernova simulations. Astron. Astrophys., **396**, 361–392 (2002)
4. Thompson, T.A., Burrows, A., Pinto, P.A.: Shock breakout in core-collapse supernovae and its neutrino signature. Astrophys. J., **592**, 434–456 (2003)
5. Liebendörfer, M., Messer, O.E.B., Mezzacappa, A., Bruenn, S.W., Cardall, C.Y., Thielemann, F.K.: A finite difference representation of neutrino radiation hydrodynamics in spherically symmetric general relativistic spacetime. Astrophys. J. Supp. Ser., **150**, 263–316 (2004)
6. Janka, H.T., Mueller, E.: Neutrino heating, convection, and the mechanism of Type-II supernova explosions. Astron. Astrophys., **306**, 167–198 (1996)
7. Lindquist, R.W.: Relativistic transport theory. Ann. Phys. (NY), **37**, 487–518 (1966)
8. Ehlers, J.: General relativity and kinetic theory. In Sachs, R.K. (ed), Proceedings of the International School of Physics "Enrico Fermi" Course XLVII: General Relativity and Cosmology, New York, Academic Press (1971)
9. Mezzacappa, A., Matzner, R.A.: Computer simulation of time-dependent, spherically symmetric spacetimes containing radiating fluids - Formalism and code tests. Astrophys. J., **343**, 853–873 (1989)
10. Cardall, C.Y., Mezzacappa, A.: Conservative formulations of general relativistic kinetic theory. Phys. Rev. D, **68**, 023006 1–26 (2003)

A Finite Element Method for the Even-Parity Radiative Transfer Equation Using the P_N Approximation

Stephen Wright, Simon Arridge, and Martin Schweiger

Department of Computer Science, University College London, Gower Street, London, WC1E 6BT

1 Introduction

The concept of using optical radiation to penetrate highly scattering media, combined with image reconstruction methods to recover optical parameters inside the media, has been a recurrent idea for over a century. However it has received great attention in the last decade due to advances both in measurement technology and in theoretical and practical understanding of the nature of the image reconstruction problem. This field has come to be known as *Diffuse Optical Tomography* (DOT); for recent reviews see [1–5].

The term "Diffuse" is employed since the usual conditions being investigated are where the medium is so highly scattering that its propagation is nearly completely described by a Diffusion Approximation (DA). However, it is well known that under certain conditions, the DA is no longer valid. In particular the presence of non-scattering (void) regions, such as occur in the Cerebro-Spinal Fluid (CSF) filled ventricles in the brain, represent a situation for which the DA is clearly inadequate. Under these circumstances, more advanced methods are required [6–9].

A general model of light transport in scattering media, but one that ignores polarisation and coherence effects, is the Boltzmann Equation. This equation has been extensively studied in the field of Neutron Transport [10–14] and in Radiation transfer [15,16] where it is known as the Radiative Transfer Equation (RTE). In this paper, we describe the RTE in its second order form, and discuss the development of a finite element method for its solution. Related methods are described in [17–19].

2 The Radiative Transfer Equation

We will discuss only the *single-group* Radiative Transfer Equation which in the steady state is written

$$\left(\hat{\mathbf{s}}\cdot\nabla + \mu_{\text{tr}}(\mathbf{r}) + \frac{i\omega}{c}\right)\phi(\mathbf{r},\hat{\mathbf{s}};\omega) = \mu_{\text{s}}(\mathbf{r})\int_{S^2}\Theta(\hat{\mathbf{s}},\hat{\mathbf{s}}')\phi(\mathbf{r},\hat{\mathbf{s}}';\omega)\,d\hat{\mathbf{s}}' + q(\mathbf{r},\hat{\mathbf{s}};\omega) \quad (1)$$

Here $\mu_{\text{tr}}(\mathbf{r}) = \mu_{\text{s}}(\mathbf{r}) + \mu_{\text{a}}(\mathbf{r})$ (units of inverse length) is the attenuation coefficient at position $\mathbf{r}$, with $\mu_{\text{s}}(\mathbf{r})$ the scattering coefficient and $\mu_{\text{a}}(\mathbf{r})$ the absorption coefficient. $\phi(\mathbf{r},\hat{\mathbf{s}};\omega)$ (units of inverse length cubed per steradian) is the number of photons per unit volume at position $\mathbf{r}$ with velocity in angular direction $\hat{\mathbf{s}}$, with $q(\mathbf{r},\hat{\mathbf{s}};\omega)$ the number of source photons, and ω the modulation frequency. $\Theta(\hat{\mathbf{s}},\hat{\mathbf{s}}')$ is the normalised phase function representing the probability of scattering from direction $\hat{\mathbf{s}}'$ to direction $\hat{\mathbf{s}}$.

In general, the phase function depends on the absolute angle $\hat{\mathbf{s}}$ and leads to anisotropic effects [20], but in this paper we will make the usual assumption of directional independent scattering $\Theta(\hat{\mathbf{s}},\hat{\mathbf{s}}') \equiv \Theta(\hat{\mathbf{s}}\cdot\hat{\mathbf{s}}') = \Theta(\cos\tau)$, whereupon the integral operator can be interpreted as a convolution on S^2 and defined

$$\mathcal{S}[\phi](\hat{\mathbf{s}}) := \int_{S^2}\Theta(\hat{\mathbf{s}}\cdot\hat{\mathbf{s}}')\phi(\hat{\mathbf{s}}')\,d\hat{\mathbf{s}}' = \Theta\odot\phi \quad (2)$$

and we define the complex attenuation coefficient

$$\tilde{\mu}_{\text{tr}}(\omega) := \mu_{\text{tr}} + \frac{i\omega}{c} = \mu_{\text{a}} + \mu_{\text{s}} + \frac{i\omega}{c} \quad (3)$$

and the combined attenuation and inscatter operator

$$\mathcal{C} := \tilde{\mu}_{\text{tr}}(\omega) - \mu_{\text{s}}\mathcal{S} = \mu_{\text{a}} + \frac{i\omega}{c} + \mu_{\text{s}}(\mathcal{I} - \mathcal{S}) \quad (4)$$

This allows us to write (1) as

$$\hat{\mathbf{s}}\cdot\nabla\phi(\mathbf{r},\hat{\mathbf{s}};\omega) + \mathcal{C}\phi(\mathbf{r},\hat{\mathbf{s}};\omega) = q(\mathbf{r},\hat{\mathbf{s}};\omega) \quad (5)$$

2.1 Even Parity Transport Equation

For the use of numerical methods based on variational principles, it is convenient to work with a second-order, self-adjoint operator. To derive the even-parity Radiative Transfer Equation, we define even parity radiance and source terms

$$\phi^{\pm}(\hat{\mathbf{s}}) := \frac{1}{2}\left(\phi(\hat{\mathbf{s}}) \pm \phi(-\hat{\mathbf{s}})\right), \quad q^{\pm}(\hat{\mathbf{s}}) := \frac{1}{2}\left(q(\hat{\mathbf{s}}) \pm q(-\hat{\mathbf{s}})\right).$$

We may also split the kernel of the convolution operator

$$\Theta^{\pm}(\cos\tau) = \frac{1}{2}\left(\Theta(\cos\tau) \pm \Theta(-\cos\tau)\right)$$

Then (2) becomes

$$\mathcal{S}[\phi] = \mathcal{S}^{+}[\phi^{+}] + \mathcal{S}^{-}[\phi^{-}] \quad (6)$$

where we have defined

$$\mathcal{S}^\pm[\phi^\pm](\hat{\mathbf{s}}) := \int_{S^2} \Theta^\pm(\hat{\mathbf{s}}\cdot\hat{\mathbf{s}}')\phi^\pm(\hat{\mathbf{s}}')\,\mathrm{d}\hat{\mathbf{s}}' = \Theta^\pm \odot \phi^\pm. \tag{7}$$

Finally we can define

$$\mathcal{C}^\pm := \tilde{\mu}_{\mathrm{tr}}(\omega) - \mu_s \mathcal{S}^\pm = \mu_a + \frac{i\omega}{c} + \mu_s\left(\mathcal{I} - \mathcal{S}^\pm\right). \tag{8}$$

The derivation of the even parity transport equation is now straightforward. We write

$$\hat{\mathbf{s}}\cdot\nabla\left(\phi^+(\hat{\mathbf{s}}) + \phi^-(\hat{\mathbf{s}})\right) + \mathcal{C}^+\phi^+(\hat{\mathbf{s}}) + \mathcal{C}^-\phi^-(\hat{\mathbf{s}}) = q^+(\hat{\mathbf{s}}) + q^-(\hat{\mathbf{s}}). \tag{9}$$

Now substituting $\hat{\mathbf{s}} \to -\hat{\mathbf{s}}$ we have

$$-\hat{\mathbf{s}}\cdot\nabla\left(\phi^+(\hat{\mathbf{s}}) - \phi^-(\hat{\mathbf{s}})\right) + \mathcal{C}^+\phi^+(\hat{\mathbf{s}}) - \mathcal{C}^-\phi^-(\hat{\mathbf{s}}) = q^+(\hat{\mathbf{s}}) - q^-(\hat{\mathbf{s}}) \tag{10}$$

Adding and subtracting (9) and (10) we get

$$\hat{\mathbf{s}}\cdot\nabla\phi^-(\hat{\mathbf{s}}) + \mathcal{C}^+\phi^+(\hat{\mathbf{s}}) = q^+(\hat{\mathbf{s}}), \tag{11}$$
$$\hat{\mathbf{s}}\cdot\nabla\phi^+(\hat{\mathbf{s}}) + \mathcal{C}^-\phi^-(\hat{\mathbf{s}}) = q^-(\hat{\mathbf{s}}). \tag{12}$$

We use (12) to define the odd parity radiance in terms of the even parity radiance as

$$\phi^-(\hat{\mathbf{s}}) = \mathcal{D}\left[q^-(\hat{\mathbf{s}}) - \hat{\mathbf{s}}\cdot\nabla\phi^+(\hat{\mathbf{s}})\right] \tag{13}$$

where the generalised diffusion operator is defined $\mathcal{D} := (\mathcal{C}^-)^{-1}$. Substituting (13) into (11) we arrive at

$$\left(\mathcal{C}^+ - \hat{\mathbf{s}}\cdot\nabla(\mathcal{D}\hat{\mathbf{s}}\cdot\nabla)\right)\phi^+(\mathbf{r},\hat{\mathbf{s}};\omega) = q^+(\mathbf{r},\hat{\mathbf{s}};\omega) - \hat{\mathbf{s}}\cdot\nabla\left(\mathcal{D}q^-(\mathbf{r},\hat{\mathbf{s}};\omega)\right) \tag{14}$$

which is the even parity Radiative Transfer Equation.

2.2 The Weak Formulation

In the weak formulation of the even parity Radiative Transfer Equation, the (even-parity) radiance is represented in a finite-dimensional space χ^{h+}

$$\phi^+(\mathbf{r},\hat{\mathbf{s}}) \simeq \phi^{h+}(\mathbf{r},\hat{\mathbf{s}}) = \sum_{k=1}^K \phi_k^+ f_k(\mathbf{r},\hat{\mathbf{s}}) \tag{15}$$

where $\{f_k; k = 1,\ldots,K\}$ are a set of basis functions for χ^{h+}. Let us assume that the basis is developed separately for the spatial and angular terms

$$f_k(\mathbf{r},\hat{\mathbf{s}}) = u_i(\mathbf{r})\theta_j(\hat{\mathbf{s}}) \quad i = 1,\ldots D\,; j = 1,\ldots S\,; K = D \times S \tag{16}$$

Since we require even parity for ϕ^{h+} it is natural to choose θ_j to have even parity. The space χ^{h+} is equipped with a norm

$$\langle \psi, \phi \rangle := \int_{S^2} \int_{\Omega} \psi(\mathbf{r}, \hat{\mathbf{s}}) \phi(\mathbf{r}, \hat{\mathbf{s}}) \, d\mathbf{r} \, d\hat{\mathbf{s}} \tag{17}$$

Since ϕ^{h+} is an approximation to ϕ^+ it does not satisfy (14) exactly, but rather

$$\left(\mathcal{C}^+ - \hat{\mathbf{s}} \cdot \nabla \left(\mathcal{D}\hat{\mathbf{s}} \cdot \nabla \right) \right) \phi^{h+}(\mathbf{r}, \hat{\mathbf{s}}; \omega) - q^{h+}(\mathbf{r}, \hat{\mathbf{s}}; \omega) + \eta^{h+}(\mathbf{r}, \hat{\mathbf{s}}; \omega) = e(\mathbf{r}, \hat{\mathbf{s}}; \omega) \tag{18}$$

where q^{h+} and η^{h+} are the projection into χ^{h+} of q^+ and $\hat{\mathbf{s}} \cdot \nabla \left(\mathcal{D}q^-(\mathbf{r}, \hat{\mathbf{s}}; \omega) \right)$ respectively.

The principle of the weak (Galerkin) approximation is that the error term $e(\mathbf{r}, \hat{\mathbf{s}}; \omega)$ be orthogonal to the space χ^{h+}, i.e. that

$$\langle f_k, e \rangle = 0 \quad \forall k = 1, \ldots, K \tag{19}$$

which leads to a discrete matrix equation for the K unknowns ϕ_k^+

$$\mathsf{M}\phi^+ = \mathbf{q}^+ \tag{20}$$

3 Boundary Conditions

3.1 Vacuum Boundary Conditions

We are considering the Radiative Transfer Equation in domain Ω with boundary $\partial \Omega$, with outward directed normal $\hat{\nu}$. The natural boundary condition for the Radiative Transfer Equation is that there is no incoming energy flux crossing the boundary

$$\hat{\nu} \cdot \hat{\mathbf{s}} \phi(\mathbf{r}, \hat{\mathbf{s}}; \omega) = 0 \quad \mathbf{r} \in \partial \Omega, \quad \forall \hat{\mathbf{s}} \cdot \hat{\nu} < 0. \tag{21}$$

Now consider the weak version of (21):

$$\int_{\hat{\mathbf{s}} \cdot \hat{\nu} < 0} (\hat{\nu} \cdot \hat{\mathbf{s}}) \theta_j(\hat{\mathbf{s}}) \phi(\mathbf{r}, \hat{\mathbf{s}}; \omega) d\hat{\mathbf{s}} = 0 \quad j = 1, \ldots S. \tag{22}$$

Let us assume that $q = 0|_{\partial \Omega}$ then we have

$$\phi = \phi^+ + \phi^- = \phi^+ - \mathcal{D}\hat{\mathbf{s}} \cdot \nabla \phi^+,$$

so that (22) becomes

$$\int_{\hat{\mathbf{s}} \cdot \hat{\nu} < 0} (\hat{\nu} \cdot \hat{\mathbf{s}}) \theta_j(\hat{\mathbf{s}}) \phi^+(\mathbf{r}, \hat{\mathbf{s}}; \omega) d\hat{\mathbf{s}} = \int_{\hat{\mathbf{s}} \cdot \hat{\nu} < 0} (\hat{\nu} \cdot \hat{\mathbf{s}}) \theta_j(\hat{\mathbf{s}}) \mathcal{D}\hat{\mathbf{s}} \cdot \nabla \phi^+(\mathbf{r}, \hat{\mathbf{s}}; \omega) d\hat{\mathbf{s}} \quad j = 1, \ldots S. \tag{23}$$

Note that the right hand side of (23) is even and so we can state

$$\int_{\hat{\mathbf{s}} \cdot \hat{\nu} < 0} (\hat{\nu} \cdot \hat{\mathbf{s}}) \theta_j(\hat{\mathbf{s}}) \mathcal{D}\hat{\mathbf{s}} \cdot \nabla \phi^+(\mathbf{r}, \hat{\mathbf{s}}; \omega) d\hat{\mathbf{s}} = \frac{1}{2} \int_{S^2} (\hat{\nu} \cdot \hat{\mathbf{s}}) \theta_j(\hat{\mathbf{s}}) \mathcal{D}\hat{\mathbf{s}} \cdot \nabla \phi^+(\mathbf{r}, \hat{\mathbf{s}}; \omega) \, d\hat{\mathbf{s}}. \tag{24}$$

3.2 A Generalised Divergence Theorem

Going back to the Galerkin formulation, we can write (19) as

$$\langle u_i \theta_j, (\mathcal{C}^+ - \hat{\mathbf{s}} \cdot \nabla \mathcal{D} \hat{\mathbf{s}} \cdot \nabla) \phi^{h+} \rangle = \langle u_i \theta_j, q^+ \rangle \tag{25}$$

Applying the Divergence Theorem ($\int_\Omega \hat{\mathbf{s}} \cdot \nabla g = \int_{\partial\Omega} \hat{\mathbf{s}} \cdot \hat{\nu} g$) we get

$$\langle u_i \theta_j, \mathcal{C}^+ \phi^{h+} \rangle + \langle \hat{\mathbf{s}} \cdot \nabla u_i \theta_j, \mathcal{D} \hat{\mathbf{s}} \cdot \nabla \phi^{h+} \rangle +$$
$$\int_{S^2} (\hat{\nu} \cdot \hat{\mathbf{s}}) \theta_j(\hat{\mathbf{s}}) \int_{\partial\Omega} u_i(\mathbf{r}) \mathcal{D} \hat{\mathbf{s}} \cdot \nabla \phi^+(\mathbf{r}, \hat{\mathbf{s}}; \omega) \mathrm{d}\mathbf{r}\, \mathrm{d}\hat{\mathbf{s}} = \langle u_i \theta_j, q^+ \rangle , \tag{26}$$

and making use of (23) and (24) we get

$$\langle u_i \theta_j, \mathcal{C}^+ \phi^{h+} \rangle + \langle \hat{\mathbf{s}} \cdot \nabla u_i \theta_j, \mathcal{D} \hat{\mathbf{s}} \cdot \nabla \phi^{h+} \rangle +$$
$$\int_{S^2} |\hat{\nu} \cdot \hat{\mathbf{s}}| \theta_j(\hat{\mathbf{s}}) \int_{\partial\Omega} u_i(\mathbf{r}) \phi^+(\mathbf{r}, \hat{\mathbf{s}}; \omega) \mathrm{d}\mathbf{r}\, \mathrm{d}\hat{\mathbf{s}} = \langle u_i \theta_j, q^+ \rangle . \tag{27}$$

4 P_N Approximation

In principle we may use any basis functions $\theta_j(\hat{\mathbf{s}})$ for the angular variable. However if we use spherical harmonics we obtain the so-called P_N approximations. We express the quantities in (1) as

$$\phi(\mathbf{r}, \hat{\mathbf{s}}; \omega) = \sum_{l}^{\infty} \sum_{m=-l}^{l} \left(\frac{2l+1}{4\pi} \right)^{\frac{1}{2}} \phi_{l,m}(\mathbf{r}; \omega) Y_{l,m}(\hat{\mathbf{s}}) \tag{28}$$

$$q(\mathbf{r}, \hat{\mathbf{s}}; \omega) = \sum_{l}^{\infty} \sum_{m=-l}^{l} \left(\frac{2l+1}{4\pi} \right)^{\frac{1}{2}} q_{l,m}(\mathbf{r}; \omega) Y_{l,m}(\hat{\mathbf{s}}) \tag{29}$$

where $Y_{l,m}(\hat{\mathbf{s}})$ is a spherical harmonic of order l degree m, and the normalisation factor $((2l+1)/4\pi)^{1/2}$ is introduced for convenience. If the phase function is assumed to be independent of the explicit angle $\hat{\mathbf{s}}$ and is written $\Theta(\hat{\mathbf{s}} \cdot \hat{\mathbf{s}}')$ then

$$\Theta(\hat{\mathbf{s}} \cdot \hat{\mathbf{s}}') = \sum_{l}^{\infty} \sum_{m=-l}^{l} \Theta_l \overline{Y}_{l,m}(\hat{\mathbf{s}}') Y_{l,m}(\hat{\mathbf{s}})$$

and (1) can be expressed as an infinite set of coupled first order equations. When these are truncated by assuming $\phi_{l,m} = 0; l > N$ for some N the result is a set of $(N+1)^2$ (in 3D) first order equations known as the P_N approximation [4, 10, 12].

The $\mathcal{P}_N$ operator has a special form

$$\mathcal{P}_{\mathrm{N}} = \begin{bmatrix} \mathcal{C}_0 & \mathcal{A}_0 & 0 & 0 & \cdots \\ \mathcal{A}_0^{\mathrm{T}} & 3\mathcal{C}_1 & 3\mathcal{A}_1 & 0 & \cdots \\ 0 & 3\mathcal{A}_1^{\mathrm{T}} & 5\mathcal{C}_2 & 5\mathcal{A}_2 & \cdots \\ \vdots & \vdots & \vdots & \vdots & \vdots \end{bmatrix} \tag{30}$$

The operator $\mathcal{C}_l$ is diagonal for angularly independent scattering and is given by

$$\mathcal{C}_l = \left(\mu_{\mathrm{a}} + (1 - \Theta_l)\mu_{\mathrm{s}} + \frac{i\omega}{c}\right)\mathcal{I}_l$$

where $\mathcal{I}_l$ is the $(2l+1) \times (2l+1)$ identity operator. The operator $\mathcal{A}_l$ is a generalisation of the divergence operator and has the following form

$$\mathcal{A}_l = \begin{bmatrix} \alpha_{l,-l}\mathcal{D}_\xi^+ & \beta_{l,-l}\frac{\partial}{\partial z} & \gamma_{l,-l}\mathcal{D}_\xi^- & 0 & \cdots \\ 0 & \alpha_{l,1-l}\mathcal{D}_\xi^+ & \beta_{l,1-l}\frac{\partial}{\partial z} & \gamma_{l,1-l}\mathcal{D}_\xi^- & \cdots \\ \vdots & \vdots & \vdots & \vdots & \vdots \\ \cdots & 0 & \alpha_{l,l}\mathcal{D}_\xi^+ & \beta_{l,l}\frac{\partial}{\partial z} & \gamma_{l,l}\mathcal{D}_\xi^- \end{bmatrix} \tag{31}$$

where

$$\alpha_{l,m} = \frac{\sqrt{(l-m+2)(l-m+1)}}{2l+1} \quad \beta_{l,m} = \frac{\sqrt{(l+1-m)(l+1+m)}}{2l+1}$$
$$\gamma_{l,m} = \frac{\sqrt{(l+m+1)(l+m+2)}}{2l+1} \quad \mathcal{D}_\xi^+ = \frac{-1}{2}\left(\frac{\partial}{\partial x} + i\frac{\partial}{\partial y}\right) \quad \mathcal{D}_\xi^- = \frac{-1}{2}\left(\frac{\partial}{\partial x} - i\frac{\partial}{\partial y}\right)$$

Applying the representation (30) to the even-parity form of the Radiative Transfer Equation leads to the system

$$\begin{bmatrix} \mathcal{P}_{0,0}^+ & \mathcal{P}_{0,2}^+ & 0 & 0 & \cdots \\ \mathcal{P}_{2,0}^+ & \mathcal{P}_{2,2}^+ & \mathcal{P}_{2,4}^+ & 0 & \cdots \\ 0 & \mathcal{P}_{4,2}^+ & \mathcal{P}_{4,4}^+ & \mathcal{P}_{4,6}^+ & \cdots \\ \vdots & \vdots & \vdots & \vdots & \vdots \end{bmatrix} \begin{bmatrix} \phi_{0,0}^+ \\ \phi_{2,m\in[-2,2]}^+ \\ \phi_{4,m\in[-4,4]}^+ \\ \vdots \end{bmatrix} = \begin{bmatrix} q_{0,0}^+ \\ q_{2,m\in[-2,2]}^+ \\ q_{4,m\in[-4,4]}^+ \\ \vdots \end{bmatrix} \tag{32}$$

which we write as

$$\mathcal{P}_{\mathrm{N}}^+ \phi^+ = \mathbf{q}^+$$

where

$$\mathcal{P}_{0,0}^+ = \mathcal{C}_0 - \tfrac{1}{3}\mathcal{A}_0\mathcal{C}_1^{-1}\mathcal{A}_0^{\mathrm{T}} \qquad \mathcal{P}_{0,2}^+ = -\mathcal{A}_0\mathcal{C}_1^{-1}\mathcal{A}_1$$
$$\mathcal{P}_{2,0}^+ = -\mathcal{A}_1^{\mathrm{T}}\mathcal{C}_1^{-1}\mathcal{A}_0^{\mathrm{T}} \qquad \mathcal{P}_{2,2}^+ = 5\mathcal{C}_2 - 3\mathcal{A}_1^{\mathrm{T}}\mathcal{C}_1^{-1}\mathcal{A}_1 - \tfrac{25}{7}\mathcal{A}_2\mathcal{C}_3^{-1}\mathcal{A}_2^{\mathrm{T}}$$
$$\ldots$$

5 Implementation

The Finite Element Method implementation of (32) is obtained by specifying that the domain Ω is divided into P elements, joined at D vertex nodes and the basis functions $\{u_i(\mathbf{r}); i = 1\ldots D\}$ are chosen to have limited support. The

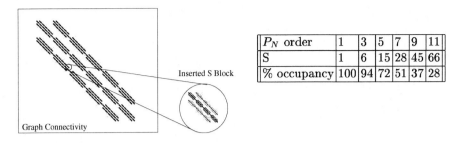

Fig. 1. Block structure of the FEM system matrix

problem of solving for ϕ^{h+} becomes one of sparse matrix inversion of (20), for which standard methods are readily available.

Because of the separability of the spatial and angular basis functions, the matrix M in (20) has the structure of a $D \times D$ graph representing the connectivity of the FEM mesh. For the general P_N equations each node that is not on the boundary needs to be expanded into a $S \times S$ block whose structure has the block tridiagonal form shown in (32); this block is also sparse, with the proportion of non-zeroes decreasing with increasing P_N order, see Figure 1.

Whereas the integration of products of spherical harmonic functions on the sphere is given analytically, the implementation of the boundary conditions calls for the integration on a half-sphere of the surface integral in (27). For this a Lebedev-Skorokhodov quadrature [21] is employed.

6 Results

As a test example we considered a cubic domain of dimension $20 \times 20 \times 20$ mm, represented as a mesh of voxel elements of size 0 mm^3. The mesh contained $D = 21^3$ nodes and the graph had 226981 non-zeros (0.26% occupancy). A source was placed at position (0,0,-8). Both an isotropic source and a directed source were considered. The domain had a uniform attenuation of $\mu_a = 0.01$ mm^{-1} and two cases were considered: a highly scattering case with $\mu_s = 10$ mm^{-1} and a low scattering case with $\mu_s = 0.1$ mm^{-1}. The former represents a case where the Diffusion Approximation can be expected to be quite accurate, whereas the latter is one where it can be expected to fail. In all examples an isotropic scattering phase function was used: $\Theta = \frac{1}{4\pi}$ and no harmonic modulation of the source was employed, i.e. $\omega = 0$.

Figure 2, row 1 shows the photon density for $\mu_s = 10.0$ mm^{-1} with a source that is isotropic in the angular variable. Here the P_1 (Diffusion Approximation) and the higher order P_7 solutions produce results that are indistinguishable. This is to be expected as within this regime the Radiative Transfer Equation can be shown to be approximated by the Diffusion Approximation.

46   Stephen Wright et al.

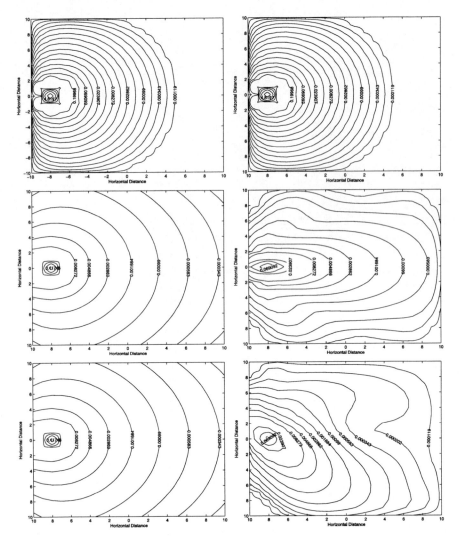

Fig. 2. The photon density fields $\phi^+_{0,0}$ for three different cases and two levels of approximation. Left column: the P_1 solution, which is identical to the diffusion approximation; right column the P_7 solution. First Row (Case 1): $\mu_a = 0.01$ mm^{-1}, $\mu_s = 10$ mm^{-1} isotropic source. Second Row (Case 2): $\mu_a = 0.01$ mm^{-1}, $\mu_s = 0.1$ mm^{-1} directed source perpendicular to boundary. Third Row (Case 3): $\mu_a = 0.01$ mm^{-1}, $\mu_s = 0.1$ mm^{-1} directed source, at 45 degrees to the boundary.

Figure 2 row 2, shows the photon density with a directed source and $\mu'_s = 0.1$ mm^{-1}. In this regime one would expect the Diffusion Approximation to break down. This figure shows that the P_1 solution is unable to take into account the directional nature of the source and the photon density distribu-

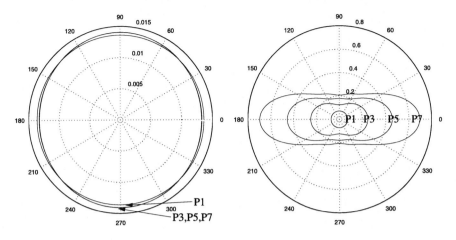

Fig. 3. Angular distribution of the even-parity radiance 5 mm from the source for the P_1, P_3, P_5, P_7 approximations. Left: Case 1 of figure (2), right Case 2 of figure (2).

tion is diffuse. However the P_7 solution produces a less diffuse photon density distribution and the directionality of the source can be see.

Further in Figure 2, row 3 the source is now directed at 45 degrees to the boundary and again the P_1 solution is unable to take this into account. The P_7 solution produces a similar photon density distribution as shown in row 2 but now rotated through 45 degrees.

Figure 3 shows the angular distribution of the even-parity radiance at a point 5 mm away from the source along the center line. For $\mu_s = 10$ mm^{-1}, i.e. the diffusion regime, the even parity radiance has little dependence on the angular variable and is independent of the order of solution. This is to be expected for the diffusion regime. For $\mu_s = 0.1$ mm^{-1} however there is a marked difference in the even parity radiance field between the various levels of approximation. For the P_1 solution again no angular dependence is possible, whereas in the P_7 approximation there is a stark angular dependence with a strong bias in the forward (and backward due to even parity) direction, i.e $\hat{s} = [0, 0, 1]^T$. There is also an increase in intensity as the P_1 solution diffuses photons whereas the higher order approximations are able to advect them as well.

7 Conclusions

In this paper we presented the derivation of a finite element method for the second order even-parity Radiative Transfer Equation, using the P_N approximations. This leads to a large sparse matrix system with a simple structure. We have shown initial results that suggest the model behaves as expected. We plan to test the model more extensively.

References

1. A. Yodh and B. Chance. Spectroscopy and imaging with diffusing light. *Phys. Today*, 48:38–40, 1995.
2. J.C. Hebden, S.R. Arridge, and D.T. Delpy. Optical imaging in medicine: I. Experimental techniques. *Phys. Med. Biol.*, 42:825–840, 1997.
3. S.R. Arridge and J.C. Hebden. Optical imaging in medicine: II. Modelling and reconstruction. *Phys. Med. Biol.*, 42:841–853, 1997.
4. S.R. Arridge. Optical tomography in medical imaging. *Inverse Problems*, 15(2):R41–R93, 1999.
5. D.A. Boas, D.H. Brooks, E.L. Miller, C.A. DiMarzio, M. Kilmer, R.J. Gaudette, and Q. Zhang. Imaging the body with diffuse optical tomography. *IEEE Sig. Proc. Magazine*, 18(6):57–75, 2001.
6. M. Firbank, S.R. Arridge, M. Schweiger, and D.T. Delpy. An investigation of light transport through scattering bodies with non-scattering regions. *Phys. Med. Biol.*, 41:767–783, 1996.
7. O. Dorn. A transport-backtransport method for optical tomography. *Inverse Problems*, 14(5):1107–1130, 1998.
8. A.H. Hielscher, R.E. Alcouffe, and R.L. Barbour. Comparison of finite-difference transport and diffusion calculations for photon migration in homogeneous and heterogeneous tissue. *Phys. Med. Biol.*, 43:1285–1302, 1998.
9. S.R. Arridge, H. Dehghani, M. Schweiger, and E. Okada. The finite element model for the propagation of light in scattering media: A direct method for domains with non-scattering regions. *Med. Phys.*, 27(1):252–264, 2000.
10. B. Davison. *Neutron Transport Theory*. Oxford University Press, 1957.
11. A.M. Weinberg and E.P. Wigner. *The Physical Theory of Neutron Chain Reactors*. University of Chicago Press, 1958.
12. M.C. Case and P.F. Zweifel. *Linear Transport Theory*. Addison-Wesley, New York, 1967.
13. J.J. Duderstadt and W.R. Martin. *Transport Theory*. John Wiley & and Sons, 1979.
14. R.T. Ackroyd. *Finite Element Methods for Particle Transport: Applications to Reactor and Radiation Physics*. Research Studies Press Ltd., Taunton, 1997.
15. S. Chandrasekhar. *Radiative Transfer*. Oxford University Press, London, 1950.
16. A. Ishimaru. *Wave Propagation and Scattering in Random Media*, volume 1. Academic, New York, 1978.
17. C.R.E. de Oliveira. An arbitrary geometry finite element method for multigroup neutron transport with anisotropic scattering. *Prog. Nucl. Energy*, 18:227–236, 1986.
18. E.D. Aydin, C.R.E. de Oliveira, and A.J.H. Goddard. A comparison between transport and diffusion calculations using a finite element-spherical harmonics radiation transport method. *Med. Phys.*, 2(9):2013–2023, 2002.
19. G.S. Abdoulaev and A.H. Hielscher. Three-dimensional optical tomography with the equation of radiative transfer. *Journal of Electronic Imaging*, 12(4):594–601, 2003.
20. J. Heino, S.R. Arridge, J. Sikora, and E. Somersalo. Anisotropic effects in highly scattering media. *Physical Review E*, 68:Article number 31908, 2003.
21. V.I. Lebedev and A.L. Skorokhodov. Quadrature formulas of orders 41,47 and 53 for the sphere. *Russian Acad. Sci. Dokl. Math.*, 45(3), 2002.

Solution of Radiative Transfer Problems with Finite Elements

Guido Kanschat

Institut für Angewandte Mathematik
Universität Heidelberg
69120 Heidelberg
Germany
kanschat@dealii.org

Summary. Mathematical modeling for monochromous radiative transfer problems is reviewed. Suitable boundary conditions for well-posed problems are introduced. Finite element discretizations for the integral operator in the angular variable and the transport operator in space are discussed. Adaptive algorithms and error estimates are explained. The structure of the resulting discrete linear system is analyzed and solution methods are suggested.

1 Introduction

We review the information leading to the design of the monochromous radiative transfer code used in [12–18, 27, 33, 34]. This code has been extended to the simulation of frequency redistribution due to scattering and Doppler shifts. For details on this topic confer to [25] in this volume.

First, we review main features of the mathematical model. In particular, the correct choice of boundary conditions is discussed, since this is crucial for the simulation. Then, discretization techniques for the integral operator are discussed. Since a general finite element discretization does not seem advisable due to computational overhead, we show how a simple discretization can be derived.

Two finite element methods for the discretization of the advection operator in space are discussed in Section 4, namely the streamline diffusion method (SDFEM), also known as streamline-upwind Petrov-Galerkin (SUPG), and the discontinuous Galerkin method (DGFEM). Both methods exhibit optimal convergence properties.

Based on these discretization methods, the question of optimal mesh design by adaptive algorithms is discussed in Section 5. Several grid generation techniques and mesh refinement strategies are presented. A major part of this section is devoted to the discussion of error estimators for the optimal reproduction of measured values by the simulation. A more detailed account of these methods can be found in [2,4].

Finally, we analyze the basic structure of the resulting linear system of equations and its consequences for the solution process. In particular, traditional iterations and preconditioners are discussed and the reasons for their failure in the scattering dominant case are explained. We propose the use of Krylov-space methods as a simple remedy (see [28] for thorough introduction to linear solvers). More complex preconditioners based on moment equations as presented in [19] are discussed in [25] in this volume.

2 Mathematical Model

The transfer of radiation in the approximation of geometrical objects is modeled by an equation of Boltzmann type. These equations are posed for the density function of the particles of a stochastic gas. Usually, this density function depends on both variables of the phase space, the location x in space and the velocity vector v of the particles. In our model, the quantity of interest is the density of photons, or its macroscopic analogue, the radiation intensity u. Since the speed of photons is always light speed c, the velocity vector is decomposed in the directional component, the solid angle ϑ, and the wave length λ. Here, we will focus on the monochromatic case and neglect the wave length dependence.

Assuming that relativistic effects do not play a role (see [25] in this volume for the inclusion of special relativistic effects), light will travel on straight lines, and the intensity function is a solution to the equation

$$\frac{1}{c}\frac{d}{dt}u(t,x,\vartheta) + \frac{d}{ds}u(t,x,\vartheta) = S(t,x,\vartheta,u).$$

Here, ds is the infinitesimal distance in direction ϑ, therefore $d/ds\, u(t,x,\vartheta) = \vartheta \cdot \nabla_x u(t,x,\vartheta)$, where ∇_x denotes the gradient with respect to the space coordinates. S is the collision term on which we will make some assumptions now. First, we assume the superposition principle of light, that is, photons do not interact with other photons. Furthermore, we assume that the optical properties of the area the light travels through are independent of the radiation. Then, the collision term S is linear[1] in u. Furthermore, we assume for simplicity that material properties are isotropic, that is, do not depend on the solid angle ϑ. Then, S consists of an extinction component $-\chi u$ and a scattering component $\int Pu$:

$$S(t,x,\vartheta,u) = -\chi(x)u(t,x,\vartheta) + \sigma(x)\int_{S^2} P(\widetilde{\vartheta},\vartheta)u(t,x,\widetilde{\vartheta})\,d\widetilde{\vartheta}.$$

---
[1] If the collision term was nonlinear, the solution process would involve Newton's method or some other nonlinear iteration, where in each step a linear radiative transfer problem must be solved

Finally, we assume that the extinction coefficient χ is the sum of the scattering coefficient σ and the absorption coefficient κ, both nonnegative. σ is chosen in such a way that P is normalized to

$$\int_{S^2} P(\widetilde{\vartheta}, \vartheta) \, d\widetilde{\vartheta} = 1. \tag{1}$$

This condition implies energy conservation of the scattering, that is,

$$\int_{S^2} \left(u(t, x, \vartheta) - \int_{S^2} P(\widetilde{\vartheta}, \vartheta) \, d\widetilde{\vartheta} \right) d\vartheta = 0. \tag{2}$$

By the principle of reversibility of light, we have the additional property

$$P(\widetilde{\vartheta}, \vartheta) = P(-\vartheta, -\widetilde{\vartheta}). \tag{3}$$

Furthermore, we assume that the scattering kernel P is strictly positive and that it is symmetric in its arguments.

Then, the stationary monochromatic radiative transfer equation reads

$$\vartheta \cdot \nabla_x u + (\kappa + \sigma) u = \sigma \int_{S^2} P(\widetilde{\vartheta}, \vartheta) u(x, \widetilde{\vartheta}) \, d\widetilde{\vartheta} + \kappa B(\lambda, T(x)), \tag{4}$$

where B is a (thermal) radiation source. This equation is solved for x in a convex spatial domain Ω in $\mathbb{R}^d$ and ϑ the angles of the unit sphere S^{d-1} of $\mathbb{R}^d$; here and in the following, ϑ will always be the unit vector in the direction of the angle and thus a vector in $\mathbb{R}^d$. The space dimension d may be one, two or three.

The solution of equation (4) is a key step in the simulation of more complex models, including frequency redistribution due to material properties and Doppler shifts and transient problems. Furthermore, it is of interest itself, if the physical problem is well described by the *grey approximation* or if just a single wavelength is of interest.

Before we discuss discretization and solution techniques for (4), we discuss the model a bit further. First, we describe boundary conditions for bounded domains. Then, we advert to the question of well-posedness and regularity of solutions.

2.1 Boundary Conditions

The model in the previous subsection was developed in the whole space $\mathbb{R}^d$. In order to be able to discretize equation (4), we have to cut out a bounded domain $\Omega \subset \mathbb{R}^d$. This implies the introduction of artificial boundaries and boundary conditions for the integro-differential equation. We will always assume that Ω is convex and that its boundary $\partial \Omega$ is a finite union of C^∞ curves of finite length. In most cases, a rectangle or rectangular brick will be sufficient.

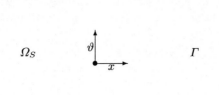

Fig. 1. Computational domain Ω_S and boundary Γ in one dimension

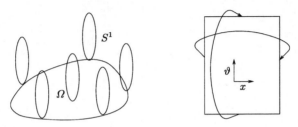

Fig. 2. Computational domain Ω_S and boundary Γ (unrolled) in two dimensions

First, we attempt a description of the computational domain

$$\Omega_S = \Omega \times S^{d-1}, \qquad d = 1, 2, 3 \tag{5}$$

and its boundary. Since Ω_S is a $(2d-1)$-dimensional object embedded into $\mathbb{R}^{2d}$, this will not be very intuitive even in two dimensions. In one dimension, the domain Ω is a bounded open interval of $\mathbb{R}^1$ and S^0 consists of the points -1 and 1.

We define the boundary manifold Γ as

$$\Gamma = \partial\Omega \times S^{d-1}. \tag{6}$$

Since S^d is a closed manifold, this Γ is the boundary of Ω_S. The one-dimensional situation is displayed in Figure 1.

In two dimensions, the Ω_S can be considered an open piece of the Euclidean plane with a circle attached to each point (see Figure 2). By unrolling $\partial\Omega$ and S^1 onto a line segment each, we can view Γ as a rectangle with connected sides (a topological 2d torus).

The extension to the three-dimensional case is obvious and we do not attempt a picture for that.

By $\mathbf{n}_\Gamma(x)$ we denote the outer normal vector to the boundary $\partial\Omega$ at x. With this vector we define

$$\Gamma_- = \{(x, \vartheta) \in \Gamma \mid \mathbf{n}_\Gamma(x) \cdot \vartheta < 0\}, \tag{7}$$
$$\Gamma_+ = \{(x, \vartheta) \in \Gamma \mid \mathbf{n}_\Gamma(x) \cdot \vartheta > 0\}. \tag{8}$$

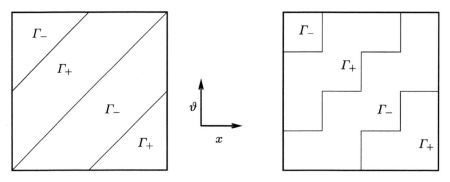

Fig. 3. Inflow (Γ_-) and outflow (Γ_+) boundary in two dimensions (left circle, right square)

The remaining part $\Gamma_0 = \Gamma \backslash (\Gamma_- \cup \Gamma_+)$ is a set of measure zero with respect to the measure on Γ and will be ignored. In the one-dimensional example of Figure 1 on the facing page, Γ_- consists of the upper right an lower left points. Figure 3 shows these parts on the boundary for the two-dimensional case in the unrolled version on the right of Figure 2. In correspondence to standard advection equations these sets will be called inflow and outflow boundary, respectively.

For any given $\vartheta \in S^{d-1}$ we will denote by Γ_-^ϑ (Γ_+^ϑ) the part of $\partial\Omega$ for which $\mathfrak{n}_\Gamma(x) \cdot \vartheta < 0 \; (> 0)$.

In order to obtain a well-posed radiative transfer problem, boundary conditions must be prescribed on Γ_-. These boundary conditions come in two flavors.

The so called vacuum boundary condition assumes that radiation will not reenter the domain Ω after it has left it. Therefore, incident radiation is independent of the solution in Ω and the boundary condition takes the simple form

$$u(x, \vartheta) = g(x, \vartheta) \quad \text{for } (x, \vartheta) \in \Gamma_-, \tag{9}$$

with g a prescribed function modeling incident radiation due to external sources. Vacuum boundary conditions are typical if the whole simulated object is embedded in the domain Ω and σ vanishes at the boundary of Ω.

Alternatively, the boundary can be reflecting. Typically, cuts through the domain at symmetry planes have this property. In technical applications, a solid wall is modeled by reflecting boundary conditions, too. They are modeled by a term similar to the scattering operator, namely

$$u(x, \vartheta) = \int_{\widetilde{\vartheta} \cdot \mathbf{n} > 0} R(x, \widetilde{\vartheta}, \vartheta) u(x, \widetilde{\vartheta}) \, d\widetilde{\vartheta} \quad \text{for } (x, \vartheta) \in \Gamma_-. \tag{10}$$

A particular case is specular reflection by a mirror, where $R(\widetilde{\vartheta}, \vartheta) = \delta(\widetilde{\vartheta}_r - \vartheta)$ and $\widetilde{\vartheta}_r$ is the reflected angle $\widetilde{\vartheta}_r = \widetilde{\vartheta} - 2(\widetilde{\vartheta} \cdot \mathbf{n}_\Gamma(x))\mathbf{n}_\Gamma(x)$. Here, $\mathbf{n}_\Gamma$ again is the outward normal vector.

2.2 Solvability and Regularity of Solutions

A very detailed account on analytical properties of equation (4) can be found in [6]. Here, we summarize just the most important results and restrict ourselves to $\mathbb{R}^3$ for simplicity.

Most modern analysis of differential equations starts with the proper statement of a so-called weak formulation. Assuming that the coefficients are in L^∞ of the corresponding domains, this weak formulation reads: find $u \in L^2(\Omega \times S^2)$ such that for any $\varphi \in L^2(\Omega \times S^2)$ holds:

$$\iint_{\Omega\, S^2} \left(\vartheta \cdot \nabla_x u(x,\vartheta)\varphi(x,\vartheta) + (\kappa(x) + \sigma(x))u(x,\vartheta)\varphi(x,\vartheta) \right.$$
$$\left. - \iint_{S^2} P(\widetilde{\vartheta},\vartheta)u(x,\widetilde{\vartheta})\varphi(x,\vartheta)\,d\widetilde{\vartheta} \right) dx\,d\vartheta + \iint_{\Gamma_-} u(x,\vartheta)\varphi(x,\vartheta)|\mathbf{n}\cdot\vartheta|\,ds\,d\vartheta$$
$$= \iint_{\Omega\, S^2} B(\lambda, T(x))\varphi(x,\vartheta)\,dx\,d\vartheta. + \iint_{\Gamma_-} g(x,\vartheta)\varphi(x,\vartheta)|\mathbf{n}\cdot\vartheta|\,ds\,d\vartheta \quad (11)$$

For this weak formulation holds:

Theorem 1. *Assume additionally to the assumptions made on P that κ and σ are nonnegative almost everywhere. Furthermore, the domain Ω on which we solve is bounded and convex, and we impose vacuum boundary conditions (9) everywhere on Γ_-. Then, the radiative transfer problem is well-posed and the weak problem (11) admits a unique solution.*

Remark 1. The same theorem holds in two dimensions, provided a genuine twodimensional model is used. It does *not* hold in cylinder geometry, where the angles of S^2 are projected onto S^1. Then, strict positivity of κ is one way to ensure well-posedness.

Remark 2. Similarly, strict positivity of κ at least in parts of the domain Ω is required for well-posedness if the reflection boundary conditions (10) is imposed on the whole inflow boundary Γ_-.

While the question of well-posedness is solved satisfactorily by above theorem, the regularity of solutions is rather weak. By the method of characteristics, we obtain that $u(.,\vartheta)$ is differentiable in direction ϑ for right hand sides in $L^2(\Omega)$. On the other hand, there is no gain in regularity for any other direction. Therefore, L^2 right hand sides will also only produce an L^2 solution. Similarly, a jump in the boundary condition will be transported through the domain, being reduced in size by scattering, but not being smeared out like in advection-diffusion problems.

3 The Integral Operator

First, we consider the discretization of integral equations of type

$$\chi u(\vartheta) - \sigma \int P(\vartheta, \widetilde{\vartheta}) u(\widetilde{\vartheta}) \, d\widetilde{\vartheta} = f(\vartheta). \tag{12}$$

This equation captures the main features of equation (4) considered in the angular variable only. The domain of integration will be the unit sphere in $\mathbb{R}^2$ or $\mathbb{R}^3$ in our application. The schemes considered here either replace the integral operator by a suitable quadrature formula or are obtained by testing the integral equation by finite element test functions.

We remark here that both methods in their standard form will produce good results only in the case that the intensity is a smooth variable of the angle. In particular, the problem of a searchlight beam in an optically thin medium poses a problem that has to be handled with care. While Nyström's method may fail to see this beam at all, the Galerkin scheme ideally will always catch it, but will smear it out into adjacent directions. Therefore, the methods must be adapted if there is a dominant direction of radiation, for instance by making this direction one of the quadrature points of Nyström's method.

3.1 Bounds on the Operator

The analysis of the finite element method is usually carried out in the Hilbert space of square integrable functions. In this space, energy conservation is not sufficient to obtain the bounds on the scattering operator needed. Therefore, we will need some further insight into the properties of the scattering operator Σ. The results here are formulated for general L^p-spaces and included for readers with a stronger mathematical interest. Since scattering only operates on the angular variable, we introduce the operator $\Sigma_\vartheta : L^p(S^{d-1}) \to L^p(S^{d-1})$ defined by

$$\Sigma_\vartheta u(\vartheta) = \int_{S^{d-1}} P(\widetilde{\vartheta}, \vartheta) u(\vartheta) \, d\vartheta. \tag{13}$$

First, we consider two-dimensional scattering.

Lemma 1. *If a two-dimensional scattering material is isotropic[2], then the phase function $P(\widetilde{\vartheta}, \vartheta)$ has a representation as*

$$P(\widetilde{\vartheta}, \vartheta) = \sum_{i=0}^{\infty} a_i \cos(i\varphi), \tag{14}$$

where $\varphi = \widetilde{\vartheta} - \vartheta$.

---
[2] This does not imply isotropic scattering, i.e., $P \equiv const$

Proof. Isotropy of material yields

$$P(\tilde{\vartheta}, \vartheta) = P(\varphi, 0),$$

which is a periodic function on the unit circle. By Fourier series expansion, we obtain

$$P(\varphi, 0) = \sum_{k=0}^{\infty} \bigl(a_k \cos(k\varphi) + b_k \sin(k\varphi)\bigr).$$

Since isotropy implies symmetry with respect to φ, the coefficients b_k in front of the sine functions must be zero.

Definition 1. *We will say that scattering is absolutely forward dominant, if all coefficients a_k in (14) are non-negative.*

We give a short comment on idea of this definition: all cosines are unity for $\varphi = 0$. Therefore, $P(\vartheta, \vartheta) = \sum a_k$. For all other angles $\varphi \neq 0$, we have $P(\varphi, 0) \leq \sum a_k$. Remark that isotropic scattering as well as Thompson and Rayleigh scattering are absolutely forward dominant.

Lemma 2. *Assume isotropic material and absolutely forward dominant scattering. Then, the scattering operator Σ_t is a positive semi-definite operator in $L^2(S^1)$.*

Proof. We have to show

$$\int_{S^1} \Sigma_\vartheta u u \, d\vartheta \geq 0.$$

By the expansion (14) and basic trigonometric function calculations, this is

$$\iint_{S^1} P(\tilde{\vartheta}, \vartheta) u(\tilde{\vartheta}) u(\vartheta) \, d\tilde{\vartheta} \, d\vartheta = \iint_{S^1} P(\tilde{\vartheta} - \vartheta, 0) u(\tilde{\vartheta}) u(\vartheta) \, d\tilde{\vartheta} \, d\vartheta$$

$$= \iint_{S^1} \sum_{i=0}^{\infty} a_i \cos\bigl(i(\tilde{\vartheta} - \vartheta)\bigr) u(\tilde{\vartheta}) u(\vartheta) \, d\tilde{\vartheta} \, d\vartheta$$

$$= \sum_{i=0}^{\infty} a_i \left\{ \left(\int \cos(i\vartheta) u(\vartheta) \, d\vartheta\right)^2 + \left(\int \sin(i\vartheta) u(\vartheta) \, d\vartheta\right)^2 \right\},$$

which is non-negative.

The three-dimensional case is slightly more involved, since we first have to reduce three-dimensional angles to two-dimensional ones. Again, the isotropy of the material is crucial:

Lemma 3. *Assume isotropic material and absolutely forward dominant scattering. Then, the scattering operator Σ_t is a positive semi-definite operator in $L^2(S^2)$.*

3 The Integral Operator   57

Proof. Again, we have to show that

$$\int_{S^2} \Sigma_\vartheta uu \, d\vartheta \geq 0.$$

For each pair of three-dimensional angles $\tilde\vartheta$ and ϑ, there is a rotation $\varrho_{\tilde\vartheta,\vartheta} \in SO(3)$, such that the rotated angles $\tilde\varphi = \varrho_{\tilde\vartheta,\vartheta}\tilde\vartheta$ and $\varphi = \varrho_{\tilde\vartheta,\vartheta}\vartheta$ are in the xy-plane and are therefore plane angles.

By isotropy we obtain

$$\iint_{S^2} P(\tilde\vartheta,\vartheta)u(\tilde\vartheta)u(\vartheta)\, d\tilde\vartheta \, d\vartheta = \iint_{S^2} P(\tilde\varphi,\varphi)u(\tilde\vartheta)u(\vartheta)\, d\tilde\vartheta \, d\vartheta$$
$$= \iint_{S^1} P(\tilde\varphi - \varphi, 0)u(\tilde\vartheta)u(\vartheta)\, d\tilde\vartheta \, d\vartheta.$$

Now, the proof continues as in Lemma 2 on the preceding page.

Lemma 4. *Provided that the scattering material is isotropic and absolutely forward dominant and the kernel P is bounded, positive and smooth, the scattering operator Σ_ϑ is a compact, positive semi-definite, self-adjoint operator from $L^2(S^{d-1})$ onto itself.*

Proof. By Fubini's theorem and symmetry of $P(.,.)$ with respect to its arguments

$$\int_{S^{d-1}} \Sigma_\vartheta uv \, d\vartheta = \iint_{S^{d-1}} P(\tilde\vartheta,\vartheta)u(\tilde\vartheta)v(\vartheta)\, d\tilde\vartheta \, d\vartheta = \int_{S^{d-1}} u\Sigma_\vartheta v \, d\vartheta.$$

Since we assumed that P is continuous and bounded, we clearly have

$$\iint_{S^{d-1}} P(\tilde\vartheta,\vartheta)\, d\tilde\vartheta \, d\vartheta < \infty,$$

and therefore a Hilbert-Schmidt kernel. Then, the operator Σ_ϑ is compact (see [35], X.2).

Due to isotropy, there is a rotation ϱ such that $\varrho\tilde\vartheta$ and $\varrho\vartheta$ are plane angles. Therefore, letting $\alpha = \tilde\vartheta - \vartheta$, there is a function $\varphi : \mathbb{R} \to \mathbb{R}$, such that $P(\tilde\vartheta,\vartheta) = \varphi(\alpha)$. Due to the symmetry with respect to $\tilde\vartheta$ and ϑ, φ is well defined and, by Fourier expansion, can only depend on powers of the cosine of α:

$$P(\tilde\vartheta,\vartheta) = \sum_{i=0}^\infty \cos(\varrho\tilde\vartheta - \varrho\vartheta) = \sum_{i=0}^\infty \cos \varrho\tilde\vartheta \, \cos \varrho\vartheta + \sin \varrho\tilde\vartheta \, \sin \varrho\vartheta.$$

The positive semi-definiteness follows from Lemma 2 on the facing page and Lemma 2 on the preceding page, respectively

Lemma 5. *If the energy conservation* (1) *and the reversibility condition* (3) *hold, the operator norm of* Σ_ϑ *is*

$$\|\Sigma\|_{L^p \to L^p} = 1, \tag{15}$$

for $1 \leq p \leq \infty$.

Proof. Equations (1) and (3) yield

$$\int_{S^{d-1}} P(\widetilde{\vartheta}, \vartheta) \, d\widetilde{\vartheta} = \int_{S^{d-1}} P(\widetilde{\vartheta}, \vartheta) \, d\vartheta = 1.$$

In the case $\Sigma : L^1 \to L^1$, we have

$$\|\Sigma u\|_{L^1} = \iint_{S^{d-1}} |P(\widetilde{\vartheta}, \vartheta)| |u(\widetilde{\vartheta})| \, d\widetilde{\vartheta} \, d\vartheta$$

$$= \int_{S^{d-1}} |u(\widetilde{\vartheta})| \int_{S^{d-1}} P(\widetilde{\vartheta}, \vartheta) \, d\vartheta \, d\widetilde{\vartheta} = \int_{S^{d-1}} |u(\widetilde{\vartheta})| \, d\widetilde{\vartheta} = \|u\|_{L^1}.$$

The L^∞-estimate is obtained by observing that

$$\Sigma u(\vartheta) = \int_{S^{d-1}} P(\widetilde{\vartheta}, \vartheta) u(\widetilde{\vartheta}) \, d\widetilde{\vartheta} \leq \|u\|_{L^\infty} \int_{S^{d-1}} P(\widetilde{\vartheta}, \vartheta) \, d\widetilde{\vartheta} = \|u\|_{L^\infty}.$$

For $1 < p < \infty$ and $1/p + 1/q = 1$ we derive the estimate by:

$$\|\Sigma u\|_{L^p}^p = \int_{S^{d-1}} \left(\int_{S^{d-1}} P(\widetilde{\vartheta}, \vartheta) u(\widetilde{\vartheta}) \, d\widetilde{\vartheta} \right)^p d\vartheta$$

$$\leq \left(\int_{S^{d-1}} P(\widetilde{\vartheta}, \vartheta)^{q/q} \, d\widetilde{\vartheta} \right)^{p/q} \int_{S^{d-1}} P(\widetilde{\vartheta}, \vartheta)^{p/p} u(\widetilde{\vartheta})^p \, d\widetilde{\vartheta} \, d\vartheta$$

$$= \int_{S^{d-1}} \int_{S^{d-1}} P(\widetilde{\vartheta}, \vartheta) u(\widetilde{\vartheta})^p \, d\vartheta \, d\widetilde{\vartheta}$$

$$= \|u\|_{L^p}^p.$$

It remains to prove that $\|\Sigma\| \geq 1$. Applying the operator to the function

$$u_0(\vartheta) \equiv 1 / \left|S^{d-1}\right|^{1/p} \tag{16}$$

if $p < \infty$ and $u_0(\vartheta) \equiv 1$ if $p = \infty$ yields

$$\|\Sigma u_0\|_{L^1} = \frac{1}{|S^{d-1}|} \iint P(\widetilde{\vartheta}, \vartheta) \, d\widetilde{\vartheta} \, d\vartheta = 1$$

$$\|\Sigma u_0\|_{L^\infty} = 1 \int P(\widetilde{\vartheta}, \vartheta) \, d\vartheta = 1$$

$$\|\Sigma u_0\|_{L^p} = \left(\int \frac{1}{|S^{d-1}|} \left(\int P(\widetilde{\vartheta}, \vartheta) \, d\widetilde{\vartheta} \right)^p d\vartheta \right)^{1/p} = 1.$$

3 The Integral Operator 59

Corollary 1. *The operator Σ_ϑ of an absolutely forward dominantly scattering isotropic material has a discrete spectrum contained in the interval $[0,1]$ and 1 is an isolated eigenvalue in the sense that there is a positive constant γ such that*

$$1 - \lambda_i \geq \gamma,$$

holds for any $\lambda_i \neq 1$ from the spectrum of Σ_ϑ.

Proof. It is a well-known result that the spectrum of compact operators is at most a countable set of points with no accumulation except at zero (cf. [35] XI.9 Theorem 1). Then, Lemma 4 implies that all eigenvalues are real and positive.

By Lemma 5 on the facing page, the largest eigenvalue equals to 1 and γ is the distance to the second largest eigenvalue.

Lemma 6. *Let the scattering material be isotropic and scattering be absolutely forward dominant. Then, the operator $\sigma - \Sigma^\sigma$ is positive semi-definite. Its null-space $\mathcal{N}$ is not empty and there exists a constant $\lambda_1 > 0$ such that*

$$\iint_{\Omega_S} (\sigma - \Sigma^\sigma) v v \, d\vartheta \, dx \geq \lambda_1 \|\sqrt{\sigma} v\|^2,$$

for any function v from the orthogonal complement of $\mathcal{N}$.

Proof. Corollary 1 implies that there is a function $v \in W$, such that $v(x,.)$ is in the eigenspace of the eigenvalue 1 of Σ_ϑ almost everywhere in Ω. Obviously, this function is in the null-space of $\sigma - \Sigma^\sigma = \sigma(1 - \Sigma_\vartheta)$. Furthermore, since 1 is an isolated eigenvalue of Σ_ϑ, 0 is an isolated eigenvalue of $1 - \Sigma_\vartheta$ and the estimate follows.

Remark 3. Our analysis in this section involved the condition that scattering must be absolutely forward dominant. This condition is not implied by the physical modeling leading to equation (4), even if it holds for isotropic and Rayleigh scattering. It is not necessary for the analysis of the continuous equation as for instance in [6]. The reason is, that it can be shown that the radiative transfer operator is a mapping of the cone of positive functions into itself.

Unfortunately, simple computations show that this is not true anymore for the discrete radiative transfer operator, where the solutions may have negative values due to over-shooting induced by the space dicretization.

If we consider scattering on the whole space $L^2(S^{d-1})$, a simple example shows the necessity of absolutely forward dominance:

In the Fourier expansion of equation (14), let $a_0 = \frac{1}{2}$ and $a_1 = -\frac{1}{2}$, for $i > 1$ let $a_i = 0$. The scattering kernel $P(\tilde{\vartheta}, \vartheta)$ is non-negative everywhere. Therefore, it maps positive functions to positive functions. But, for $u(\vartheta) = \cos(\vartheta)$, we obtain by orthogonality of the Fourier basis

$$\int_{S^1} \Sigma_\vartheta u u \, d\vartheta = a_1 \left(\int \cos(\vartheta)^2 \, \delta\vartheta \right)^2 < 0$$

3.2 Nyström's Method

A classical method for discretizing integral operators like in equation (12) is the application of numerical quadrature, also called Nyström's method.

For our two-dimensional computations—i. e. one ordinate dimension—any Newton-Coates- or Gauß-formula may be used. We use the iterated midpoint rule on the unit circle with m equidistributed points:

$$\vartheta_i = \frac{2\pi}{m} i$$
$$w_i = \frac{1}{m} \qquad (17)$$

This formula allows an efficient implementation and is of high accuracy. In fact, by the Euler-McLaurin summation formula, its approximation order depends on the regularity of the solution only.

For integration over S^2 the first question arising is the construction of (nearly) equidistributed quadrature points to avoid discretization artifacts. For instance, using a parameterization of S^2 by two angles from the rectangle $Q = [0, 2\pi) \times [0, \pi)$ and using a grid on this rectangle (a longitude–latitude mesh) produces solutions showing distinct features in polar directions, introducing a nonphysical axis of symmetry. This is due to the fact that the quadrilaterals close to the poles degenerate and this becomes even worse using finer grids.

On the other hand, it is well known that there are only five regular polyhedra. Our approach uses the triangles of S^2 obtained by successive subdivision of an icosahedron (see Figure 4). Quadrature points are the cell centers projected on S^2 and spherical cell volumes serve as weights. The result is a scheme converging with second order.

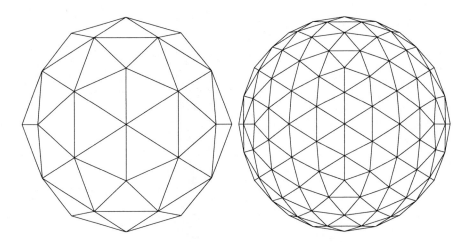

Fig. 4. Refined icosahedron (80 and 320 cells)

If we consider the number of cells lying on a great circle to be a measure for the quality of such a subdivision, we can compare our triangulation with the longitude-latitude mesh. Using for instance 40 cells on a great circle, we need a total number of 320 cells, whereas the parameterization approach needs 800.

The discretization by Nyström's method converts equation (12) into a system of equations of the form

$$\chi u_i - \sigma \sum_{k=1}^{m} w_k P(\theta_i, \theta_k) u_k = f_i \qquad i = 1, \ldots, m.$$

In order to symmetrize this linear system, we multiply each row by w_i, obtaining the equivalent system

$$\chi w_i u_i - \sigma \sum_{k=1}^{m} w_i w_k P(\vartheta_i, \vartheta_k) u_k = w_i f_i \qquad i = 1, \ldots, m. \qquad (18)$$

The matrix associated with this system of equations is

$$\begin{pmatrix} w_1 & & \\ & \ddots & \\ & & w_m \end{pmatrix} - \begin{pmatrix} w_1^2 P(\vartheta_1, \vartheta_1) & \cdots & w_1 w_m P(\vartheta_1, \vartheta_m) \\ \vdots & & \vdots \\ w_m w_1 P(\vartheta_m, \vartheta_1) & \cdots & w_m^2 P(\vartheta_m, \vartheta_m) \end{pmatrix}, \qquad (19)$$

which obviously inherits symmetry from the phase function P.

Remark 4. Since energy conservation of the scattering is very decisive for the character of the solution of the discrete problem, it should always be asserted, that (1) also holds for the discretized operator.

3.3 Galerkin Discretization

Having in mind the development of more refined error estimates, it is advisable to look at Galerkin discretizations for the integral operator. The integral equation (12) will then be used in its weak formulation: Search $u \in X = L^2(S^{d-1})$ such that

$$\int_{S^{d-1}} \chi u(\vartheta) v(\vartheta) d\vartheta - \iint_{S^{d-1}} \sigma P(\widetilde{\vartheta}, \vartheta) u(\widetilde{\vartheta}) v(\vartheta) d\widetilde{\vartheta} \, d\vartheta$$

$$= \int_{S^{d-1}} f(\vartheta) v(\vartheta) d\vartheta \quad \forall \, v \in X. \qquad (20)$$

Replacing X by a finite element space X_h, the computation of matrix elements requires evaluation of an integral sum of the type

$$\sum_{i,k} \int_{T_i} \int_{T_j} P(\widetilde{\vartheta}, \vartheta) \varphi_k(\widetilde{\vartheta}) \varphi_l(\vartheta) . \, d\widetilde{\vartheta} \, d\vartheta$$

The sum extends over all cells $T_i \cap \operatorname{supp} \varphi_k \neq \emptyset$ and $T_j \cap \operatorname{supp} \varphi_l \neq \emptyset$[3]. The integration on the cells is then accomplished by a quadrature formula. Unlike differential equations, where two basis functions are coupled only locally, this integration has to be summed up over all possible pairs of cells. This is already a fourdimensional integration for the scattering operator alone and leads to an approximately sevendimensional one for the full problem and therefore unaffordable. Since no simplification can be made for general trial functions and accuracy in the ordinate space is not considered as crucial as in the spatial variable, we restrict ourselves in the following to the simplest case.

This integration becomes much simpler in the case of piecewise constant polynomials, also known as DG(0) method. On each cell of $\mathbb{T}_{S^{d-1}}$, the finite element functions $v_h(x, \vartheta)$ (and their derivatives with respect to x in the full problem) are constant with respect to the angular variable ϑ. Then, $\chi u_h v_h$ and $\Sigma_\vartheta u_h v_h$ can be integrated independent of u_h and v_h on each cell. This property leads to the choice of our finite element space

$$X_h = \{ v \mid v|_T = c_T, \forall T \in \mathbb{T}_{S^{d-1}} \}, \qquad (21)$$

where $\mathbb{T}_{S^{d-1}}$ is a subdivision of S^{d-1}. We obtain $\mathbb{T}_{S^{d-1}}$ by projecting polyhedra as in Figure 4 on page 60 onto the unit sphere. Independent of the mesh width h, the interior angles α in these triangulations are limited by

$$54° \leq \alpha \leq 72°$$

and areas of the triangles differ at most by a factor of two. So this triangulation is nearly uniform and does not produce the artifacts at the poles known from longitude-latitude meshes.

In the case of isotropic scattering ($P(\widetilde{\vartheta}, \vartheta) = 1/|S^{d-1}|$), the DG(0)-discretization is equivalent to Nyström's method with the mid-point rule. For non-isotropic scattering phase functions it will be principally the same, but with modified quadrature weights. An important feature of this method is though, that the scattering matrix in equation (19) can be computed once and for all for each phase function and angular discretization. Therefore, the angular discretization does not affect the program structure of the radiative transfer simulation tool.

We are now ready to state convergence of our discretization.

Theorem 2. *The physical assumptions of Section 2 given, the DG(0)-discretization of (12) is of first order and the error is limited by*

$$\| u - u_h \|_{L^2} \leq Ch \frac{\chi}{\kappa} \| u \|_{H^1}$$

with C an interpolation constant. Furthermore, in the barycenters x_T of the triangulation cells T, we have the superconvergence estimate

$$| u(x_T) - u_h(x_T) | \leq Ch^2 \frac{\chi}{\kappa} \| u \|_{W^{2,\infty}}$$

---

[3] supp f is the support of the function f, i.e. the closure of the set of points where $f \neq 0$

4 Space Discretization

In this section, we describe two possible stable and accurate finite element discretization schemes for the linear advection equation

$$\vartheta \cdot \nabla_x u(x) + \chi(x) u(x) = f(x), \tag{22}$$

which covers the dependence of equation (4) on the space variable x. The standard finite element method may produce spurious oscillations in upwind direction due to a lack of stability and the resulting linear systems are hard to solve. Therefore, the method has to be improved. The two solutions presented here consist either in weakening the continuity across cell boundaries, yielding discontinuous Galerkin schemes, or in stabilizing the streamline derivative by a diffusion term, resulting in the so called streamline diffusion (SDFEM) or also called streamline upwind Petrov-Galerkin (SUPG) method.

The finite element method always sets out from a weak formulation of the differential equation. Similarly to the Lagrange formalism in theoretical physics, this weak formulation is derived by multiplying the equation with a suitable test function and integrating the product over the domain of computation. The next subsections will be devoted to establishing such weak formulations suitable for radiative transfer problems.

Finite difference schemes are beyond the scope of this article; in general, it can be stated that the derivation of higher order schemes and in particular of schemes allowing for mesh adaptation is considerably more difficult for finite differences.

4.1 The Streamline Diffusion Finite Element Method

The streamline diffusion finite element method (SDFEM) sets out from a weak formulation of (22) incorporating the inflow boundary condition (9) in weak form like (11). In its first version, it reads: for given functions $f \in L^2(\Omega)$ and $g \in V$ find $u \in V$ such that

$$\int_\Omega \vartheta \cdot \nabla_x u + \chi u \varphi \, dx + \int_{\Gamma_-^\vartheta} u \varphi \mathbf{n} \cdot \vartheta \, ds = \int_\Omega f \varphi \, dx + \int_{\Gamma_-^\vartheta} g \varphi \mathbf{n} \cdot \vartheta \, ds \quad \forall \varphi \in V, \tag{23}$$

where V is the space of functions u in $L^2(\Omega)$ with derivatives $\vartheta \cdot \nabla_x u$ in $L^2(\Omega)$, such that (23) is well-defined on this space. Remark the bilinear form on the left as well as the linear form on the right consist of two parts, one for the differential equation in the interior of Ω and one enforcing the inflow boundary condition on Γ_-^ϑ.

Originally, this weak formulation of boundary conditions was introduced for the sake of the analysis. The method works as well with strong boundary conditions. Nevertheless, the weak form has two advantages: incorporation of the reflexion boundary condition (10) and formulation of the dual problem used later for error estimation and mesh refinement are straight forward.

The second step of the derivation of a finite element method is replacing the function space V for the solution as well as the test functions in (23) by a finite dimensional subspace V_h. This subspace is constructed by the following steps: first, a mesh consisting of tetrahedra or hexahedra (triangles or quadrilaterals in two dimensions) is used to cover the domain Ω. Then, a simple, finite dimensional function space is chosen on each mesh cell. Typically, this is a polynomial space on triangles and tetrahedra. Finally, these piecewise polynomial functions are glued together at the surfaces between grid cells, for instance by requiring continuity of the resulting function.

We use quadrilaterals and hexahedra for our meshes, therefore, the construction of the local spaces on each cell is a bit more involved. First, on the reference cube $[-1,1]^3$, we define the space of trilinear[4] polynomials

$$Q_1 = \{p(\xi,\eta,\zeta) = (a_1\xi + b_1)(a_2\eta + b_2)(a_3\zeta + b_3)\}.$$

Then, it is possible to construct a mapping Ψ_T, mapping the reference cube to any hexahedron T of the mesh (possibly with curved faces), using only polynomials in Q_1 for the x-, y- and z-component of Ψ_T. Finally, the finite element space on T is the space of functions in Q_1, mapped by Ψ_T, that is,

$$Q_1(T) = \{p(x) = \tilde{p}(\Psi_T^{-1}(x)) | \tilde{p} \in Q_1\}. \tag{24}$$

Since a bilinear function on a face of the reference cube is determined uniquely by its value in the four corner points, the mapped mesh cells will indeed touch each other at their common faces and continuity of the finite element function is enforced easily by requiring continuity in the grid points.

If the space V_h constructed in this way is used in the weak formulation (23), the results are not satisfactorily. In particular if the right hand side f is not continuous, spurious oscillations may occur in the finite element solution. These oscillations unphysically spreading against the transport direction are due to a lack of stability of this weak form on V_h.

In order to increase stability an additional diffusion term in transport direction is added [5,10,11], yielding the SDFEM (or SUPG) method

$$\int_\Omega (\vartheta \cdot \nabla_x u + \chi u)\varphi \, dx + \int_\Omega (\vartheta \cdot \nabla_x u + \chi u)\delta\vartheta \cdot \nabla_x \varphi \, dx + \int_{\Gamma^\vartheta_-} u\varphi \mathbf{n}\cdot\vartheta \, ds$$
$$= \int_\Omega f\varphi \, dx + \int_\Omega f\delta\vartheta \cdot \nabla_x \varphi \, dx + \int_{\Gamma^\vartheta_-} g\varphi \mathbf{n}\cdot\vartheta \, ds \quad \forall \varphi \in V. \tag{25}$$

The stabilization parameter δ has to be chosen according to the size of χ and proportional to the local mesh width to obtain the optimal balance between accuracy and stability. As a rule of thumb, this should be

---

[4] "trilinear", because they are linear in each of the three coordinate directions. The same construction is possible with bilinear polynomials in two dimensions or with polynomials of degree higher than one.

$$\delta \approx \begin{cases} 0.3h & \text{if } \chi < 1 \\ 0 & \text{if } \chi \gg 1, \end{cases} \qquad (26)$$

and preferably smooth in the transition region.

Remark that the solution u of the original problem (22) is a solution of (25), too. Therefore, SDFEM introduces a *fully consistent* stabilization.

We apply the streamline diffusion method to functions from the space V_h defined above.

The streamline diffusion method is similar to a weighted form of the least-squares discretization

$$\int_\Omega (\vartheta \cdot \nabla_x u + \chi u)(-\vartheta \cdot \nabla_x \varphi + \chi \varphi \, dx) = \int_\Omega f(-\vartheta \cdot \nabla_x \varphi + \chi \varphi \, dx) \qquad (27)$$

to the standard scheme. These discretizations have the advantage, that their bilinear form (and therefore the associated matrix) is symmetric and positive definite, simplifying the analysis as well as the solution process. On the other hand, the differential operator in equation (27) is of second order in the direction ϑ. Therefore, the physical boundary condition of (22) has to be modified carefully to preserve the physical meaning. By Lemma 7 below follows that the streamline diffusion scheme does not have this drawback. A solution method for the radiative transfer equation with a least-squares discretization is proposed by Ressel in [26] for one space dimension.

The approximation order of the streamline diffusion method for linear finite elements has been proven to be $3/2$ (and $k+\frac{1}{2}$ for Q_k-elements) on general grids, but there is evidence for second order convergence in the L^2-norm on nearly all computationally interesting grids. Unfortunately, a mathematical proof for second order is available only on regular grids. In [37] the construction of grids with lesser convergence rate is discussed. Those grids are constructed purposely to show optimality of the theoretical results and do not occur in real calculations due to our grid generation techniques.

Instead of the L^2-norm we define the stronger problem adapted norm

$$\|u\|_\delta^2 = \|\sqrt{\delta}\vartheta \cdot \nabla_x u\|_2^2 + \|\sqrt{\chi}u\|_2^2 + \|u\|_\Gamma^2, \qquad (28)$$

where

$$\|u\|_\Gamma^2 = \|u\|_{\vartheta,\Gamma}^2 = \frac{1}{2} \int_\Gamma u^2(x) \, |\mathbf{n}_\Gamma(x) \cdot \vartheta| \, dx. \qquad (29)$$

A crucial ingredient to the analysis of the streamline diffusion method is the strong stability of the method, that is, the bilinear form on the left of equation (25) admits the estimate

$$\int_\Omega (\vartheta \cdot \nabla_x u + \chi u) u \, dx + \int_\Omega (\vartheta \cdot \nabla_x u + \chi u) \delta \vartheta \cdot \nabla_x u \, dx + \int_{\Gamma_-^\vartheta} uu \mathbf{n} \cdot \vartheta \, ds \geq C \|u\|_\delta,$$

as soon as $\chi \in L^\infty(\Omega)$ and
$$\delta \leq \frac{1}{\chi}, \tag{30}$$
almost everywhere, where $\chi \neq 0$.

Since boundedness of the operator is obtained by the choice of the spaces, this stability estimate ensures convergence of the method. The weighted second order term in direction ϑ suffices to suppress oscillations in upwind direction occurring with the standard Galerkin scheme. This is supported by the following lemma due to [11], which establishes that the streamline diffusion operator mimics physical transport phenomena.

Lemma 7. *Let u_h solution of (25). For an arbitrary sub-domain $\Omega_1 \subset \Omega$ as in Figure 5 choose Ω_2 with $\Omega_1 \subset \Omega_2 \subset \Omega$ such that*

- *for all $x \in \Omega_2$, there holds $y = x - s \cdot \vartheta \in \Omega_2$ for $s > 0$ and $y \in \Omega$, i. e. all upstream points of x belong to Ω_2.*
- *For arbitrary points $x \in \Omega_1$ and $y \in \Omega - \Omega_2$ we split the difference vector $r = y - x$ in parts $r^\perp$ and $r^\parallel$ orthogonal and parallel to ϑ respectively. These parts are to fulfill $r^\parallel > ch \log \frac{1}{h}$ and $|r^\perp| > c\sqrt{h} \log \frac{1}{h}$.*

With these assumptions, the estimate
$$\|u - u_h\|_{L^2(\Omega_1)} \leq Ch^{k+1/2}\big(\|u\|_{H^{k+1}(\Omega_2)} + \|f\|_{L^2(\Omega)}\big)$$
holds.

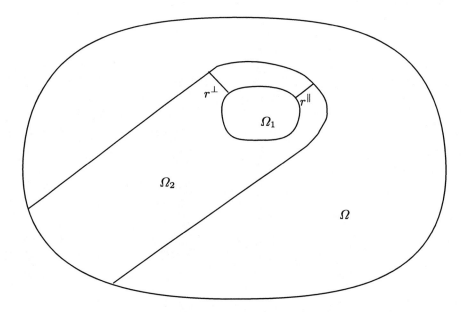

Fig. 5. Domain of dependency

Rannacher and Zhou proved a similar estimate for the maximum norm in [38]. It follows that influence of errors at point y on the solution at point x with $|x - y| = d$ decays exponentially with $\sqrt{d}$ in crosswind and with d in upwind direction. This result corresponds to the fact, that information is transported mostly in streamline direction mimicking the physical model. Due to this, we can assume the same inflow and outflow boundary conditions as in the original equation (25).

The main advantage of the streamline diffusion finite element method is its being a consistent discretization. This implies Galerkin orthogonality for the approximate solution. We can therefore apply the a posteriori error control techniques developed in Section 5.

4.2 The Discontinuous Galerkin Finite Element Method

The discontinuous Galerkin finite element method (DGFEM) sets out from the weak formulation (23), but it applies the weak boundary conditions not only on the inflow boundary Γ_-^ϑ, but on the inflow boundary of each grid cell. In detail, let T be a grid cell and ∂T_-^ϑ be the inflow part of ∂T as defined in (7) for Ω. Then, we require that

$$\int_T (\vartheta \cdot \nabla_x u + \chi u)v \, dx + \int_{\partial T_-^\vartheta} uv \, ds = \int_T fv \, dx + \int_{\partial T_-^\vartheta} u^\uparrow v \, ds \qquad (31)$$

for any test function $v \in C^1(T)$. Here, $u^\uparrow(x)$ is either the inflow boundary data $g(x)$ if $x \in \Gamma$ or the value of u on the cell $T^\uparrow$ on the other side of dT_-^ϑ. Then, the continuity requirement on the space V_h is dropped[5], such that the finite element functions on two different grid cells are completely independent.

It turns out that the discontinuous bilinear form has nearly the same stability properties as the streamline diffusion form in (25), and therefore the discontinuous Galerkin method converges with the same asymptotic behavior. We remark here that for linear problems the discretization error produced by the discontinuous Galerkin method is about the same as for the continuous method with the same number of grid cells. We should point out here, that this means a loss of a factor eight in number of degrees of freedom for trilinear finite elements, a number becoming smaller if higher order polynomials are used. Nevertheless, the discontinuous Galerkin method offers some advantages from the point of view of implementation. Furthermore, applied to advection or radiative transfer problems, it does not have a parameter which must be tuned.

In [19], we showed that the DGFEM for radiative transfer can be combined very efficiently with a DGFEM for the diffusion approximation in order to obtain an efficient preconditioner (see also [25] in this volume).

---

[5] In fact, the form in (31) is equal to the one in (23) if the functions in V_h are continuous

One important feature of the DGFEM becomes evident by a closer look at equation (31): since the coupling between grid cells is achieved exclusively through the boundary integrals, there is no flow of information in upwind direction at all, a property exhibited by the original problem, too. As a consequence, the resulting linear system can be solved in a single weep over all grid cells, starting at the inflow boundary Γ_-^ϑ.

5 Adaptive Mesh Refinement

Numerical computation of solutions to partial differential equations often requires a huge amount of memory and computation time. For economic reasons—considering radiative transfer problems even for the ability to compute at reasonable accuracy—it is inevitable to reduce the size of the discrete system to be solved. A well-known method to reach this goal is the exploitation of symmetry to reduce the dimension of a problem, but many problems need computation of a two-dimensional or the full three-dimensional model. Here, adaptive methods provide a flexible means to reduce costs of computation and—combined with a posteriori error estimates—still produce reliable and accurate solutions. This way, even radiative transfer problems can be solved on a workstation (at least in the two-dimensional case). Three-dimensional radiative transfer needs sophisticated grid adaption even on supercomputers to be able to compute with decent accuracy.

The adaptive solution of partial differential equations involves three components: first, a local error indicator determining areas of the grid where large contributions to the overall error are produced. Ideally, these indicators are tightly linked to an error estimate such that the method can produce a guaranteed accuracy on an efficient grid. It is important to understand that mesh refinement is *not* required where large errors occur, but where the error indicators are large. This accounts for error propagation. The error indicators are usually coupled to the residual of the discrete solution in the differential equation and are described in detail in subsection 5.3.

The second ingredient is a mesh generating algorithm transforming the output of the error indicator to a new mesh. Several options are discussed in subsection 5.2. Since we only implement the technique of local refinement, the treatment of so called hanging nodes is described in this subsection, too.

Finally, error indicators and mesh generator are coupled by an iterative procedure, the adaptive iteration itself. Since these three aspects are fairly independent of each other, we will describe them in the following subsections in detail, starting with the adaptive iteration as the governing process. Before going into details, we want to put this whole process into the general framework of optimization problems.

There are two aims in simulating models of physical problems: first, the display of the qualitative behavior of the physical quantity approximated. This means that visualizations of the results are produced showing the whole domain or parts of it. The norm in which the error should be measured is essentially the optical difference of these visualizations. Since only qualitative behavior is important a refinement indicator not tightly linked to an error estimate may be sufficient. It must be able to produce meshes capturing essential features of the solution economically. Indicators based on estimates for the so called energy error or the mean quadratic error are good candidates for reaching this goal.

The second aim is the comparison with real measurements. In this case, the numerical solution is post-processed by an operator imitating the physical measuring process. We denote this quantity of interest by the measurement functional $J(u)$. The goal of the simulation is the reproduction of this measurement with the error below a certain, prescribed tolerance. It may be written in the abstract form

$$\left|J(u-u_h)\right| \leq \text{TOL}. \tag{32}$$

Since the error $u-u_h$ is not known it will be replaced by an estimate $\eta_J(h, u_h)$ in Section 5.3.

The solution should be obtained with as few as possible resources, i. e. on a nearly optimal mesh in the finite element context. Using the function $h(x)$ denoting the local mesh width, this task reads as the following optimization process:

$$\begin{aligned} h(x) &= \max \quad \forall\, x \in \Omega \\ \eta_J(h, u_h) &\leq \text{TOL}. \end{aligned} \tag{33}$$

The iterative process applied to determine the optimal mesh function h must converge rapidly, since each step requires the solution of the discretized differential equation. In the first subsection, we present several heuristic optimization strategies for this problem.

5.1 Adaptive Algorithms

The iterative process for solving a partial differential equation essentially reads like the algorithm in Figure 6 on the following page. In the first step, we solve the problem $Au = f$ on a start grid (lines 9–11). Then, in each adaptive step, we adapt the grid according to the approximate solution u (5–7) and transfer u to the new mesh (8) to be used as start value of the iterative solution (11). This step can also be eliminated by starting with 0 on each mesh; using Krylov-space solvers, the loss in efficiency is not crucial. If the global estimate η is lower than a given tolerance (12–13) we stop the loop and do the post-processing.

There are two criteria for the efficiency of this adaptive algorithm:

1. optimality of the final grid and
2. speed of convergence.

```
1   Triangulation  tr := start_triangulation
2   Vector  u := 0;
3   for step :=0 to maxsteps
4       if (step ≠ 0)
5           adapt(tr, u);
6           tr.coarse();
7           tr.refine();
8           tr.interpolate(u);
9       Vector  b := tr.rhs(f);
10      Matrix  A := tr.matrix();
11      u := A⁻¹ b;
12      double  η := tr.estimate(u);
13      if (η ≤ TOL) break;
14  tr.postprocess(u);
```

Fig. 6. Adaptive Iteration

If a multi-grid method is used to solve the discrete linear system in each adaptive step, the weight is put on the first criterion. Indeed, Becker has shown in [3], that a tightly coupled adaptive multi-grid method is much faster than solving to a given tolerance and refining alternatingly.

As we are restricted to use iterative methods based on orthogonal sequences of vectors, it is just the other way round. Each refinement step means a restart of the iteration, since only the start vector is interpolated from the previous grid, not the whole iteration history. Due to the change in the operator, the orthogonality of other vectors is lost. Usually, the first steps of Krylov-space iterations are fast (cf. Figure 17 on page 95). Therefore, we restart the iteration from scratch in each adaptive step.

So we have to look for an refinement strategy, which is a good compromise between grid optimality and convergence rate, putting stress on the latter.

Given a local refinement indicator η_T indicating the relative error contribution of cell T, we consider the following strategies:

1. Refine cell T if $\eta_T \cdot \lambda > \gamma \text{TOL}$ with $\lambda = \#T$ the overall number of cells and γ a constant slightly below unity.
2. Refine cell T if $\eta_T > \alpha \hat{\eta}$. The comparison value $\hat{\eta}$ and relaxation parameter α offer additional choices:
   a) Comparison with the maximal error: $\hat{\eta} = \max_T \eta_T$ and $\alpha \in]0, 1[$.
   b) Comparison with the mean error: $\hat{\eta} = \sum \eta_T / \lambda$ and α about unity.
3. Refine the ν cells ("fixed number") or $\mu\lambda$ cells ("fixed fraction") having the largest error indicators η_T.
4. Refine the cells with largest error indicators until for the refined cells $\sum_T \eta_k \geq \alpha \eta$ holds with $\alpha \in (0, 1)$ ("fixed reduction").

Strategy 1 compares each local refinement indicator with the global tolerance. The weights are chosen such that local refinement stops automatically if the global criterion is reached. Indeed, we can replace the check of the global error by this local criterion. If this method is applied using λ of the existing grid, the number of cells may be wrong by a factor of up to 4 (8 in the three-dimensional case). Therefore, the cells should be sorted according to the size of the criteria and λ updated after each refined cell. The parameter γ ensures, that convergence is not asymptotically to TOL, but that the estimate reaches TOL in a few steps. As opposed to the other strategies, this one will enforce global refinement on the coarse grids. Only, when the estimate gets closer to the tolerance, local refinement will start. The computational results below show, that this indeed is a fast converging method, but the intermediate grids may be inefficient with respect to the estimated value.

Refinement strategies 2 to 4 try to equidistribute the error over the domain. Since the work load for each cell is the same, this should be a good heuristic for optimal grids. In fact, there is an optimality proof under certain conditions (cf. [9]). These methods attempt a sequence of "optimal" meshes for the estimated value in the sense, that each mesh is quasi-optimal for the accuracy obtained.

Method 3 is especially valuable, if a computation "as accurate as possible" is desired. Then, the parameter ν has to be determined by the remaining memory resources.

There has been a lot of discussion on the proof of convergence of adaptive algorithms, see e.g. [7], where this proof is achieved for the fixed reduction strategy and a local error indicator related to energy norm estimates. Since these proofs are of minor importance for the application, we prefer to point out the behavior of these methods for a concrete problem in radiative transfer simulation.

By introducing additional parameters, these strategies are generalized to double refinement for cells with especially large local error indicators and coarsening of cells with a small indicator. This way, a faster speed of convergence can be achieved and refinement that has not proven necessary can be undone. This is especially important for time dependent problems, but even with stationary equations the residual in one place can be reduced by better resolution in another.

We compare these refinement strategies applied to the dust cloud example in Figure 10. Parameter values are $\gamma = 0.8$ for strategy 1 and $\mu = 1/3$ for the fixed fraction strategy (strat3). The parameter α in strategy 2a is $\alpha = 0.01$. The results shown in Figures 7 and 8 are all produced with mesh sizes where the results still have not converged to their final value (the estimated relative error in the last step is about 1/6). We observe in Figure 7 on the following page, that all three methods generate grids of nearly the same efficiency; only the efficiency of the grids generated by strategy 1 is lower, but then this strategy needs only 5 steps to reach the final grid instead of 8-9.

The choice of the parameter α in Strategy 2 is important as may be seen in Figure 8 on the next page. While e.g. Verfürth in [32] proposes $\alpha \approx 0.5$ for Poisson's equation, a sufficient refinement rate is not achieved even for

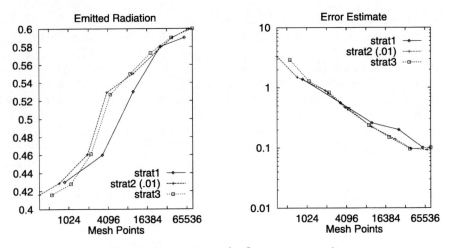

Fig. 7. Comparison of refinement strategies

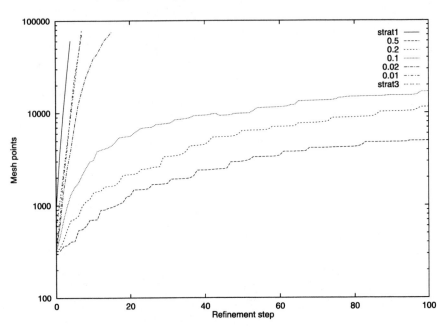

Fig. 8. Parameter dependence of the second strategy

$\alpha \in \{0.1, 0.2\}$ in our case; fewer and fewer cells are refined while the adaptive iteration advances. Only parameter values of $\alpha \in \{0.01, 0.02\}$ in our case yield the necessary convergence speed.

The convergence of the second strategy depends strongly on the parameter α and the structure of the problem. The dust cloud model problem has very localized features and the error contributions are limited to a very small portion of the domain. The result are very small refinement regions, often just a single cell if $\alpha > 0.1$. The necessary fine tuning of α makes this algorithm inadequate if we want to cover a wide range of applications. With the last strategy, we can easily control the growth rate of meshes and therefore the memory consumption and convergence speed by varying the parameter ν. Since the action of ν is obvious, this allows experimenting with different convergence rates, especially, if memory is insufficient to reach the desired accuracy. In this case, we can approximate the best possible value with the second strategy. Provided computing resources are sufficient and an optimal error estimate exists, the first strategy is the best to reach the prescribed tolerance, since it is very fast and it does not need parameter tuning.

5.2 Grid Generation

An important topic in designing adaptive algorithms is the generation of computational meshes. We can interpret the output of a local error indicator as a mesh function $h : \Omega \to \mathbb{R}$, describing the desired local mesh width. Different grid generators have been developed to produce meshes approximating this function. We discuss three prototypes in detail.

1. Execute the following three steps to compute a new mesh:
   a) Randomly distribute points in the domain Ω with local density $h^d(x)$.
   b) Connect these points using a so-called Delaunay algorithm to get a simplicial (triangular/tetrahedral) subdivision.
   c) Apply a smoothing method to avoid degenerate mesh cells.

   Algorithms of this kind have their particular advantage in being able to approximate a given mesh function quite accurately. Furthermore, a high regularity of the mesh cells can be achieved.

   On the other hand, transfer between meshes is cumbersome. In order to preserve accuracy, L^2-projection is required, involving the construction of quadrature rules for the required scalar products and the ability to locate quadrature points of one of the meshes in the other.

2. The "advancing front" generator begins with a triangulation of the boundary. It consecutively constructs layers of cells protruding to the interior and thus filling the whole domain. This method is known to have severe topological problems, if the domain is not convex or the boundary surface has rapidly changing curvature. Otherwise, it shares the properties of the previous method.

3. Finally, adaptive meshes can be constructed by successive local refinement of a very coarse starting triangulation. Curved boundaries are approximated by pulling division points of those boundary edges (faces in 3D) on the desired curve (surface). This generator is the only one, which generates the structures needed by multi-level methods. It is this last property, which made us decide for successive refinement, in particular in view of the method proposed in [19].

Table 1. Comparison of grid generation methods

distributed points	advancing front	local refinement
\multicolumn{3}{c}{Grid regularity}		
good, depends on the smoothing algorithm and is therefore a tradeoff to the approximation of $h(x)$	good if topologic problems of overlapping cells are solved	perfect for straight boundaries, tends to degenerate cells in critical boundary regions
\multicolumn{3}{c}{Approximation of mesh function $h(x)$}		
best approximation	good approximation	mediocre approximation, since the mesh width can only jump by factors of two between adjacent cells.
\multicolumn{3}{c}{Multi-grid}		
grid hierarchies have to be constructed a posteriori	same as left entry	grids are constructed as hierarchy, so multi-grid is inherent
\multicolumn{3}{c}{Transfer between adaptive grids}		
Needs point search algorithms and interpolation in cells. This often causes loss of accuracy.	same as left entry	uses multi-grid prolongation and restriction operators, therefore is highly accurate

The generation of non-uniform grids by using tensor products of arbitrary one-dimensional meshes is a strategy widely used. We do not consider its application, since the grids obtained are restricted too much to approximate the mesh function h. Additionally, this method often produces cells with a very high (10^4 and more) aspect ratio. High aspect ratios on the other hand impose difficulties for finding an optimal stabilization parameter for the streamline diffusion method. Furthermore, the resulting matrices are ill-conditioned and linear solvers may be slowed down.

"Hanging Nodes"

Refining a grid locally, there are edges between different levels of refinement and the problem of "hanging nodes" arises; see Figure 9 for an example. Using special interface cells not only disturbs the topology of the grid hierarchy, but causes difficulties implementing accurate grid transfers. We decided to choose numerical treatment of these nodes. Remark that the discontinuous Galerkin formulation automatically takes care of nonmatching cells sharing a common boundary as soon as a suitable quadrature rule on the boundary is chosen. Therefore, the following only applies to the continuous elements of subsection 4.1. It can be applied to elements with weaker continuity properties like the rotated bilinear element as well (see [22]). For theoretical treatment of our methods, we use the concept of "hierarchical bases" developed in [36].

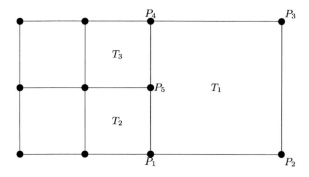

Fig. 9. A hanging node

For sake of simplicity, we will refer only to bilinear elements in the generic situation of Figure 9. It is obvious from the construction, that the method applies to all conforming shape functions.

Cell T_1 needs five shape functions, which are denoted in the hierarchical basis notation

- the 4 usual bilinear shape functions $\varphi_1 \ldots \varphi_4$ on a quadrilateral cell and
- 1 additional function φ_5 satisfying
  - the Lagrange interpolation condition $\varphi_5(P_i) = \delta_{5i}$ for all points of T_1,
  - $\varphi_5 = 0$ on the non-divided edges of T_1 and
  - φ_5 is linear on the edges $T_1 \cap T_2$ and $T_1 \cap T_3$.

We remark, that we do not require any knowledge of the behavior of φ_5 in the interior of T_1, because it will be eliminated later anyway. It should be chosen to offer good interpolating properties. Outside T_1, on T_2 and T_3, φ_5 has the shape of a standard bilinear finite element function.

While the hierarchical basis approach is natural for T_1, from the point of view of cells T_2 and T_3, the shape functions in P_1, P_4 and P_5 should be in nodal representation. We need a transformation between the hierarchical and nodal base functions $\{\varphi_i\}$ and $\{\psi_i\}$. For the construction, we evaluate a function $u_h = \sum u_i^H \varphi_i = \sum u_i^N \psi_i$ in the mesh points:

$$
\begin{aligned}
u(P_1) &= u_1^N = u_1^H \\
u(P_4) &= u_4^N = u_4^H \\
u(P_5) &= u_5^N = u_5^H + \tfrac{1}{2} u_1^H + \tfrac{1}{2} u_4^H
\end{aligned}
\tag{34}
$$

This results in the matrix C of basis exchange (only considering the interesting points)

$$C : u^H \mapsto u^N \tag{35}$$

$$\begin{pmatrix} u_1^N \\ u_4^N \\ u_5^N \end{pmatrix} = \begin{pmatrix} 1 & 0 & 0 \\ 0 & 1 & 0 \\ \tfrac{1}{2} & \tfrac{1}{2} & 1 \end{pmatrix} \begin{pmatrix} u_1^H \\ u_4^H \\ u_5^H \end{pmatrix} \tag{36}$$

At this point we have to decide, whether there should be a degree of freedom at point P_5. First, we consider the case of adding this degree of freedom. Interpolation estimates for T_2 and T_3 stay the same, while they are not worse (compared to the standard case) for T_1. As we use the complete matrix in nodal representation, the element matrices of T_2 and T_3 have the correct form. We have to modify the element matrix of cell T_1. We build it as usual

$$a_{ij}^H = a(\varphi_i, \varphi_j)_{T_1} \qquad i,j = 1,\ldots,5$$

The implementation of this algorithm is simplified by the following remark:

The hierarchical basis matrix with additional hanging nodes consists of two parts. A sub-matrix corresponding to the usual nodes coincides with the standard element matrix without hanging nodes. For each hanging node, one row and column are added due to the extra basis function.

To get the nodal representation required to build the global matrix, we apply the basis transfer

$$A^N = C A^H C^T \qquad (37)$$

Alternatively, the node P_5 may not carry a degree of freedom. This has algorithmic advantages in the use of element matrices, but we have to deal with the problem of node P_5 occurring in the matrices of T_2 and T_3 but not in T_1. It is of algorithmic advantage to let point P_5 be part of the mesh, so we search a treatment on matrix level. The hierarchical representation of $u(P_5)$ is a natural choice. The desired behavior is achieved by setting $u_5^H = 0$ and thus omitting base function φ_5. This way of canceling node values at P_5 clearly involves no modifications of the right hand side, as would be necessary in nodal representation. Accordingly, we have to apply the inverse base transformation as in (37), where C is deprived of the third row. (36):

$$\begin{pmatrix} u_1^H \\ u_4^H \end{pmatrix} = \begin{pmatrix} 1 & 0 & \frac{1}{2} \\ 0 & 1 & \frac{1}{2} \end{pmatrix} \begin{pmatrix} u_1^N \\ u_4^N \\ u_5^N \end{pmatrix}$$

This conforms to first averaging u_1^N and u_4^N to u_5^H, then applying the operator A and finally distributing the value of u_5^H to the neighboring points. The corresponding changes of base have to be applied to the right hand side generated by finite element integration:

$$f_1^H = f_1^N + f_5$$
$$f_4^H = f_4^N + f_5.$$

5.3 Error Estimates and Local Error Indicators

In this section, we discuss several methods of controlling the local mesh refinement by local error indicators. Here, the commonly used term *error indicator*

5 Adaptive Mesh Refinement

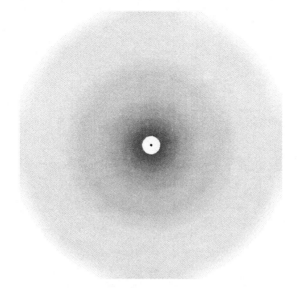

Fig. 10. Model example of a dust enshrouded star

is slightly inaccurate, since we consider this term as an indicator for necessary refinement only. In particular, it does *not* estimate the local error on a grid cell. As will be shown later in this section, sometimes these local indicators can be building blocks of an error estimate. In this case, they measure the *error production* of a grid cell, but not the local error itself. This difference is due to the fact that the error in a certain grid cell may propagate to other cells.

In order to illustrate the different effects of the methods considered, we introduce a model example of a dust enshrouded star in two dimension (see Figure 10). A small star (size 1/100 of the diameter of the domain) is surrounded by a whole of diameter 1/10. Then follows a scattering cloud with density distribution $\widehat{\chi}/r^2$.

A Priori Information

Linear problems like the radiative transfer equation in particular offer the chance to produce reasonably efficient meshes by exploiting information on the coefficients and the source function in the equation. For instance jumps in the source function or the density χ may be indicators that accurate discretization requires fine grid cells in a particular area.

A grid generation strategy could ascertain that the optical depth $h_T \chi$ of a mesh cell is limited, requiring that for each cell $h_T \leq c/\max_T \chi$. In applications we have in mind, this requirement usually would produce meshes mush too fine to be tractable on available computers. Therefore, it is not feasible. In fact, the solution is usually most smooth in areas of high optical depth and it can be approximated on rather coarse meshes.

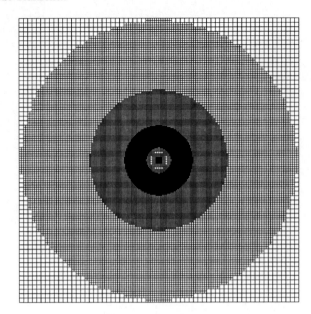

Fig. 11. A priori refined grid for the dust enshrouded star in Figure 10

A priori refinement strategies have the advantage that no iterative solution process is necessary for mesh construction. Therefore, they are quite cheap and it is a good idea to use such a method as starting mesh for the adaptive schemes below.

Using this approach of controlling grid refinement by χ, a grid for the enshrouded star problem may look like in Figure 11. In particular, all features of the data are recognizable (the empty whole around the star is refined since we do not allow a jump of more than one refinement level at a single edge).

Roughness Estimates

Setting out from the observation, that smooth functions can be approximated well on coarse meshes, measuring the "roughness" of the discrete solution on a given grid can indicate where refinement may be required.

Considering linear finite elements, these refinement indicators are based on the following heuristics: since globally linear functions can be represented exactly by the method, a large second derivative of the true solution u indicates that a fine mesh is required. Now, the only thing left is to do is to find an approximation to this second derivative only involving the finite element solution u_h.

In case of the streamline diffusion method with Q_1-elements, such an approximation may be a second difference quotient. Let T_1 and T_2 be two adjacent mesh cells, x_1 and x_2 their barycenters and h the distance vector between these two points. Then, the gradients $\nabla u_h(x_1)$ and $\nabla u_h(x_2)$ are computable and in many cases[6],

$$(h \cdot \nabla)\nabla u \approx \nabla u(x_1) - \nabla u(x_2),$$

which has to be divided by $|h|$ to get a directional derivative. Under the reasonable assumption that the distance vectors from the barycenter of one grid cell to those of its neighbors are linearly independent, the maximum of the norms of these vectors is an approximation to the maximum norm of the tensor $\nabla^2 u$.

This scheme can be modified to involve jumps of the normal derivatives at the edges. In the case of discontinuous elements, the jumps of the function itself are also an indicator.

If the roughness of a function has been determined for each cell, there is still the question of the weighting of cells of different size remaining. This can be answered heuristically again. The general approximation theory tells us that we loose one power of h for each derivative. Therefore, if we want to balance the error in L^2, we should use h^2 multiplied by an approximate second derivative.

Estimating Errors

In this section, we develop a posteriori error estimates for the Galerkin discretizations described above. The technique presented follows essentially the method proposed by Johnson et al. in [8] and by Rannacher et al. in [3,4,13,29]. Consider a solution u to the radiative transfer equation and u_h to its discretization. Let $e = u - u_h$ be the error function. Computing quantitative results for comparison with measurements, the desired accuracy is prescribed typically in the form

$$|J(e)| \leq \text{TOL}. \tag{38}$$

Here $J(.)$ is a functional describing the value to be computed by the simulation, e.g. a weighted mean value, the value at one point. A situation prototypical for astrophysical applications is displayed in Figure 12 on the next page. The object observed is far away and even in a strong telescope appears as a single dot only. As a result, only a single intensity value is measured at a certain wave length. This value is the integral over the part of the boundary facing the observer of the intensity of light in direction of the observer

$$J(u) = \int_{\Gamma_+(\vartheta_{\text{Obs}})} u(x, \vartheta_{\text{Obs}}) \, \mathbf{n}_\Gamma \cdot \vartheta_{\text{Obs}} \, ds. \tag{39}$$

Together with such a functional J we choose r_J such that

$$J(e) = \iint_{\Omega_S} e r_J \, d\vartheta \, dx. \tag{40}$$

---

[6] Here, some superconvergence is required, which in many cases holds at least in the interior of regular grids

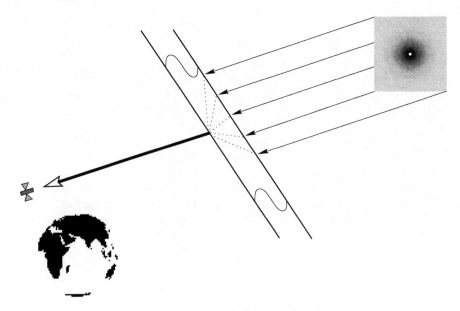

Fig. 12. A typical observer situation

Examples for r_J are

$$J(e) = \|e\|_{L^2} \qquad r_J = \frac{e}{\|e\|} \qquad (41)$$

$$J(e) = \frac{1}{|M|} \int_M e\, d\mu(x) \qquad r_J = \chi(M) \qquad (42)$$

$$J(e) = e(x_0) \qquad r_J = \delta_{x_0}. \qquad (43)$$

Here $\mu(x)$ denotes a suitable measure on M and $\chi(M)$ is the characteristic function of the set M defined as a distribution by

$$\int_\Omega \chi(M)\varphi\, dx = \frac{1}{|M|} \int_M \varphi\, d\mu(x) \qquad \forall\, \varphi \in C^\infty(\Omega)$$

and $\delta_{x_0} = \chi(\{x_0\})$ the Dirac functional.

In order to simplify the notation in the following paragraphs, we introduce some operator notation for the components of the radiative transfer equation. These operators are defined by the weak conditions

$$\iint_{\Omega_S} Tu(x,\vartheta)\varphi(x,\vartheta)\, d\vartheta\, dx = \iint_{\Omega_S} \vartheta\cdot\nabla_x u(x,\vartheta)\varphi(x,\vartheta)\, d\vartheta\, dx$$
$$+ \iint_{\Gamma_-} u(x,\vartheta)\varphi(x,\vartheta)|\mathbf{n}_\Gamma\cdot\vartheta|\, d\vartheta\, dx,$$

$$\iint_{\Omega_S} \Sigma u(x,\vartheta)\varphi(x,\vartheta)\, d\vartheta\, dx = \iint_{\Omega_S} \int_{S^2} P(\widetilde{\vartheta},\vartheta)u(x,\widetilde{\vartheta})\, d\widetilde{\vartheta}\varphi(x,\vartheta)\, d\vartheta\, dx.$$

With these abbreviations, the radiative transfer equation (4) can be rewritten as

$$\int\int_{\Omega S^2} \Big(T + \chi(x) + \Sigma\Big) u(x,\vartheta)\varphi(x,\vartheta)\, dx\, d\vartheta + \int\int_{\Gamma_-} u(x,\vartheta)\varphi(x,\vartheta)|\mathbf{n}\cdot\vartheta|\, ds\, d\vartheta$$

and its weak formulation analogue to (11) is

$$= \int\int_{\Omega S^2} B(\lambda, T(x))\varphi(x,\vartheta)\, dx\, d\vartheta + \int\int_{\Gamma_-} g(x,\vartheta)\varphi(x,\vartheta)|\mathbf{n}\cdot\vartheta|\, ds\, d\vartheta$$

For the error estimation, we need a so-called dual problem, which originates from the weak formulation by shuffling the operators onto the test function φ. Since we assumed the scattering phase function P symmetric in its arguments, we obtain by partial integration on the left hand side

$$\int\int_{\Omega_S} \Big(-T + \chi(x) + \Sigma\Big)\varphi(x,\vartheta) u(x,\vartheta)\, d\vartheta\, dx + \int\int_{\Gamma_+} u(x,\vartheta)\varphi(x,\vartheta)|\mathbf{n}\cdot\vartheta|\, ds\, d\vartheta.$$

Replacing φ by the dual solution z and u by a dual test function ψ, we obtain the dual problem

$$-Tz + \chi z - \Sigma z = r_J,$$

with boundary condition $z(x,\vartheta) = 0$ on Γ_+ (unless the evaluation functional is located on the outflow boundary, see below).

Neglecting boundary terms for simplicity, we represent the error in (38) in terms of z and the residual $\mathcal{R}(u_h) = (T + \chi - \Sigma)u_h - \kappa f$ by

$$J(e) = \int\int_{\Omega_S} e(x,\vartheta) r_J(x,\vartheta)\, d\vartheta\, dx \tag{45}$$

$$= \int\int_{\Omega_S} e(-T + \chi - \Sigma) z\, d\vartheta\, dx$$

$$= \int\int_{\Omega_S} (T + \chi - \Sigma) e z\, d\vartheta\, dx$$

$$= \int\int_{\Omega_S} (T + \chi - \Sigma) e (z - z_h)\, d\vartheta\, dx \tag{46}$$

$$= \int\int_{\Omega_S} (f - (T + \chi - \Sigma) u_h)(z - z_h)\, d\vartheta\, dx$$

$$= \int\int_{\Omega_S} \mathcal{R}(u_h)(z - z_h)\, d\vartheta\, dx, \tag{47}$$

In equation (46), we inserted an additional function z_h in the finite element space using a characteristic feature of finite element methods, the Galerkin orthogonality

$$\iint_{\Omega_S} (T + \chi - \Sigma)(u - u_h) w_h \, d\vartheta \, dx = 0 \qquad \forall \, w_h \in V_h. \qquad (48)$$

This function will allow us to use an additional approximation argument in order to obtain an optimal estimate.

We apply the formalism developed so far to the radiative transfer equation and summarize

Lemma 8. *Let u be solution to the radiative transfer equation (4) and u_h to a Galerkin discretization. Then, the error $J(u - u_h)$ admits the representation*

$$J(u - u_h) = \iint_{\Omega_S} (Tu_h + \chi u_h - \Sigma u_h - \kappa f)(z - z_h) \, d\vartheta \, dx \qquad (49)$$

where z is solution to the dual problem

$$-Tz + \chi z - \Sigma z = r_J \qquad \text{in } \Omega \qquad (50)$$
$$z = 0 \qquad \text{on } \Gamma_+. \qquad (51)$$

Now, the dual solution z is as unknown as the solution u of the original problem. Therefore, the error representation (49) can be used only if a reasonable approximation to $z - z_h$ can be found. Several possibilities exist, which fall in two classes: the first uses analytical results for the dual problem in order to obtain an estimate for $z - z_h$. This is the original approach by Eriksson and Johnson in [8] and we will present it in the subsection on the mean quadratic error below.

The second class actually computes an approximation of z and then uses this instead of z. Since the resulting estimator consists of local residuals weighted by factors computed by a dual problem, it is also called the dual weighted residual (DWR) method (see [2,4]). Again, several options exist and are explained here with the example of linear and d-linear polynomials

1. Computation of z with the same discretization. In this case, z is already in the finite element space and $z - z_h$ is zero in the optimal case. Therefore, some other means has to be found. It is obtained by the local approximation estimate

$$\|z - z_h\|_T \le C \|\nabla^2 z\|_T$$

   on each grid cell. Even if the second derivatives of a linear function are zero, they can be approximated fairly well by using difference quotients of first derivatives on neighboring cells.

2. Alternatively, z can be computed using a more accurate finite element space, usually by applying higher order polynomials, but using a finer grid is also possible. In this case, a projection of z into the original finite element space actually can be computed, such that we obtain a higher order approximation for (49). This approach was applied for instance in [1,20]. While it provides very good estimates, it suffers from the fact that the solution of the dual problem requires considerably *more* effort than the solution of the original radiative transfer problem.

3. These two approaches can be combined by first computing the dual solution with the same discretization and then using extrapolation techniques to produce a higher order approximation.

Boundary Integral Error

Now we consider a situation like in Figure 12 on page 80. The value desired from the computation is given by the functional in equation (39). Since the support of this functional is the outflow boundary Γ_+, the dual problem (50), (51) is replaced by

$$-Tz + \chi z - \Sigma z = 0 \qquad \text{in } \Omega$$
$$z = |\mathbf{n} \cdot \vartheta| \qquad \text{on } \Gamma_+.$$

Now we can continue at (47) to get an efficient a posteriori estimate (see also [4]) for this special value:

$$|J(e)| = \left| \iint_{\Omega_S} \mathcal{R}(u_h)(z - z_h) \, d\vartheta \, dx \right|$$
$$= \left| \sum_T \int_T \mathcal{R}(u_h)(z - z_h) \, dx \right| \qquad (52)$$
$$\leq C_{\text{Sec}} C_i \sum_T h_T^2 \|\mathcal{R}(u_h)\|_T \|\nabla^2 z\|_T,$$

with the solution z of the dual problem not depending on u_h. The constant C_i only depends on the interpolation estimate (58) for the finite element space. It is independent of the equation solved and depends on the polynomial degree and the shape of the mesh cells only.

Following the outline of the first method for assessing z described above, we compute z with the same discretization as u_h, obtaining a cellwise d-linear approximation $\tilde{z}$.

The first derivatives of finite element functions are bounded functions with discontinuities at cell boundaries. As a result, they are not in $H^1(\Omega)$, but only of bounded variation. The norm $\|\nabla^2 \tilde{z}\|_{L^2}$ is not defined for these functions. On the other hand, taking the norms over each cell separately destroys important information, which is obvious in the case of linear elements.

The second derivative can be viewed as a measure $d\mu$ with support on the cell interfaces. Let $\mathbb{E}$ be the union of all cell interfaces and $\mathfrak{n}_F$ be the normal vector on the face F. For each point $x \in F$ we define the jump $[\![.]\!]$ of a discontinuous function by

$$\mathfrak{n}_F [\![v_h]\!] = \mathfrak{n}_F \lim_{s \searrow 0} \big(v_h(x + s\mathfrak{n}_F) - v_h(x - s\mathfrak{n}_F)\big). \qquad (53)$$

We remark that $\mathfrak{n}_F [\![v_h]\!]$ is uniquely determined, even if there is an arbitrary choice of two normal vectors. The following investigation holds strictly for linear elements on simplicial cells only. It is accurate up to a constant for d-linear elements, though.

The second derivative $\partial_i \partial_j v_h$ of a piece-wise linear function understood as a measure on $C^0(\bar{\Omega})$ is

$$\int_\Omega \varphi\, d(\partial_i \partial_j v_h) = \int_\mathbb{E} \varphi\, [\![\partial_i v_h]\!]\, \mathbf{n}_F \cdot \mathbf{e}_j\, ds \quad \forall \varphi \in C^0(\bar{\Omega}),$$

where $\mathbf{e}_j$ denotes the unit vector of $\mathbb{R}^d$ in coordinate direction j. Restricting ourselves to the union of faces $\mathbb{E}$, we define a norm of the second derivative by

$$\|\nabla^2 v_h\|_\mathbb{E}^2 = \sum_{i,j=1}^d \int_\mathbb{E} |[\![\partial_i v_h]\!]\, \mathbf{n}_F \cdot \mathbf{e}_j|^2\, ds.$$

This second derivative applied to the computed dual solution $\tilde{z}$ will be approximated by the cell-wise constant difference quotient

$$|D_h^2(\tilde{z})|_T = \left(\sum_{F \in \partial T} \left| \frac{\partial \tilde{z}_i - \partial \tilde{z}_e}{h_F} \right|^2 \right)^{1/2}. \tag{54}$$

h_F denotes the distance between the cell centers left and right of an edge. In the case of even-sided triangles, these values coincide. Otherwise, they differ by a constant factor, which can be predetermined an computed for different cell shapes.

These considerations result in

Lemma 9. *The discretization error $J(u - u_h)$ is limited by*

$$|J(e)| \leq \eta = \sum_{T \in \mathcal{T}} \eta(T), \tag{55}$$

where $\eta(T)$ denotes the local error indicator

$$\eta(T) = C_{Sec} C_i h_T^2 \|\mathcal{R}(u_h)\|_T \, \|D_h^2 z\|_T. \tag{56}$$

The constant C_{Sec} has to be chosen such that errors replacing z by $\tilde{z}$ the approximating second derivatives by difference quotients are covered.

When we apply this concept to the dust enshrouded star problem, we obtain the grids in Figure 13, where we show several steps of the adaptive iteration. These meshes are not symmetric anymore like the one in Figure 11, because the asymmetric dual problem enters (ϑ_{Obs} is chosen pointing WSW).

The dual solution for the problem is shown in Figure 14. The four plots show the dual solution for four representative directions, where the spacing of

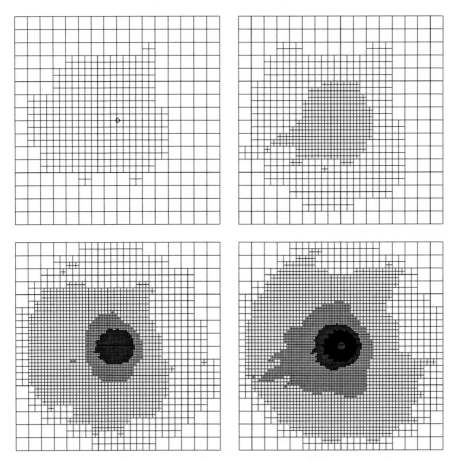

Fig. 13. Refinement history for the dust cloud (steps 1, 2, 4 and 6)

the isolines is the same for all of them. First, we see that the component for the observer direction ϑ_{Obs} (WSW) is larger by orders of magnitude, causing the stronger refinement in this direction. Still, the other components get scattered "light" from the cloud. Additionally, we see that all components, even the one facing in opposite direction of ϑ_{Obs}, are very small behind the star (seen from the observer). Therefore, refinement there is driven by the large residual of the solution u_h.

Remark 5. By exchanging each radiation direction with its opposite, the dual problem corresponds to the radiative transfer problem of a cloud (with a negligible dark star in its center) illuminated by an external light source infinitely distant in direction ϑ_{Obs}.

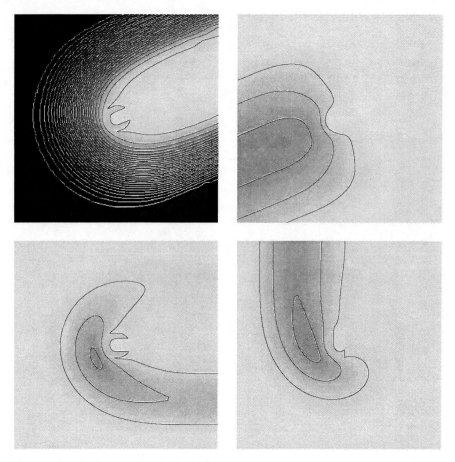

Fig. 14. Dual solutions for the boundary integral estimate. Four different radiation directions, WSW, ENE, west and south

Mean Quadratic Error

In this subsection we discuss estimates for the mean quadratic error in the space variable. We consider u to be the solution of a semi-discrete radiative transfer equation with a fixed number of ordinates. In this case, the solution of the problem is a mapping from Ω into $\mathbb{R}^m$, where m is the number of discrete angles.

The appropriate error functional for L^2-error estimates is $r_J = e/\|e\|$. We cannot solve the dual problem with this right hand side, since e is the unknown quantity. We therefore follow [8] and would like to try to use a stability estimate of the form

$$\|z\|_{H^\alpha(\Omega;\mathbb{R}^m)} \leq C_s \|e\|_{L^2(\Omega;\mathbb{R}^m)}. \tag{57}$$

Here, the norm in $H^\alpha(\Omega;\mathbb{R}^m)$ denotes the norm of some derivatives of order $\alpha > 0$. Then, we could exploit an approximation estimate of the form

$$\|z - I_h z\|_{L^2(\Omega;\mathbb{R}^m)} \leq C_i h^\alpha \|z\|_{H^\alpha(\Omega;\mathbb{R}^m)}, \tag{58}$$

with $\alpha > 0$.

Lemma 10. *Assume estimates (57) and (58) hold for the solution z of the dual problem and some $\alpha \geq 0$. Then, we obtain a residual based error estimate*

$$\|u\|_{L^2(\Omega;\mathbb{R}^m)} \leq C \|h^\alpha \mathcal{R}(u_h)\|_{L^2(\Omega;\mathbb{R}^m)}. \tag{59}$$

Proof. We continue (47) using Hölder inequality:

$$\begin{aligned}
\|u\|_{L^2(\Omega;\mathbb{R}^m)}^2 &= \int_\Omega ee\, dx = \int_\Omega \mathcal{R}(u_h)(z - z_h)\, dx \\
&\leq \sum_{T \in \mathbb{T}} \|\mathcal{R}(u_h)\|_{L^2(T;\mathbb{R}^m)} \|z - z_h\|_{L^2(T;\mathbb{R}^m)} \\
&\leq C_i \sum_{T \in \mathbb{T}} h_T^\alpha \|\mathcal{R}(u_h)\|_{L^2(T;\mathbb{R}^m)} |z|_{H^\alpha(T;\mathbb{R}^m)} \\
&\leq C_i \|h^\alpha \mathcal{R}(u_h)\|_{L^2(\Omega;\mathbb{R}^m)} |z|_{H^\alpha(\Omega;\mathbb{R}^m)} \\
&\leq C_i C_s \|h^\alpha \mathcal{R}(u_h)\|_{L^2(\Omega;\mathbb{R}^m)} \|u\|_{L^2(\Omega;\mathbb{R}^m)}
\end{aligned}$$

Dividing by the L^2-error yields the result.

If the solution u is smooth, the optimal a priori error estimate for general grids is $h^{3/2}$. The residual converges to zero with first order. To obtain optimal bounds, we would like to have $\alpha = 1/2$, yielding the error estimate

$$\|e\| \leq C \|\sqrt{h} \mathcal{R}(u_h)\|. \tag{60}$$

This estimate requires the boundedness of derivatives of z in (57), but it holds that the operator on the left hand side of (50) allows control only of $\|z\|_{\Omega \times S^2}$ and $\|\vartheta \cdot \nabla_x z\|_{\Omega \times S^2}$, but not of the gradient ∇z. Consequently, standard approximation theory is not very helpful here. This is the more true considering that the analytical constant C_s in equation (57) becomes very large in the presence of varying coefficients.

Due to these problems, we do not attempt to estimate the mean quadratic error, but we still use grids that are optimal for an estimate of this error by a heuristic argument. First, we observe that the dual solution u is usually smooth in most of the domain. Additionally, if the grids show at least some regularity, the error will be asymptotically $\|e\| = \mathcal{O}(h^2)$ (see e.g. [24,37,38]). Therefore, we choose α in (59) to obtain an estimate of order h^2. Since the residual converges of first order, if the solution is smooth, this leads to $\alpha = 1$.

Using this method yields efficient grids for getting a general overview over the main features of the solution. Since these grids resemble the one in Figure 11 very much, we do not display them here.

Table 2. Comparison between indicators based L^2-error and boundary integral error control

L^2-indicator		boundary-indicator		estimates	
points	value	points	value	η	$\frac{C_{\text{Sec}}\eta}{e}$
564	0.181	576	0.417	3.1695	23.77
1105	0.210	1146	0.429	1.0804	8.62
2169	0.311	2264	0.461	0.7398	7.11
4329	0.405	4506	0.508	0.2861	3.94
8582	0.460	9018	0.555	0.1375	3.33
17202	0.488	18857	0.584	0.0526	2.39
34562	0.537	39571	0.599	0.0211	1.76
68066	0.551	82494	0.608	0.0084	1.40

Comparison

As soon as a functional of the solution is to be approximated, the approach using the computed dual solution should be preferred. We show this at the dust enshrouded star example of Figure 10, using the observer functional (39). In Table 2, we display the value of the output functional on a series of grids, obtained by the same refinement strategy, but with the two different estimators. Clearly, the weighted residual estimator for the boundary integral yields a sequence converging much faster to a limit value. In the last columns, we display the value of this estimator and its efficiency index. Here, the error was determined by extrapolating the sequence of values. The security constant C_{Sec} was chosen to be $3/2$.

In particular, we observe that accuracy on a grid with 8,500 points generated with the dual weighted residual estimator is the same as on a grid with 68,000 points with the L^2-indicator. Consequently, we can save about a factor of eight in memory and about four in computation time (since two problems have to be solved for the DWR estimator). These are considerable savings, in particular since the grids in Figure 13 show that the refinement is mostly driven by the residual and the dual solution introduces just a small anisotropy.

By these results it becomes evident, that heuristic refinement criteria may be helpful when a qualitative study is conducted. As soon as simulated and measured values are to be compared quantitatively, the DWR approach is much superior.

6 Solution of the Discrete Systems

6.1 Properties of the Discrete System

We give a short description of the matrices resulting from the various discretizations introduced in this article. From a linear algebra point of view, the

6 Solution of the Discrete Systems

discrete system has the form

$$Ax = b \tag{61}$$

where $x, b \in X = \mathbb{R}^{nm}$ and $A : X \to X$, n being the number of degrees of freedom of the space discretization and m the number of ordinates.

From the radiative transfer equation (4), we derive the splitting

$$A \equiv T_h + M_h(\kappa) + S_h = b, \tag{62}$$

with T_h the discretization of the advection operator $\vartheta \cdot \nabla_x$., a suitable discretization $M_h(\kappa)$ of the multiplication with the extinction coefficient $\kappa(x)$ depending on the space variable and S_h the discrete scattering operator.

For our tensor product discretizations, these operators may be split up further, assuming an ordering of the degrees of freedom by the angular variable:

$$T_h = \text{diag}(w_1 \mathfrak{T}_1, \ldots, w_m \mathfrak{T}_m)$$
$$M_h(\kappa) = \text{diag}(w_1 \mathfrak{M}_1(\kappa), \ldots, w_m \mathfrak{M}_m(\kappa))$$
$$S_h = \begin{pmatrix} w_1 \mathfrak{M}_1(\sigma) & & \\ & \ddots & \\ & & \mathfrak{M}_m(\sigma) \end{pmatrix} - \begin{pmatrix} w_{11} \mathfrak{M}_1(\sigma) & \cdots & w_{1m} \mathfrak{M}_1(\sigma) \\ \vdots & & \vdots \\ w_{m1} \mathfrak{M}_m(\sigma) & \cdots & w_{mm} \mathfrak{M}_m(\sigma) \end{pmatrix}. \tag{63}$$

The splitting between $M_h(\kappa)$ and S_h may seem quite odd (instead of using $M_h(\chi)$), but it elucidates the eigenvalue structure of the problem much better. In fact, S_h is a discrete energy preserving scattering operator analogous to equation (2). In order for the numerical scheme to reflect the features of the equation properly, it is of highest importance that S_h is energy preserving, that is, has a zero eigenvalue (see below).

With the finite element streamline diffusion tensor product discretization, the entries of these matrices are defined through

$$\mathfrak{T}_i^{jk} = \int_\Omega (\varphi_j + \delta \vartheta_i \cdot \nabla_x \varphi_j) \vartheta_i \cdot \nabla_x \varphi_k \, dx \tag{64}$$
$$+ \int_{\Gamma_-^{\vartheta_i}} \varphi_j \varphi_k |\mathbf{n}_\Gamma \cdot \vartheta_i| \, ds$$

$$\mathfrak{M}_i^{jk}(\chi) = \int_\Omega \chi(\varphi_j + \delta \vartheta_i \cdot \nabla_x \varphi_j) \varphi_k \, dx \tag{65}$$

$$\mathfrak{M}_i^{jk}(\sigma) = \int_\Omega \sigma(\varphi_j + \delta \vartheta_i \cdot \nabla_x \varphi_j) \varphi_k \, dx. \tag{66}$$

For the discontinuous Galerkin discretization, this structure is even simpler, since the mass matrix does not involve the directional stability. Therefore, let T be the support of the test function φ_j and

$$\mathfrak{T}_i^{jk} = \int_T \varphi_j \vartheta_i \cdot \nabla_x \varphi_k \, dx + \int_{\partial T_-^\vartheta} \varphi_j |\beta \cdot \mathbf{n}| \varphi_k \, ds \qquad (67)$$

$$\mathfrak{M}_i^{jk}(\chi) = \mathfrak{M}^{jk}(\chi) = \int_\Omega \chi \varphi_j \varphi_k \, dx_T \qquad (68)$$

$$\mathfrak{M}_i^{jk}(\sigma) = \mathfrak{M}^{jk}(\sigma) = \int_\Omega \sigma \varphi_j \varphi_k \, dx_T. \qquad (69)$$

We now investigate the eigenvalue structure of these matrices. We use the example of the discontinuous Galerkin method, since its mass matrices are symmetric and thus yield a simpler structure of the spectrum. Still, similar results hold for the SDFEM, if the penalty parameter is chosen according to (26).

For any continuous function χ, the maximal and minimal eigenvalues of $\mathfrak{M}(\chi)$ are

$$\mu_{\min}(\chi) \approx \min_{T \in \mathbb{T}_\Omega} |T| \min_{x \in T} \chi(x) \approx \min_{T \in \mathbb{T}_\Omega} h_T^d \min_{x \in T} \chi(x)$$

$$\mu_{\max}(\chi) \approx \max_{T \in \mathbb{T}_\Omega} |T| \max_{x \in T} \chi(x) \approx \max_{T \in \mathbb{T}_\Omega} h_T^d \max_{x \in T} \chi(x)$$

Similar, more refined estimates hold in the case of discontinuous coefficients. We denote by $\nu(\kappa)$ the eigenvalues of $T_h + M_h(\kappa)$ and obtain for their real parts

$$\Re \nu_{\min}(\kappa) \geq \mathcal{O}(h) + \mu_{\min}(\kappa),$$
$$\Re \nu_{\max}(\kappa) \leq \mathcal{O}(1) + \mu_{\max}(\kappa).$$

In order to simplify the next steps, we assume that the phase function P is constant, yielding $w_i = 1/(4\pi m)$ (or $1/(2\pi m)$ in two dimensions).

$$S_h = \begin{pmatrix} (1 - \frac{1}{m})\mathfrak{M}(\sigma) & -\frac{1}{m}\mathfrak{M}(\sigma) & \cdots & -\frac{1}{m}\mathfrak{M}(\sigma) \\ -\frac{1}{m}\mathfrak{M}(\sigma) & \ddots & & \vdots \\ \vdots & & \ddots & \vdots \\ -\frac{1}{m}\mathfrak{M}(\sigma) & \cdots & -\frac{1}{m}\mathfrak{M}(\sigma) & (1 - \frac{1}{m})\mathfrak{M}(\sigma) \end{pmatrix} \qquad (70)$$

Let v be an arbitrary vector in $\mathbb{R}^n$. Then, $(v, \ldots, v)^T$ is an eigenvector of S_h with eigenvalue zero. If on the other hand $(\alpha_1, \ldots, \alpha_m)^T$ is a vector with mean value zero, then

$$S_h \begin{pmatrix} \alpha_1 v \\ \vdots \\ \alpha_m v \end{pmatrix} = \begin{pmatrix} \mathfrak{M}(\sigma)v \\ \vdots \\ \mathfrak{M}(\sigma)v \end{pmatrix}. \qquad (71)$$

we see by choosing $(\alpha_1 v, \ldots, \alpha_m v)^T$, that we have another $(m-1)n$ eigenvalues in the interval $[m\mu_{\min}(\sigma), m\mu_{\max}(\sigma)]$.

We conclude that the real parts of the eigenvalues λ of the radiative transfer matrix A are within two (possibly overlapping) intervals of the positive real axis, namely

$$\lambda \in [\Re\nu_{\min}(\kappa), \Re\nu_{\max}(\kappa)] \cup [m\mu_{\min}(\sigma) + \Re\nu_{\min}, m\mu_{\max}(\sigma) + \Re\nu_{\max}]. \quad (72)$$

From this analysis, we deduce some important properties of the linear system. Here, we restrict ourselves to the cases where the coefficients κ and σ and the mesh size h are of the same order of magnitude everywhere in the domain. The case of strongly varying coefficients is not attainable to such simple arguments.

1. If σ is small everywhere, then the spectrum of A is similar to the spectrum of $T_h + M_h(\kappa)$ and the whole system can be solved efficiently by the Gauß-Seidel method.
2. If σ and κ are large[7], the spectrum consists of two clusters of eigenvalues, one around $mh^d\kappa$ and one around $mh^d\sigma$. This is a situation perfectly suited for Krylov-space methods like Bicgstab or GMRES (cf. [28]).
3. If σ is large and κ is small, the interesting case from the point of view of applications, then, the real parts of the eigenvalues are between $\mathcal{O}(h)$ and $mh^d\sigma$. Here, the lower cluster of eigenvalues is close to zero, a situation not immediately tractable by Krylov-space methods. In fact, it turns out that while considerable improvements compared to the Λ-iteration can be made by the techniques in the following subsections (e.g. reducing the number of steps from $\mathcal{O}(\sigma^2)$ to $\mathcal{O}(\sigma)$), only more involved preconditioners can avoid hundreds of iteration steps for high opacities.

6.2 Iterative Methods

The Λ-Iteration

Common practice in solving system (62) is an alternation between solving the transport part of the system and applying the integral operator, the so called Λ-iteration in Figure 15. In numerical analysis of integral equations, it is known as Picard-iteration, while it is called source iteration in application fields.

$$(T_h + M_h(\chi))x^{(k+1)} = (-S_h - M_h(\sigma))x^{(k)} + b. \quad (73)$$

It is clear, that a limit to this iteration, where $x^{(k+1)} = x^{(k)}$, is a solution to (62). It is also evident that the right hand side does not change much between two steps if σ is small. Therefore, it is expected and indeed observed that this method converges very fast for optically thin media. In optically thick regimes, this method becomes intolerably slow, such that improvements have to be found.

---

[7] large compared to the grid spacing, that is $\sigma, \kappa > 1/h$.

$u^{(0)} := u_0$
for $i := 1$ to *steps*
    for $k := 1$ to m
        $v_k^{(i-1)} = \sum_{l=1}^{N} \omega_{kl} u_l^{(i-1)}$
    for $k := 1$ to N
        $u_k^{(i)} := T_k^{-1} v_k^{(i-1)} + F_k$

Fig. 15. The Λ-Iteration

To this end, we first rewrite (73) in order to obtain a defect correction scheme. Subtracting $(T_h + M_h(\chi))x^{(k)}$ on both sides yields

$$(T_h + M_h(\chi))(x^{(k+1)} - x^{(k)}) = (-S_h - M_h(\sigma))x^{(k)} + b. \tag{74}$$

Defining the preconditioning matrix

$$C_\Lambda = \mathrm{diag}(\mathfrak{T}_i + \mathfrak{M}_i(\chi))^{-1},$$

we obtain the scheme

$$x^{(k+1)} = x^{(k)} + C_\Lambda(b - Ax^{(k)}). \tag{75}$$

Written this way, the Λ-iteration resembles the famous block-Jacobi method. The only difference consists in the preconditioner C, which would be

$$C_J = \mathrm{diag}(\mathfrak{T}_i + \mathfrak{M}_i(\chi) - \omega_{ii}\mathfrak{M}_i(\sigma))^{-1}$$

for the block-Jacobi scheme. Turek compared the preconditioners C_Λ and C_J in [30], concluding that C_J is slightly more efficient, while not causing any more numerical effort and therefore should be preferred.

It is well known, that the convergence of iterative schemes in the form (75), if applied to symmetric positive definite matrices, is limited by the condition number $\mathrm{cond}_2(CA) = \lambda_{\max}(CA)/\lambda_{\min}(CA)$ of the preconditioned matrix. The errors after two consecutive steps admit the estimate

$$\frac{\|x^{(k+1)} - x\|}{\|x^{(k)} - x\|} \leq \frac{\mathrm{cond}_2(CA) - 1}{\mathrm{cond}_2(CA) + 1}.$$

It is observed that in the course of the iteration the left hand side of this estimate converges to the right hand side, such that the estimate is optimal.

For non-symmetric matrices, estimates of this type are more difficult, but a lower bound for the convergence of the method can still be obtained by using the real parts of the eigenvalues in the definition of the condition number.

Both of the preconditioners above have in common that they solve the transport problem in the absence of scattering exactly. Therefore, $\mathrm{cond}_2(CA) \approx 1$ and the method converges in a single step. On the other hand, they do not even attempt to solve the scattering problem. Therefore, the condition number of CA is very large in this case according to (72). Alas, even the preconditioned iterative scheme converges only in a large number of steps.

A different approach to preconditioning exploits the fact that under certain conditions, the solution of the moment equations can be used as approximations to the solutions. The SDA and QD methods (see [23] and the literature cited there) use zero and first order moments, while the MMP method (see [19,25]) uses the zero order moment only. These preconditioners prove to be efficient even in the scattering dominated case.

Additionally to improving preconditioners, it is advisable to study faster and more robust iteration schemes, the so-called Krylov-space methods.

Krylov Space Methods

When the linear iterative scheme (75) is performed with start vector $x^{(0)} = 0$, then the residual $r^{(k)} = Ax^{(k)} - b$ evolves like

$$r^{(0)} = b,$$
$$r^{(1)} = b - ACb,$$
$$r^{(2)} = b - 2ACb + (AC)^2 b,$$
$$r^{(3)} = b - 3ACb + 3(AC)^2 b - (AC)^3 b,$$
$$\ldots$$

Therefore, the residual after the k-th step is in the affine space

$$K_k(b, A) = \mathrm{span}\{A^i b\}_{i=0,\ldots,k},$$

the k-th Krylov space of the matrix A and the vector b. In this sense, the iterates of the linear scheme are just a fixed sequence in these Krylov-spaces.

The aim of Krylov-space methods is now to actually choose an iterate $x^{(k)}$ in $K_k(b, A)$. The GMRES method (see e.g. [28]) for instant chooses $x^{(k)}$ such that the residual $r^{(k)}$ is the smallest possible in $K_k(b, A)$. By orthogonalization of the residual vectors, this minimization can be reduced from an nm-dimensional problem to a k-dimensional one, being comparably cheap. Since every vector in $K_k(b, A)$ can be represented as a polynomial in A applied to the vector b, the following estimate can be derived.

Lemma 11. *The residual norm after the k-th iteration step is estimated by*

$$\|r^{(k)}\| \leq \eta(CA, k) \|r^{(0)}\| \tag{76}$$

with η being the maximal value on the spectrum $\sigma(CA)$ of an optimal polynomial,

$$\eta(CA, \nu) = \min_{\substack{p \in P_n \\ p(0)=1}} \max_{\lambda \in \sigma(CA)} |p(\lambda)|.$$

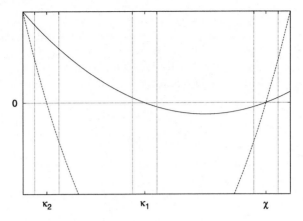

Fig. 16. Optimal polynomials for eigenvalue clusters around χ and $\kappa_1 = 0.5$ (solid) and $\kappa_2 = 0.1$ (dashed)

This lemma shows that not only the condition number of the preconditioned matrix enters into the convergence estimate, but also their distribution. For instance, take a matrix with only two eigenvalues $\lambda_{1/2} \neq 0$. Since a quadratic polynomial p can be found with $p(0) = 1$ and $p(\lambda_1) = p(\lambda_2) = 0$, the GMRES scheme converges in two steps, while the linear iterative scheme may be very slow, depending on the eigenvalues. If now the eigenvalues of the system are clustered around the two values $\lambda_{1/2} \neq 0$, the situation we are in for $\sigma, \kappa \gg 1$, then this polynomial will be small on both clusters (since the derivative of the polynomial is small), and the GMRES method yields a good error reduction already in the first two steps. This situation is explained in Figure 16: shown are clusters around the eigenvalues *chi* and κ, as well as the optimal polynomials for two configurations. The figure shows that the maximum of $|p(x)|$ is much smaller on both clusters for $\kappa_1 = 0.5$, than for the less favorable situation with $\kappa_2 = 0.1$.

Even if the eigenvalues are not clustered in a favorable way, a convergence estimate depending on the condition number in much a better way than for the linear method can be derived.

Lemma 12. *If all eigenvalues of CA are real and positive, the residual after the $k + 1$-st GMRES iteration step admits the estimate*

$$\frac{\|r^{(k+1)}\|}{\|r^{(k)}\|} \leq \frac{\sqrt{\mathrm{cond}_2(CA)} - 1}{\sqrt{\mathrm{cond}_2(CA)} + 1}.$$

The disadvantage of the GMRES method is the fact that it requires one additional auxiliary vector in each iteration step. While this may be tolerable up to a number of maybe 30, it becomes completely unbearable if hundreds of iteration steps are needed. Since radiative transfer problems are particularly demanding with respect to memory, it is advisable to look for a method not

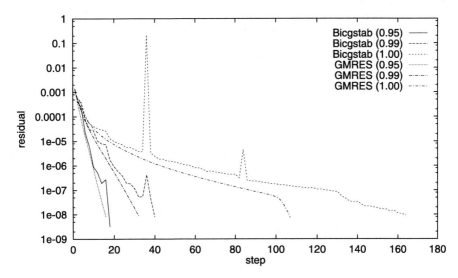

Fig. 17. Comparison of Bicgstab and GMRES with preconditioner C_J for different scattering parameters

as lavish. Such a method is the Bicgstab method of van der Vorst (cf. [31]), which uses only 8 auxiliary vectors. On the other hand, it does not minimize the residual or any other quantity over the Krylov spaces. Instead, it attempts to accelerate convergence by mutual orthogonalization of two sequences. Indeed, the Bicgstab algorithm proved rather promising in Turek's paper [30]. It usually reduces the errors much faster than stationary methods, but it shows a very irregular convergence history. We compare it to GMRES, the only method for non-symmetric systems, which really minimizes the error in each step (thus, the only one for which the estimate (76) holds). The results are shown in Figure 17 for a problem with $\chi = 100$ on a fairly fine grid. We display the norm of the residual over the number of matrix-vector-multiplications, since Bicgstab uses two of them per step. We observe, that GMRES converges faster for moderate ($\sigma = 0.95\chi$) scattering values as well as the equilibrium case $\sigma = \chi$. Furthermore, the norm of the residual vector grows by several orders of magnitude for once in a while. These outbursts are due to a near failure of the orthogonalization process. It should be mentioned here, that Bicgstab, while we have very good experiences with the algorithm, is susceptible to so-called breakdowns: due to its delicate orthogonalization process, it may happen that the next basis vector for a Krylov space becomes zero, but the system is still not solved. In this case, the method must be restarted.

On the other hand, the drawback of the GMRES method can be relieved by restarts for instance every 20–30 iteration steps. In conclusion, both methods seem to be adequate to solve radiative transfer problems and a choice should be taken according to the question whether memory or time efficiency is more important.

Acknowledgements. The results presented in this article were produced by software based on the original DEAL library developed with Franz-Theo Suttmeier [21].

References

1. W. Bangerth and R. Rannacher. Finite element approximation of the acoustic wave equation: Error control and mesh adaptation. *East-West J. Numer. Math.*, 7(4):263–282, 1999.
2. W. Bangerth and R. Rannacher. *Adaptive Finite Element Methods for Solving Differential Equations.* Birkhäuser, Basel, 2003.
3. R. Becker. *An Adaptive Finite Element Method for the Incompressible Navier-Stokes Equations on Time-Dependent Domains.* Dissertation, Universität Heidelberg, 1995.
4. R. Becker and R. Rannacher. An optimal control approach to a posteriori error estimation in finite element methods. *Acta Numerica*, 10:1–102, 2001.
5. A. Brooks and T.J.R. Hughes. Streamline upwind/Petrov-Galerkin formulation for convection dominated flows with particular emphasis on the incompressible Navier-Stokes equations. *Comput. Methods Appl. Mech. Engrg.*, 32:199–259, 1982.
6. R. Dautray and J.-L. Lions. *Mathematical Analysis and Numerical Methods for Science and Technology*, volume 6. Springer, 2000.
7. W. Dörfler. A convergent adaptive algorithm for Poisson's equation. *SIAM J. Numer. Anal.*, 33:1106–1124, 1996.
8. K. Eriksson and C. Johnson. Adaptive finite element methods for parabolic problems I: A linear model problem. *SIAM J. Numer. Anal.*, 28:43–77, 1991.
9. J.P.d.S.R. Gago, D.W. Kelly, O.C. Zienkiewicz, and I. Babuška. A posteriori error analysis and adaptive processes in the finite element method: Part II— Adaptive mesh refinement. *Internat. J. Numer. Methods Engrg.*, 19:1621–1656, 1983.
10. C. Johnson. *Finite Element Methods for Partial Differential Equations.* Studentlitteratur, Lund, 1993.
11. C. Johnson, U. Nävert, and J. Pitkäranta. Finite element methods for linear hyperbolic problems. *Comput. Methods Appl. Mech. Engrg.*, 45:285–312, 1984.
12. G. Kanschat. Parallel adaptive algorithms for radiative transfer problems. In P. Fritzson and L. Finmo, editors, *Parallel Programming and Applications*, volume 45 of *Transputer and OCCAM Engineering Series*, pages 238-243, Amsterdam, 1995. IOS Press.
13. G. Kanschat. *Parallel and Adaptive Galerkin Methods for Radiative Transfer Problems.* Dissertation, Universität Heidelberg, 1996. Preprint SFB 359, 1996-29.
14. G. Kanschat. Efficient and reliable solution of multi-dimensional radiative transfer problems. In F. Karsch, B. Monien, and H. Satz, editors, *Multiscale Phenomena and Their Simulation*, pages 245-249, Singapore, 1997. World Scientific.
15. G. Kanschat. New algorithms for radiative transfer in accretion disks and surroundings. In D.T. Wickramasinghe, G.V. Bicknell, and L. Ferrario, editors, *Accretion Phenomena and Related Outflows; IAU Colloquium 163*, pages 736-737, San Francisco, California, 1997. Astronomical Society of the Pacific.

16. G. Kanschat. A robust finite element discretization for radiative transfer problems with scattering. *East-West J. Numer. Math.*, 6(4):265–272, 1998.
17. G. Kanschat. Parallel computation of multi-dimensional neutron and photon transport in inhomogeneous media. In H.-J. Bungartz, F. Durst, and C. Zenger, editors, *High Performance Scientific and Engineering Computing*, volume 8, pages 431–440. Springer, 1999.
18. G. Kanschat. Solution of multi-dimensional radiative transfer problems on parallel computers. In P. Bjørstad and M. Luskin, editors, (eds.), *Parallel Solution of Partial Differential Equations*, volume 120 of *IMA Volumes in Mathematics and its Applications*, pages 85-96. New York, 2000. Springer.
19. G. Kanschat and E. Meinköhn. Multi-model preconditioning for radiative transfer problems. Preprint 2004-33, SFB 359, Heidelberg, 2004. submitted.
20. G. Kanschat and R. Rannacher. Local error analysis of the interior penalty discontinuous Galerkin method for second order elliptic problems. *J. Numer. Math.*, 10(4):249–274, 2002.
21. G. Kanschat and F.-T. Suttmeier. Datenstrukturen für die Methode der finiten Elemente. unpublished, Bonn-Venusberg, 1992.
22. G. Kanschat and F.-T. Suttmeier A posteriori error estimates for nonconforming finite element schemes. *Calcolo*, 36(3):129–141, 1999.
23. E.W. Larsen. Transport acceleration methods as two-level multigrid algorithms. In *Modern Mathematical Methods in Transport Theory*, pages 34–47, Basel, 1989. Birkhäuser.
24. P. LeSaint and P.-A. Raviart. On a finite element method for solving the neutron transport equation. In C. de Boor, editor, *Mathematical aspects of finite elements in partial differential equations*, pages 89–123, New York, 1974. Academic Press.
25. E. Meinköhn. A general-purpose finite element method for 3d radiative transfer problems in moving media. In this volume, 2005.
26. K.J. Ressel. *Least-Squares Finite-Element Solution of the Neutron Transport Equation in Diffusive Regimes*. PhD thesis, University of Colorado, Denver, 1994.
27. S. Richling, E. Meinköhn, N. Kryzhevoi, and G. Kanschat. Radiative transfer with finite elements I. basic method and tests. *A&A*, 380:776–788, 2001.
28. Y. Saad. *Iterative Methods for Sparse Linear Systems*. Oxford University Press, 2nd edition, 2000.
29. F.-T. Suttmeier. *Adaptive Finite Element Approximation of Problems in Elasto-Plasticity Theory*. Dissertation, Universität Heidelberg, 1996.
30. S. Turek. An efficient solution technique for the radiative transfer equation. *Imp. Comput. Sci. Engrg.*, 5:201–214, 1993.
31. H. van der Vorst. Bi-CGSTAB: A fast and smoothly converging variant of Bi-CG for the solution of nonsymmetric linear systems. *SIAM J. Sci. Stat. Comput.*, 13(2):631–644, 1992.
32. R. Verfürth. A posteriori error estimation and adaptive mesh-refinement techniques. *J. Comput. Appl. Math.*, 50:67–83, 1994.
33. R. Wehrse and G. Kanschat. Radiative fluxes and forces in non-spherical winds. In B. Wolf, O. Stahl, and A.W. Fullerton, editors, *Variable and Non-spherical Stellar Winds in Luminous Hot Stars*, pages 144–150. Springer, 1999.
34. R. Wehrse, E. Meinköhn, and G. Kanschat. A review of Heidelberg radiative transfer equation solutions. In P. Stee, editor, *Radiative Transfer and Hydrodynamics in Astrophysics*, pages 13–30. EDP Sciences, 2002.

35. K. Yosida. *Functional Analysis.* Springer, 1980.
36. H. Yserentant On the multi-level splitting of finite element spaces. *Numer. Math.*, 49:379–412, 1986.
37. G. Zhou How accurate is the streamline diffusion method? *Math. Comput.*, 66(217):31–44, 1997.
38. G. Zhou and R. Rannacher Pointwise superconvergence of the streamline diffusion finite element method. *Numer. Meth. PDE*, 12:123–145, 1996.

A General-Purpose Finite Element Method for 3D Radiative Transfer Problems

Erik Meinköhn[1,2]

[1] Institute f. Theoret. Astrophysics, University of Heidelberg, Tiergartenstr. 15, D–69121 Heidelberg, Germany
[2] Institute of Applied Mathematics, University of Heidelberg, Im Neuenheimer Feld 294/293, D–69120 Heidelberg, Germany

Summary. This paper presents a continuous finite-element method for solving the resonance-line transfer problem in moving media. The algorithm is capable of dealing with three spatial dimensions, using unstructured grids which are adaptively refined by means of an a-posteriori error indicator. Application of the method to coherent isotropic scattering and complete redistribution gives a result of matrix structure which is discussed in the paper. The solution is obtained by way of an iterative procedure, which solves a succession of quasi-monochromatic radiative transfer problems. It is therefore immediately evident that any simulation of the extended frequency-dependent model requires a solution strategy for the elementary monochromatic transfer problem, which is fast as well as accurate. The present implementation is applicable to arbitrary model configurations with optical depths up to 10^3–10^4. Additionally, a combination of a discontinuous finite-element method with a superior preconditioning method is presented, which is designed to overcome the extremely poor convergence properties of the linear solver for optically thick and highly scattering media.

1 Introduction

Under a variety of circumstances, as e.g., in combustion physics, thermonuclear fusion and astrophysics, energy transfer via radiation plays a key role. Radiation can be treated as transport of photons or neutrons, which is accompanied by absorption and emission effects depending on the particular medium, and the corresponding balance equation is generally known as the radiative transfer equation or the radiation transport equation. It takes the form of a partial integro-differential equation that turns out to be equivalent to a linear Boltzmann equation if certain simplifications can be made. These simplifications arise, for instance, from an a-priori knowledge of the temperature field or from the assumption of local thermodynamic equilibrium. But as soon as these simplifications are dropped, non-linearities occur by way of the material cross sections or opacities, which depend on the temperature or on the number density of the energy-level population.

In astrophysics, particularly, great progress resulted from sensitivity improvements of the telescope equipment, which was achieved in the 1980s. But current telescope development is characterised by advances in even greater strides, which gave an extraordinary increase in spatial resolution. These latter-day improvements resulted partly from putting telescopes on space satellites, but mainly from the use of so-called new-generation telescopes that have computer controlled active and adaptive optical components. If, especially, these new instruments are used in combination with powerful modern interferometers, observations of excellent spatial resolution are obtained, that allow celestial objects of great geometrical complexity to be revealed in fine detail. In view of this enhancement in astronomical precision, numerical codes for spherically symmetric or plane-parallel geometries, commonly used for astrophysical calculations since the early 1960s, are no longer sufficient. When modelling the requisite multidimensional transfer problems, it is indispensable to use codes that are capable of handling at least a two-dimensional, in many cases even a fully three-dimensional structure.

The present paper proposes a finite element method for solving the resonance line transfer problem in moving media. The appearance of an emission line profile is predominantly determined by complex angle-frequency coupling as well as by the macroscopic velocity field. Thus, if scattering and Doppler effects are taken into consideration, the corresponding three-dimensional radiative transfer is given by a partial integro-differential equation for the invariant radiation intensity $\mathcal{I} = \mathcal{I}(\mathbf{x}, \mathbf{n}, \nu, t)$ (cf. [MW84]), where $\mathbf{x}$ is the spatial variable, $\mathbf{n}$ is the direction of photon propagation, ν the frequency and t the time:

$$\frac{1}{c}\frac{\partial}{\partial t}\mathcal{I} + \mathbf{n} \cdot \nabla_x \mathcal{I} - \frac{\nu}{1 + \mathbf{n} \cdot \frac{\mathbf{v}}{c}} \mathbf{n} \cdot \nabla_x \left(\mathbf{n} \cdot \frac{\mathbf{v}}{c}\right) \frac{\partial \mathcal{I}}{\partial \nu} = $$
$$- (\kappa + \sigma)\mathcal{I} + \sigma \int\int_{\mathbb{R}^+ \ S^2} R\mathcal{I}d\hat{\omega}d\hat{\nu} + \kappa B. \quad (1)$$

Equation (1) depends on the following parameters:

$$\begin{aligned}
S^2 &: \quad \text{unit sphere of } \mathbb{R}^3 \\
\omega &: \quad \text{solid angle} \\
c &: \quad \text{speed of light} \\
\mathbf{v} &: \quad \text{velocity field} \\
\kappa = \kappa(\mathbf{x}, \nu) &: \quad \text{absorption coefficient} \\
\sigma = \sigma(\mathbf{x}, \nu) &: \quad \text{scattering coefficient} \\
R = R(\hat{\nu}, \hat{\mathbf{n}}; \nu, \mathbf{n}) &: \quad \text{redistribution function} \\
B = B(\nu, T(x)) &: \quad \text{Planck function} \\
T(x) &: \quad \text{temperature field}
\end{aligned}$$

If we assume that the velocity field causes no change of sign of the Doppler term, i.e. the term including the frequency derivative, (1) is generally solved with the constraint that the initial values of both the time and the frequency variable are fixed, and that the intensity of the incident radiation originating from outside the spatial domain is a given function at the boundary. More complex velocity fields give rise to a frequency boundary value problem rather than an initial value problem.

A finite element method is proposed which solves the resonance line transfer problem in moving media. The algorithm works in three spatial dimensions on hierarchically structured grids which are locally refined by means of duality-based a-posteriori error estimates ("DRW method"; see [BR01]). The solution is obtained by using an iterative approach which solves a succession of quasi-monochromatic radiative transfer problems. Efficient simulation of the extended frequency-dependent model therefore crucially depends on the solution strategy for the monochromatic transfer problem, which needs to be as fast and as accurate as possible (for more details see [Kan96, RM01, MR02, Mei02]).

The paper is organised as follows. Sect. 2 gives a detailed overview of recent numerical methods for multidimensional radiative transfer problems. A solution strategy which is based on a detailed account of the monochromatic 3D radiative transfer problem is derived in Sect. 3. In Sect. 4 an algorithm for the solution of the frequency-dependent line transfer problem is presented. The code allows for arbitrary velocity fields, with velocities up to 10% of the speed of light, and complete redistribution. The latter condition is critical for the correct modelling of Lyα line profiles in optically thick media. A sample of test runs of the model and a comparison of the monochromatic problem with the line transfer problem is presented in Sect. 5, which intends to illustrate the underlying physics. More realistic 3D configurations, which then are of greater complexity, are investigated in Sect. 6. Finally, in Sec. 7 a preconditioner is presented which accelerates the solution of monochromatic radiative transfer problems in optically thick and highly scattering media. The preconditioner makes use of a smoothing operator for specific intensities and is based on a multi-level method for the diffusion approximation of the mean intensity. Sec. 7 ends by demonstrating that the method actually brings about a fast solution of the discrete linear problems in those cases where scattering dominates as well as when transition layers occur.

2 Overview: Numerical Methods

The development of advanced discretisation methods is of fundamental importance for numerical treatments of the radiation transport equation, since it is still extremely challenging to model increasingly complex multi-dimensional problems while demanding at the same time strong increases in accuracy:

- The radiation intensity turns out to be a function of seven variables, if polarisation is neglected: three variables $\mathbf{x} = (x, y, z)$ giving the spatial position, two angular variables describing the direction $\mathbf{n}$ of photon propagation, one frequency variable ν, and the time variable t. Choosing a degree of discretisation which is still fairly moderate, namely 10^2 grid points for each independent variable, already results in a huge discrete problem with a collection of 10^{14} unknowns. Handling such an amount of data not only requires computers which are extremely powerful, but also efficient numerical methods designed to reduce the memory and CPU requirements. An appropriately designed discretisation method is of fundamental importance, if this latter point is to be satisfied.
- The intensity is usually a rapidly changing function of the spatial, angular and frequency variables. This gives jumps in the intensity itself or in its derivatives across thin layers of the corresponding computational domain. For many discretisation methods these jumps cause a considerable loss in accuracy.
- For different ranges of the coefficients, the linear Boltzmann equation can have a totally different character: in regions devoid of matter it corresponds to a hyperbolic wave equation; in optically thick media where scattering is dominant, it corresponds to an elliptic diffusion equation for states which remain steady, whereas it is a parabolic evolution equation for states which vary in time; in regions where scattering is highly peaked in the forward direction, it corresponds to a parabolic equation. It is extremely difficult to find a discretisation method universally capable of dealing efficiently with all of these different types of behaviour.

Up into the early sixties of last century the main focus was on the investigation of the radiative transfer equation for spherically symmetric or plane-parallel systems, for which analytical solutions exist [Cha60, Kou63]. An efficient numerical treatment of these transport problems was impossible before the advent of computers of sufficient capacity. In an activity which continued up to recent times, these 1D radiative transfer equations were treated by a host of numerical methods of increasing ingenuity(cf. [Mih78, Kal87]), which gave rise to numerical results of increasing precision, so that static and moving systems for spherically symmetric and plane-parallel configurations are now known in great detail. Numerical algorithms solving multidimensional radiative transfer problems started to appear in the late 1980s and gave rise to large a number of publications. When classified, these algorithms roughly fall into three categories: Monte-Carlo methods, Discrete-Ordinate methods and Angle-Moment methods.

The stochastic approach of *Monte-Carlo* codes is extremely flexible, since it is based on the concept of tracking photon packages, a principle which is applicable to arbitrary multidimensional configurations, as, e.g., in ultraviolet continuum transfer [Spa96], in optical and infrared continuum transport [WHS99], in molecular line transfer [Hog98, Juv97, PH98], and in Compton scattering [PSS79, HM91]. If the optical depth of the configuration is not too large, a Monte-Carlo method converges reasonably fast, but in the case of large optical depths, its drawback makes itself felt, namely that it is extraordinarily time-consuming. For the latter case, however, certain *Angle-Moment* methods are applicable, if the so-called flux-limited diffusion approximation is adopted (see [SPY95, MCK94] and references therein). Such an approach has been applied extensively, but it lacks the accuracy of consistent solutions of the radiative transfer problem in optically thin or semi-transparent regions or close to an open boundary. More sophisticated Angle-Moment methods, e.g., the variable Eddington factor method as described in [RA02], avoid such inconsistencies. For these improved methods, the coupled set consisting of the Boltzmann equation and the first two moment equations is iterated to convergence. Unfortunately, the number of moment equations involved is less than the number of unknowns being generated. To overcome this difficulty, variable Eddington factor methods are used for closing the set of moment equations, with the closure relationships obtained from the solution of the Boltzmann equation. As the emissivity and the opacity depend in general on the angular moments of the specific intensity, a nonlinear integro-differential equation set results. However, this latter aggravation is mitigated if zeroth and first order angular moments are assumed to be given. This allows the derivation of a formal solution of the Boltzmann equation, which is then used to compute the next iterate of the Eddington factors. The variable Eddington factor method performs very well for 1D geometries, but when modelling transfer problems which are multi-dimensional, this method is found to be inefficient due to the strong nonlinear coupling between the Boltzmann equation and the two moment equations (see [RA02]). Another class of Angle-Moment methods, the so-called P_N or *Differential-Moment* methods, were first introduced in [Kro55]. To overcome the "closure" problem of the moment equations as described above, the intensity is approximated by a series expansion in spherical harmonics that is truncated after N terms. In the asymptotic limit of large "N" the P_N method yields the solution of the actual Boltzmann equation. This approximation is in fact a generalisation of the Milne-Eddington method as described in [Mih78]. For complicated multi-dimensional problems it has a major drawback in that a large number of complex equations must be solved. Furthermore, in the P_N approximation the positivity of the radiation field cannot be guaranteed. This led to the formulation of the simplified P_N, i.e. the SP_N, method. The SP_N approximation is based on a coupled system of diffusion equations which are independent of the angular variable (see [BL00, LMM96, LTK02] for the theoretical foundation and the asymptotic derivation of the SP_N equations). The number of SP_N equations, and their complexity, are considerably reduced in comparison to the full set of P_N equations. Formally, and given that the circumstances are suitably ideal, approximations of the radiative transfer equation result which are of an order which increases in proportion as the value of N is increased. However, it cannot be guaranteed that in the limit of large "N" the SP_N method yields the solution of the actual transport equation, because two conditions may interfere with the accuracy of the asymptotic expansion: these conditions arise from the spatial non-constancy of the opacity coefficients and from complications due to particular boundary conditions. In practice only lower moments are used, and a wealth of results has been obtained from an application of the SP_1 and SP_3 methods (e.g., see [BL00, LMM96, LTK02]). The SP_1 approximation, for instance, yields the familiar radiation diffusion approximation, which removes all angular dependence from the radiation field. The SP_3 method retains some angular dependence in the radiation field and gives a reasonable approximation of the Boltzmann problem in spatial boundary layers. The SP_3 equation, therefore, gives results which are good enough even when the regime is not strongly diffusive, while it is clear that in optically thin media or in vacuum this approximation is bound to break down. Another deterministic discretisation method is the higher-order S_N approximation, which retains a much higher level of angular dependence in the radiation field. The *Discrete-Ordinate* (S_N) method for radiative transfer was first formulated in [Cha50]. For each S_N approximation, the angular dependence in the transport equation is discretised along a set of $N(N+2)$ ordinates. In order to avoid numerical artifacts, the ordinate domain must be subdivided in a uniform way. Numerous approaches and formulations have been devised by previous authors for solving the S_N transport equations (e.g., see [Kan96, Tur93, SSW91, DT00, Bal01]), which proved to have a number of attractive aspects. However, situations exist which still turn out to be problematic, especially when modelling transfer problems in optically thick and highly scattering media, where convergence is slow for the S_N approaches, which then need to be complemented by methods which ensure convergence to be accelerated. A discussion of how to achieve this acceleration is given below.

Spatial discretisation, i.e., discretisation of the transport operator of the radiative transfer equation, is often performed by using approaches based on finite differences [SSW91] or by using the method of characteristics [DT00]. These methods usually employ structured grids, that yield algebraic systems for which solutions are rapidly obtained for homogeneous media and smooth data, but which fail in the case of complex geometries or steep gradients of the solution or of the coefficients. The resulting system is by far too large to be solved even on today's supercomputers, since very high resolution is needed to produce accurate solutions if steep gradients occur. For many radiative transfer problems steep gradients of the coefficients or the solution are confined to thin layers of the computational domain. High resolution is then only needed within these layers whereas the rest of the spatial domain can be treated with moderate resolution, because of the smoothness of photon transport. Unstructured grids, when they are solution-adapted, permit to achieve a high resolution of rapidly changing solutions over small domains, whereas in regions of smooth transport the grid remains coarse. Unfortunately, the generation of adaptively-refined grids by way of a finite difference method or, similarly, by way of the method of characteristics is extremely difficult. But fortunately there is an alternative, namely the finite element method (FEM), which is exceptionally suited to applications of adaptively refined grids and which delivers good approximations when the geometries are complex or when there are steep gradients of the coefficients and of the solution. A FEM with upwind discretisation on pre-refined grids is presented in [Tur93]. Kanschat then stabilises this FEM discretisation by introducing streamline diffusion, and obtains adaptively refined and therefore problem-adapted grids by means of an a-posteriori error indicator (for details see [Kan96] and [BR01]). Even though it is quite evident that FEM is superior because of its many advantages, the astronomical community rarely uses it, because code development for FEM is more complicated than for the other discretisation methods which were described above.

As mentioned earlier, discretisation of multidimensional radiative transfer problems leads to extremely large systems of linear algebraic equations, for which an application of direct solution strategies is as yet impracticable. Rather, it is standard practice to use an iterative scheme to obtain a solution of such large systems of equations (e.g., see [Hac93, Var00]). In astrophysics, particularly, the standard iterative method is the so-called Λ- or source iteration method [Mih78]. As applied to the overall discrete system, it is a Richardson method with nearly block-Jacobi preconditioning. It then follows that use of a full Jacobi preconditioner is a first step towards a better rate of convergence [Tur93]. Since the transport operator can be inverted explicitly, these methods converge extremely fast for transport dominated problems. If the triangular matrix structure of an upwind discretisation is exploited, inversion is indeed very cheap since it requires only one matrix-vector-multiplication operation. In the case of highly scattering, optically thick media, on the other hand, these methods, and, similarly, all other stationary iterations, fail to work, because the condition number of the iteration matrix becomes very large. Preconditioned Richardson iteration methods are characterised by a rate of convergence which depends strongly on the condition number, and therefore these methods are not suited for cases where scattering dominates. But here [Kan96] comes in handy, showing that the eigenvalue distribution of the transport matrix is clustered, with one of the eigenvalues vanishing whilst the others are either close to unity or at least bounded away from zero. Krylov space methods like GMRES or bi-CGSTAB (for details see [SF93] and references therein) are known to be particularly suited to deal with such a kind of eigenvalue distribution [Kan96]. In comparison with stationary iteration methods, Krylov-type solvers are found to give a superior performance, but they still lead to poor convergence for transfer problems where scattering dominates, so that convergence acceleration needs to be brought in. It is therefore quite fortunate that during the last decades the development of acceleration methods has progressed significantly, for instance by preconditioning Discrete-Ordinate methods in such a way that iterative convergence in these problems sped up considerably. A comprehensive review article was published recently (see [AL02]), which covers practically all of the main methods that have been discussed above. When comparing the various acceleration methods, the Diffusion Synthetic Acceleration (DSA) turns out to be a very efficient preconditioning method guaranteeing rapid convergence for all optical thicknesses and scattering intensities. DSA exploits the diffusion approximation, a well-known approximation of the Boltzmann equation for highly scattering, optically thick media, so as to establish a solution of the full radiative transfer problem in a very efficient way. Preconditioning leads to acceleration in the discretised Λ- or source iteration, which corresponds to a Richardson iteration, and the algorithm makes use of DSA efficiently as well as robustly. Discrete transport problems, for which a convergence in source iteration would ordinarily require hundreds, thousands, or even millions of iteration steps, are now solved with DSA in less than a few tens of iteration steps. Unfortunately, this method shows a loss in the effectiveness for multi-dimensional Cartesian grids in the presence of material discontinuities. But [WWM04] shows that a Krylov iteration method, preconditioned with DSA, is an effective remedy that can be used to efficiently compute a solution also for this class of problems. Results from numerical experiments indicate that transfer problems which are virtually intractable with accelerated source iteration can be solved efficiently if a preconditioned Krylov method replaces source iteration (see [WWM04]).

3 Monochromatic 3D Radiative Transfer

3.1 The Radiative Transfer Problem

The purpose of the paper is to determine the radiation field in a given material continuum which fills some finite spatial domain $\Omega \subset R^3$. Ω is assumed to be embedded in vacuum, and for the sake of simplicity it is taken to be of convex shape. Further, it is assumed that radiation leaving Ω will not re-enter. For these assumptions, the specific intensity I in Ω satisfies the monochromatic radiative transfer equation

$$\mathbf{n} \cdot \nabla_x I(\mathbf{x},\mathbf{n}) + \kappa(\mathbf{x}) I(\mathbf{x},\mathbf{n}) \\ + \sigma(\mathbf{x}) \Big(\mathrm{I}(\mathbf{x},\mathbf{n}) - \frac{1}{4\pi} \int_{S^2} P(\mathbf{n}',\mathbf{n}) \mathrm{I}(\mathbf{x},\mathbf{n}') \, d\omega' \Big) = f(\mathbf{x}), \qquad (2)$$

where $\mathbf{x} \in \Omega$ is location in space and $\mathbf{n}$ is the unit vector pointing in the direction of the solid angle $d\omega$ of the unit sphere S^2. The optical properties of the matter contained in Ω are given by the absorption coefficient $\kappa(\mathbf{x})$ and the scattering coefficient $\sigma(\mathbf{x})$. The angular phase function P in the scattering integral is normalised, such that $\frac{1}{4\pi} \int P(\mathbf{n}',\mathbf{n}) \, d\omega' = 1$. The source term

$$f(\mathbf{x}) = \kappa(\mathbf{x}) B(T(\mathbf{x})) + \epsilon(\mathbf{x}) \qquad (3)$$

comprises the combined effect of thermal emission arising from a temperature distribution $T(\mathbf{x})$ in Ω in conjunction with the Planck-Function B, and of light emission from a point source or an extended object embedded in Ω, which is taken into account via the emissivity $\epsilon(\mathbf{x})$. To be able to solve (2), boundary conditions of the form

$$I(\mathbf{x}, \mathbf{n}) = g(\mathbf{x}, \mathbf{n}) \tag{4}$$

must be imposed on the "inflow boundary"

$$\Gamma_- = \{(\mathbf{x}, \mathbf{n}) \in \Gamma \big| \mathbf{n}_\Gamma \cdot \mathbf{n} < 0\}, \tag{5}$$

where $\mathbf{n}_\Gamma$ is the unit normal of the surface Γ of Ω. $\mathbf{n}_\Gamma$ is taken to point outward, so that the sign of the product $\mathbf{n}_\Gamma \cdot \mathbf{n}$ indicates the direction of the photon flux across the boundary. If there are no light sources outside the domain Ω, the function g vanishes everywhere on Γ.

The left hand side of (2) is abbreviated by introducing an operator $\mathcal{A}$ which acts on the intensity I, giving the radiative transfer equation in compact operator form as follows

$$\mathcal{A}I(\mathbf{x}, \mathbf{n}) = f(\mathbf{x}). \tag{6}$$

This equation is five-dimensional, with $\mathbf{x}$ representing the three variables of spatial location and $\mathbf{n}$ representing two variables giving the direction of photon propagation. In addition, the numerical solution is complicated by the fact that I is rapidly changing over narrow slices of the domain, while transport is smooth over the rest of the domain, which is much larger. In the case of three spatial dimensions a reasonable resolution in space of $h = 1/1000$ and the use of 1000 directions, or *ordinates*, already gives 10^{12} unknowns. Since even state-of-the-art supercomputers cannot handle such an amount of data, efficient error estimation and grid adaption techniques must be used, if reliable quantitative results are to be obtained. Finite element methods, in particular so-called Galerkin methods, turn out to be most suited for these techniques.

3.2 Finite Element Discretisation

Equation (2) is analysed in [DL00]. The natural space for finding solutions is

$$W = \{I \in L^2(\Omega \times S^2) \,\big|\, \mathbf{n} \cdot \nabla_x I \in L^2(\Omega \times S^2)\}, \tag{7}$$

where L^2 is the Lebesgue space of the square-integrable functions. For homogeneous vacuum boundary condition, i.e. $g(\mathbf{x}, \mathbf{n}) = 0$, the solution space is

$$W_0 = \{I \in W \,\big|\, I = 0 \text{ on } \Gamma_-\}. \tag{8}$$

To apply a finite element method, we have to use a weak formulation of (2). Therefore, we multiply both sides of (2) by a trial function $\varphi(\mathbf{x}, \mathbf{n})$ and integrate over the whole domain $\Omega \times S^2$. Thus, in the weak formulation, $I \in W_0$ is to be found, such that $\forall \varphi \in W_0$

$$\iint_{\Omega\ S^2} \mathbf{n}\cdot\nabla_x I(\mathbf{x},\mathbf{n})\varphi(\mathbf{x},\mathbf{n})\,d\omega\,d^3x$$

$$+\iint_{\Omega\ S^2}(\kappa(\mathbf{x})+\sigma(\mathbf{x}))I(\mathbf{x},\mathbf{n})\varphi(\mathbf{x},\mathbf{n})\,d\omega\,d^3x$$

$$-\iiint_{\Omega\ S^2\ S^2}\sigma(\mathbf{x})P(\mathbf{n}',\mathbf{n})I(\mathbf{x},\mathbf{n}')\varphi(\mathbf{x},\mathbf{n})\,d\omega'\,d\omega\,d^3x$$

$$=\iint_{\Omega\ S^2} f(\mathbf{x})\varphi(\mathbf{x},\mathbf{n})\,d\omega\,d^3x. \tag{9}$$

To extend the definition of the L^2-scalar product we introduce the abbreviation

$$(I,\varphi)=(I,\varphi)_{\Omega\times S^2}=\iint_{\Omega\ S^2} I\varphi\,d\omega\,d^3x \tag{10}$$

so that the weak formulation of the operator form (6) is

$$(\mathcal{A}I,\varphi)=(f,\varphi)\qquad\forall\,\varphi\in W_0. \tag{11}$$

If there is no scattering, i.e. $\sigma(\mathbf{x})=0$ on Ω, the problem leads to a hyperbolic system of convection equations on Ω. If the solutions are not smooth, standard finite-element techniques applied to this type of equations are known to produce spurious oscillations. We can achieve stability by adding the streamline diffusion modification:

$$(\mathcal{A}I,\varphi+\delta\mathbf{n}\cdot\nabla_x\varphi)=(f,\varphi+\delta\,\mathbf{n}\cdot\nabla_x\varphi)\quad\forall\,\varphi\in W_0. \tag{12}$$

The cell-wise constant parameter function δ depends on the width of the local mesh and on the coefficients κ and σ. Note that the solution of (11) solves (12), too. No additional consistency error is induced by the stabilisation. In the interest of conciseness, streamline diffusion will always assumed to be contributing, but from now on the corresponding terms will be suppressed in the formulations that follow.

When applying standard Galerkin finite elements to solve (12), we choose a finite dimensional subspace W_h of W consisting of functions that are piecewise polynomial with respect to a subdivision or *triangulation* $T_h:=\mathbb{T}^\Omega\otimes\mathbb{T}^{S^2}$ of $\Omega\times S^2$:

$$\begin{aligned}W_h:=&W_h^\Omega\otimes W_h^{S^2}\\=&\Big\{v\in W\ \Big|\ v(\cdot,\mathbf{n})_{|_{K^\Omega}}\in Q^1(K^\Omega)\,\forall K^\Omega\in\mathbb{T}^\Omega,\ v_{|_{K^\Omega}}\in C^0(\Omega),\\&\quad v(\mathbf{x},\cdot)_{|_{K^{S^2}}}\in P^0(K^{S^2})\,\forall K^{S^2}\in\mathbb{T}^{S^2}\Big\}.\end{aligned} \tag{13}$$

$C^0(\Omega)$ is the space of continuous functions on the 3D domain Ω while $Q^1(K^\Omega)$ is the space of trilinear functions on the spatial grid cell $K^\Omega \in \mathbb{T}^\Omega$. $P^0(K^{S^2})$ is the space of constant functions on the ordinate grid cell $K^{S^2} \in \mathbb{T}^{S^2}$. Note, that we use continuous finite elements exclusively for spatial discretisation, whereas discontinuous elements are permitted for ordinate discretisation. The mesh size h is the piecewise constant function defined on each spatial triangulation cell K^Ω by $h|_K^\Omega = h = \text{diam}K^\Omega$. The discretised problem derived from (12) consists in finding $I_h \in W_h$, such that

$$(\mathcal{A}I_h, \varphi_h) = (f, \varphi_h) \quad \forall\, \varphi_h \in W_h. \tag{14}$$

The construction of the subspace W_h needs some further consideration (see [Kan96]). The discretised domain is a tensor product of two sets of completely different length scales: While Ω represents a domain in physical space, S^2 is the unit sphere in the Euclidean space $\mathbb{R}^3$. Therefore, we use a tensor product splitting of the five-dimensional domain $\Omega \times S^2$, such that a grid cell of the five-dimensional grid will be the tensor product of a two-dimensional cell K^{S^2} and a three-dimensional cell K^Ω. Accordingly, the mesh sizes with respect to $\mathbf{x}$ and ω will be different.

On S^2 we use a fixed discretisation based on a refined icosahedron. Quadrature points are given by projecting the centres of the triangular faces of the icosahedron on S^2. This projection method guarantees equally distributed quadrature points, each of which being associated with a solid angle $d\omega$ of the same size. Whereas other numerical quadrature methods need to impose additional symmetry conditions, these are redundant in the present approach, and discretisation artifacts at the poles are avoided. Furthermore, we use piecewise constant trial functions (cf. (13)). That way, the seven-dimensional integration of the scattering term in the weak formulation (9) can be performed very efficiently, since integration weights are not necessary. For example, the integration of the intensity over the whole unit sphere is simply replaced by a sum divided by the number of discrete ordinates M

$$\frac{1}{4\pi}\int_{S^2} I(\mathbf{n})\,d\omega \rightarrow \frac{1}{M}\sum_{j=1}^M I_j\,. \tag{15}$$

The discretisation scheme is second-order accurate in the evaluated ordinate points due to super-convergence (see [Kan96]).

For the space domain Ω we use locally refined hexahedral meshes. In the remaining sections of the paper, the mesh size with respect to the space variable will be denoted by h. Since the boundaries are arbitrary for our astrophysical application, we can choose a unit cube for Ω and do not have to worry about boundary approximations. We use continuous piecewise-trilinear trial functions in space.

3.3 Error Estimation and Adaptivity

The calculation of complex radiation fields in astrophysics often requires high resolution in parts of the computational domain, for example in regions with strong opacity gradients. In such a case, reliable error bounds are necessary to suppress numerical errors. Due to the high dimensionality of the radiative transfer problem, a well-suited method for error estimation and grid adaptation is necessary to achieve results of sufficient accuracy even if parallel computing is used.

In addition, computational goals in astrophysics are often more specific because the result of a simulation is to be compared with observations, so as to develop a physical model for the celestial object. For instance, in the case of a distant unresolved object, the total flux leaving the domain Ω in *one* particular direction is of interest. Generally, a measured quantity like the total flux can be expressed as a linear functional $\mathcal{M}(I)$. By linearity, the error of the measured quantity is $\mathcal{M}(I) - \mathcal{M}(I_h) = \mathcal{M}(e)$, where $e = I - I_h$. We will now show, that it is possible to obtain an *a-posteriori* estimate for $\mathcal{M}(e)$, even if the exact solution I is unknown.

Let us suppose that $z(\mathbf{x}, \mathbf{n})$ is the solution of the dual problem

$$\mathcal{M}(\varphi) = (\varphi, \mathcal{A}^* z) \qquad \forall\, \varphi \in W_0, \tag{16}$$

where the dual radiative transfer operator is defined by

$$\mathcal{A}^* z(\mathbf{x}, \mathbf{n}) = -\mathbf{n} \cdot \nabla_x z(\mathbf{x}, \mathbf{n}) + \big(\kappa(\mathbf{x}) + \sigma(\mathbf{x})\big) z(\mathbf{x}, \mathbf{n}) \\ - \sigma(\mathbf{x}) \int_{S2} P(\mathbf{n}', \mathbf{n}) z(\mathbf{x}, \mathbf{n}')\, d\omega'. \tag{17}$$

The boundary conditions for the dual problem are complementary to those in the primal problem, i.e. $I = 0$ on the "outflow boundary" $\Gamma_+ = \{(\mathbf{x}, \mathbf{n}) \in \Gamma \mid \mathbf{n}_\Gamma \cdot \mathbf{n} > 0\}$. Then, we get the formal error representation

$$\begin{aligned}\mathcal{M}(e) &= (e, \mathcal{A}^* z) \\ &= (\mathcal{A}e, z) \\ &= (\mathcal{A}e, z - z_i) \\ &= \sum_{K \in T_h} \big(f - \mathcal{A}I_h, z - z_i\big)_K \end{aligned} \tag{18}$$

for arbitrary $z_i \in W_h$. In the third line of (18) a characteristic feature of finite element methods is used, the so-called Galerkin orthogonality

$$(\mathcal{A}I - \mathcal{A}I_h, \varphi_h) = 0 \qquad \forall\, \varphi_h \in W_h. \tag{19}$$

Since the dual solution z is unknown, it is a usual approach to apply Hölder's inequality and standard approximation estimates of finite element spaces to obtain the estimate

$$\mathcal{M}(e) \leq \eta = \sum_{K \in T_h} \eta_K \tag{20}$$

with local error indicators for each cell $K \in T_h$

$$\eta_K = C_K h_K^2 \|\varrho\|_K \|\nabla^2 z\|_K, \tag{21}$$

where the constant C_K is determined by local approximation properties of W_h. The residual function ϱ of I_h is defined by $\varrho = f - \mathcal{A}I_h$. Since the dual solution z is not available analytically, it is usually replaced by the finite element solution z_h to the dual problem (16). This involves a second solution step of the same structure as the primal problem. It is clear, that by this replacement the error estimate (20) is not strictly true anymore. Experience shows that the additional error is small and the estimate may be used if multiplied with a small security factor larger than one.

A first approach in the development of strategies for grid refinement based on a-posteriori error estimates is the control of the global energy or L^2-error involving only local residuals of the computed solution (see [FK97]). Formally, using the functional $\mathcal{M}(\varphi) = \|e\|^{-1}(e, \varphi)$ in the dual problem (16), we obtain $\mathcal{M}(e) = \|e\|^{-1}(e, e) = \|e\|$ for the left-hand side of (18). Estimating the right hand side of (18) yields

$$\|e\| \leq \tilde{\eta} = \sqrt{\sum_{K \in T_h} \eta_{L^2}^2} \tag{22}$$

with local L^2-error indicators for each cell $K \in T_h$

$$\eta_{L^2} = C_K C_s h_K^2 \|\varrho\|_K, \tag{23}$$

where C_s only depends on the shape and size of Ω. This a-posteriori bound for the L^2-error is well-known (see e.g. [EEH95] or [Ver96]). Estimates of this kind are always sub-optimal in the case of inhomogeneous coefficients, and especially so when scattering dominates. Still, the L^2-indicators η_{L^2} provide a good refinement criterion to study the qualitative behaviour of the solution I everywhere in Ω. For an illustration of this approach, see also in [WMK99].

The grid refinement process on the basis of an a-posteriori error estimate is organised in the following way: Suppose that some error tolerance TOL is given. The aim is then to find the most economical grid T_h on which

$$|\mathcal{M}(e)| \leq \hat{\eta}(I_h) = \sqrt{\sum_{K \in T_h} \eta_K^2(I_h)} \leq TOL, \tag{24}$$

with *local refinement indicators* η_K taken from (21). A qualitative comparison between grids obtained via an adaptive refinement procedure based on different local error indicators (21) and (23) is published in [RM01]. Having computed the solution on a coarse grid, the so-called *fixed-fraction* grid refinement strategy (see [Kan96, BR01]) is applied: The cells are ordered according to the size of η_K and a fixed portion ν (say 30%) of the cells with largest η_K is refined. This guarantees, that in each refinement cycle a number of cells is refined which is sufficient large. Then, a solution is computed on the new grid and the process is continued until the prescribed tolerance is achieved. This algorithm is especially valuable, if a computation "as accurate as possible" is desired, that is, if the tolerance is not reached, whereas computer memory is exhausted. Then, the parameter ν has to be determined by the remaining memory resources.

3.4 Resulting Matrix Structure

Given a discretisation with N degrees of freedom (= number of *vertices* of the hexahedral mesh) in Ω and M ordinates in S^2, the discrete system has the form

$$\mathbf{A} \cdot \mathbf{u} = \mathbf{f}, \tag{25}$$

with the vector $\mathbf{u}$ containing the discrete specific intensities and the vector $\mathbf{f}$ the values of the source term. Both vectors have $(N \cdot M)$ components, whereas $\mathbf{A}$ is a $(N \cdot M) \times (N \cdot M)$ matrix. Applying the tensor product structure proposed above, we may write

$$\mathbf{A} = \mathbf{T} + \mathbf{K} + \mathbf{S} \tag{26}$$

with the block structure

$$\mathbf{T} = \begin{pmatrix} \mathbf{T}_1 & & 0 \\ & \ddots & \\ 0 & & \mathbf{T}_M \end{pmatrix},$$

$$\mathbf{K} = \begin{pmatrix} \mathbf{K}_1 & & 0 \\ & \ddots & \\ 0 & & \mathbf{K}_M \end{pmatrix},$$

$$\mathbf{S} = \begin{pmatrix} \omega_{11}\mathbf{S}_1 & \cdots & \omega_{1M}\mathbf{S}_1 \\ \vdots & \ddots & \vdots \\ \omega_{M1}\mathbf{S}_M & \cdots & \omega_{MM}\mathbf{S}_M \end{pmatrix},$$

where $\omega_{il} = P(\mathbf{n}_i, \mathbf{n}_l)/M$. Because of finite-element streamline-diffusion discretisation, the entries of the $N \times N$ blocks are defined by

$$\mathbf{T}_i^{jk} = (\varphi_j + \delta \mathbf{n}_i \cdot \nabla_x \varphi_j, \mathbf{n}_i \cdot \nabla_x \varphi_k)$$
$$\mathbf{K}_i^{jk} = (\varphi_j + \delta \mathbf{n}_i \cdot \nabla_x \varphi_j, (\kappa + \sigma)\varphi_k)$$
$$\mathbf{S}_i^{jk} = (\varphi_j + \delta \mathbf{n}_i \cdot \nabla_x \varphi_j, \sigma \varphi_k),$$

where $j = 1, ..., N$ and $k = 1, ..., N$.

3.5 Iterative Methods

The linear system of equations resulting from the discretisation described above is large (10^7 unknowns at least), sparse, and strongly coupled due to the integral operator. Usually, an iterative scheme is applied to solve such large systems of equations (e.g., see [Hac93, Var00]). The standard algorithm used in astrophysics in this case is the so-called Λ- or source iteration method [Mih78] which can be viewed as a form of the Richardson method with nearly block-Jacobi-preconditioning. It then follows that use of a full Jacobi preconditioner is a first step towards a better rate of convergence [Tur93]. Since the transport operator can be inverted explicitly, these methods converge extremely fast for transport dominated problems. If the triangular matrix structure of an upwind discretisation is exploited, inversion is indeed very cheap since it requires only one matrix-vector-multiplication operation.

Unfortunately, the present interest is in opaque and highly scattering media where this method – like other stationary iterations – breaks down, because the condition number of the iteration matrix becomes very large. Since the convergence rate of preconditioned Richardson iteration methods exclusively depends on the condition number, stationary iterations are not suited for the scattering dominated case. It can be shown, though, that the eigenvalue distribution of the transport matrix is clustered, with one eigenvalue located at the origin, while the others are close to unity or at least bounded away from zero (for details see [Kan96] and the contribution by G. Kanschat to this special Volume). Krylov space methods like GMRES or bi-CGSTAB described in [SF93] are known to be particularly suited to deal with such kind of eigenvalue distribution [Kan96]. While Krylov-type solvers are superior to stationary iteration methods, they still have poor convergence properties for scattering-dominated and opaque transfer problems, so that convergence acceleration methods need to be brought in. During the last several decades, significant progress has been achieved in the development of acceleration methods, i.e. preconditioning methods, which were designed to speed up the iterative convergence of discrete-ordinates approaches. A comprehensive review recently appeared in the literature (see [AL02]), covering the main methods that have been proposed. Of all the acceleration methods, the Diffusion Synthetic Acceleration (DSA) is one of the most efficient preconditioning methods, guaranteeing rapid convergence for all optical thicknesses and scattering values. The DSA method exploits the diffusion approximation, a well-known approximation of the Boltzmann equation for highly scattering, optically thick media, and solves the full radiative transfer problem efficiently. A preconditioning algorithm makes it possible to accelerate the discretised source iteration, i.e. the Richardson iteration scheme, which gives DSA its efficiency and its robustness. Discrete transport problems, for which a convergence in source iteration would ordinarily require hundreds, thousands, or even millions of iteration steps, are now solved with DSA in less than a few tens of iteration steps. Unfortunately, this method shows a loss in efficiency for multidimensional

Cartesian grids in the presence of material discontinuities. In [WWM04] it is shown that a Krylov iterative method, preconditioned with DSA, is a convenient remedy that can be used to efficiently compute the solutions for this class of problems. Results from numerical experiments show that replacing source iteration with a preconditioned Krylov method can efficiently solve transfer problems that are virtually intractable with accelerated source iteration (see [WWM04]). In Sect. 7 a DSA-type preconditioner is presented that acts like a smoothing operator for specific intensities and which corresponds to a multilevel method for the diffusion approximation of the mean intensity. The word "multilevel" or "multigrid" often connotes a sequence of increasingly coarse spatial meshes. However, for transport problems, space and angle are variables, and the method we consider involves "collapsing" the transport problem onto "coarser grids" independent of the photon direction. It is demonstrated that the method allows to arrive at a fast solution of discrete linear problems in those cases where scattering is dominant as well as when transition regions occur.

3.6 Parallelisation

Transport dominated problems differ in one specific point from elliptic problems: There is a distinct direction of information flow. This has to be taken into account when developing parallelisation strategies. While a domain decomposition for Poisson's equation should minimise the length of interior edges, such an approach does not yield an efficient method for transport equations.

In [Kan00], it was concluded that parallelisation strategies for transport equations should not divide the domain across the transport direction. The solution of the radiative transfer equation consists of a bunch of transport inversions for different directions. The construction of an efficient domain decomposition method would require a direction-dependent splitting of the domain Ω. Since this causes immense implementation problems, we decided to use ordinate parallelisation.

This strategy distributes the ordinate space S^2 of the radiative transfer equation over a number of processors. Since we use discontinuous shape functions for the ordinate variables and since there is no local coupling as there is no contribution of an integral operator, our approach results in truly non-overlapping parallelisation. Clearly, our approach also has its disadvantages: As the integral is a global operator, ordinate parallelisation involves global communication. Furthermore, the resolution in space is restricted by the per-node-memory and not by the total memory of the machine. In [RM01] memory and CPU time requirements are investigated for a test model-configuration, where the ordinate space S^2 is distributed over 1 to 16 processors. The results show that the code scales optimally, i.e. doubling the number of processors gives a reduction by half in what is required as to CPU time and memory.

4 Polychromatic 3D Line Transfer

In the previous section we discussed monochromatic 3D radiative transfer in static media and gave an account of the difficulties encountered when solving the corresponding partial integro-differential equation. The observed light originating from celestial objects contains much information beyond that which is treated in monochromatic radiative transfer. It is therefore highly desirable to extend the algorithm, so that frequency-dependent properties can also be dealt with. There are many differentially moving astronomical objects in which the radiation field is important for the energy balance. Because line profiles are highly affected by radiative transfer processes, the restriction to a static medium is not at all satisfactory. It is therefore necessary to also include arbitrary macroscopic velocity fields in our physical model.

Typical celestial systems with differentially moving media are novae, supernovae, collapsing molecular clouds, collapsing as well as expanding rotating gas clouds (halos), and accretion discs. The present paper gives an application of our model to 3D radiation fields in gas clouds from the early universe, with the particular objective to bring out the influence of varying distributions of density and velocity. In observations of high-redshift gas clouds, the Lyα transition from the first excited energy level to the ground state of the hydrogen atom is usually found to be the only prominent emission line in the entire spectrum. By a well-known assumption, highly redshifted hydrogen clouds are taken to be the precursors of present-day galaxies. Thus, an investigation of the Lyα line is of pivotal importance for the theory of galaxy formation and evolution. The observed Lyα line – or more precisely, its profile – reveals the complexity as to the spatial distribution and the kinematics of the interstellar gas, and the nature of the photon source can be derived from it.

The transfer of resonance-line photons is strongly influenced by scattering in space and frequency. Analytical (see [Neu90]) as well as early numerical methods [Ada72, HK80] were restricted to one-dimensional, static media. Only recently, codes based on the Monte Carlo method were developed which are capable to investigate the more general case of a multi-dimensional medium (see [ALL01, ALL02, ZM02]).

In the section below, a finite-element method is presented for solving the resonance line transfer problem in non-relativistic moving media. The algorithm works in three spatial dimensions on unstructured grids which are adaptively refined by means of an a-posteriori error indicator. The resulting matrix structure for coherent isotropic scattering and complete redistribution is discussed. The solution is obtained by way of an iterative procedure, where elementary monochromatic radiative transfer problems are solved successively at each step. Thus, a fast and accurate solution strategy for the elementary monochromatic transfer problem is crucial if the extended frequency-dependent model is to be simulated efficiently.

4.1 Line Transfer in Moving Media

The frequency-dependent radiation field in moving media is obtained by solving the non-relativistic radiative transfer equation in a co-moving frame. For a three-dimensional domain Ω the operator form of the equation is

$$\left(\mathcal{T} + \mathcal{D} + \mathcal{S} + \chi(\mathbf{x},\nu)\right)\mathcal{I}(\mathbf{x},\mathbf{n},\nu) = f(\mathbf{x},\nu). \tag{27}$$

$\mathcal{T}$ is the transfer operator, $\mathcal{D}$ the "Doppler" operator, and $\mathcal{S}$ the scattering operator, which are defined as follows

$$\mathcal{T}\mathcal{I}(\mathbf{x},\mathbf{n},\nu) = \mathbf{n} \cdot \nabla_x \mathcal{I}(\mathbf{x},\mathbf{n},\nu),$$

$$\mathcal{D}\mathcal{I}(\mathbf{x},\mathbf{n},\nu) = w(\mathbf{x},\mathbf{n})\,\nu\frac{\partial}{\partial \nu}\mathcal{I}(\mathbf{x},\mathbf{n},\nu),$$

$$\mathcal{S}\mathcal{I}(\mathbf{x},\mathbf{n},\nu) = -\frac{\sigma(\mathbf{x})}{4\pi}\int_0^\infty \int_{S^2} R(\hat{\mathbf{n}},\hat{\nu};\mathbf{n},\nu)\mathcal{I}(\mathbf{x},\hat{\mathbf{n}},\hat{\nu})\,d\hat{\omega}\,d\hat{\nu}.$$

The relativistic invariant specific intensity $\mathcal{I}$ depends on six variables, three of which give the spatial location $\mathbf{x}$, while two variables give the direction $\mathbf{n}$ (pointing in the direction of the solid angle $d\omega$ of the unit sphere S^2), and one variable gives the frequency ν.

The extinction coefficient $\chi(\mathbf{x},\nu) = \kappa(\mathbf{x},\nu) + \sigma(\mathbf{x},\nu)$ is the sum of the absorption coefficient $\kappa(\mathbf{x},\nu) = \kappa(\mathbf{x})\phi(\nu)$ and the scattering coefficient $\sigma(\mathbf{x},\nu) = \sigma(\mathbf{x})\phi(\nu)$. The frequency-dependence arises by way of a normalised profile function $\phi \in L^1(R^+)$. The core of the Lyα line is dominated by Doppler broadening. The effects of radiation and resonance damping may be important in the wings of the line at low column densities. Under the assumption that these mechanisms are all uncorrelated, one can account for their combined effects by a convolution procedure. The folding of the Doppler profile with the Lorentz profiles from radiation and resonance damping gives a Voigt profile, which is the general description of Lyα line profiles (see [Mih78]). For all applications presented in this paper $v_{\text{turb}} \gg v_{\text{therm}}$ is adopted which gives rise to a very broad Doppler core dominating the Lorentzian wings of the Voigt profile. We may therefore conclude that a Doppler profile of the following form

$$\phi(\nu) = \frac{1}{\sqrt{\pi}\,\Delta\nu_D}\exp\left[-\left(\frac{\nu-\nu_0}{\Delta\nu_D}\right)^2\right] \tag{28}$$

gives a reasonable approximation of a Lyα line profile. ν_0 is the frequency of the line centre. The Doppler width $\Delta\nu_D$ and the Doppler velocity v_D are determined by the thermal velocity v_{therm} of the hydrogen atoms and the macroscopic turbulent velocity v_{turb}, which gives

$$\Delta\nu_D = \frac{\nu_0}{c}v_D = \frac{\nu_0}{c}\sqrt{v_{\text{therm}}^2 + v_{\text{turb}}^2}, \tag{29}$$

where c is the speed of light.

Due to the particular makeup of the source term

$$f(\mathbf{x},\nu) = \kappa(\mathbf{x},\nu)B(T(\mathbf{x}),\nu) + \epsilon(\mathbf{x},\nu), \tag{30}$$

thermal radiation and non-thermal radiation can be investigated separately. In the case of radiation which is exclusively thermal, f is calculated from the temperature distribution $T(\mathbf{x})$, by way of the Planck function $B(T,\nu)$.

The "Doppler" operator $\mathcal{D}$ causes the Doppler shift of the photons. A derivation of the operator for non-relativistic velocities ($v/c < 0.1$) can be found in [WBW00]. In contradistinction to when transfer is fully relativistic (cf. [MW84]), for non-relativistic transfer all terms causing aberration or advection effects are neglected. The function

$$w(\mathbf{x},\mathbf{n}) = -\mathbf{n}\cdot\nabla_x\left(\mathbf{n}\cdot\frac{\mathbf{v}(\mathbf{x})}{c}\right) \tag{31}$$

is the gradient of the velocity field $\mathbf{v}(\mathbf{x})$ along the direction $\mathbf{n}$. At this point it should be recognised that velocity fields $\mathbf{v}$ exist, for which the sign of w may change.

In scattering processes, described by the operator $\mathcal{S}$ in (27), both the direction and frequency of a photon may change. These changes are described by a redistribution function

$$R(\hat{\mathbf{n}},\hat{\nu};\mathbf{n},\nu)\frac{d\hat{\omega}}{4\pi}\frac{d\omega}{4\pi}\,d\hat{\nu}\,d\nu, \tag{32}$$

which gives the joint probability that a photon will be scattered from direction $\hat{\mathbf{n}}$ in solid angle $d\hat{\omega}$ and frequency range $(\hat{\nu},\hat{\nu}+d\hat{\nu})$ into solid angle $d\omega$ in direction $\mathbf{n}$ and frequency range $(\nu,\nu+d\nu)$ (for details see [Mih78]). If we are primarily interested in redistribution in frequency and not in angle, we could assume that $\mathcal{I}(\mathbf{x},\mathbf{n},\nu)$ is nearly isotropic and replace it in (27) with the mean intensity $J(\mathbf{x},\mathbf{n})$. Then the scattering operator in (27) can be written as follows

$$\mathcal{S}\mathcal{I} = -\sigma(\mathbf{x})\int_0^\infty R(\hat{\nu},\nu)J(\mathbf{x},\hat{\nu})\,d\hat{\nu}, \tag{33}$$

where the angle-averaged redistribution function

$$R(\hat{\nu},\nu) \equiv \frac{1}{4\pi}\int_{S^2} R(\hat{\mathbf{n}},\hat{\nu};\mathbf{n},\nu)\,d\hat{\omega} \equiv \frac{1}{4\pi}\int_{S^2} R(\hat{\mathbf{n}},\hat{\nu};\mathbf{n},\nu)\,d\omega \tag{34}$$

gives the redistribution probability from $(\hat{\nu},\hat{\nu}+d\hat{\nu})$ to $(\nu,\nu+d\nu)$. Equation (33) provides an extremely useful approximation in line transfer problems. The crucial phenomenon there is the frequency diffusion of photons from the opaque line core, where they are trapped, to the line wings. The frequency shift causes an exponential decrease of the absorption profile which is in fact a Doppler profile (cf. (28)). Thus, the line wing photons may escape from regions where the assumption of an isotropic radiation field is nearly fulfilled.

We now have to investigate whether our assumption of radiation isotropy is justified. For large optical depths $\tau_{\hat{\nu}} \geq 1$ this assumption is essentially correct. But at any given frequency the radiation field will depart significantly from isotropy at points which are located away from the surface in optical depths $\tau_{\hat{\nu}}$ of order unity or less. On the other hand, line formation under the condition of non-local thermodynamic equilibrium is characterised by a source function for which the surface values are determined by photons originating from extended regions which are deeply embedded in the domain's interior, and which are of a size given in terms of a destruction length. It is important to recognise that in all of this region the radiation field is in fact isotropic (for details see [Mih78]). This may result in $\mathcal{I}(\mathbf{x}, \mathbf{n}, \nu)$ showing departures from isotropy at the surface or in optically thin regions, even though the source term $-\mathcal{S}\mathcal{I} + f(\mathbf{x}, \nu)$ in (27) is determined by processes occurring in depths where anisotropy is effectively negligible. Consequently, if photon frequencies are reshuffled, this is entirely due to the angle-averaged redistribution function (34). Further down, our test calculations will show that frequency-reshuffling is of crucial importance for any interpretation of measured line profiles, as given by astronomical observations.

The function defined by (34) is normalised such that

$$\int_0^\infty d\hat{\nu} \int_0^\infty R(\hat{\nu}, \nu) \, d\nu = 1. \tag{35}$$

Integration over all emitted photons yields the absorption profile

$$d\hat{\nu} \int_0^\infty R(\hat{\nu}, \nu) \, d\nu = \phi(\hat{\nu}) d\hat{\nu}. \tag{36}$$

In the angle-averaged approximation the emission profile is given by

$$\psi(\nu) d\nu = \frac{\int_0^\infty R(\hat{\nu}, \nu) J(\mathbf{x}, \hat{\nu}) \, d\hat{\nu}}{\int_0^\infty \phi(\hat{\nu}) J(\mathbf{x}, \hat{\nu}) \, d\hat{\nu}} \, d\nu \tag{37}$$

which shows that the distribution of emitted photons depends upon the frequency profile of the incoming radiation.

In the limiting case that the intensity is independent of frequency, we obtain natural excitation, with

$$\psi^*(\nu) d\nu = d\nu \int_0^\infty R(\hat{\nu}, \nu) \, d\hat{\nu}. \tag{38}$$

If we have $R(\hat{\nu}, \nu) = R(\nu, \hat{\nu})$, which holds in practically all cases of interest (cf. [Mih78]), then $\psi^*(\nu) = \phi(\nu)$. Natural excitation prevails, of course, in thermodynamic equilibrium, and it is usually thought of in that context. There are, however, other physical circumstances in which the result $\psi(\nu) = \phi(\nu)$ is recovered. Suppose, for instance, that atoms in their excited state are completely reshuffled in such a way that there is no correlation between the frequencies of the incoming photons and the scattered ones. In other words, there is no way of tracing back the scattered photons to their pre-scattering states. Then the frequency distribution of the absorption profiles $\phi(\nu)$ before and after completion of the scattering process are fully uncorrelated, which situation is referred to as *complete redistribution*, or *complete non-coherence*. This case is realized to a good approximation, for instance, when atoms are so strongly perturbed by collisions during the scattering process that the excited electrons are randomly redistributed over substates of the upper level. In this case, both the absorption and the emission probabilities are independent. They are proportional to the number of substates available at each frequency within the line (i.e., to $\phi(\nu)$ itself), and the joint absorption-emission probability $R(\hat{\nu}, \nu)$ is the product of these two independent distributions: $R(\hat{\nu}, \nu) = \phi(\hat{\nu})\phi(\nu)$. Then, for *complete redistribution* the scattering operator in (27) gives

$$\mathcal{S}^{\mathrm{crd}}\mathcal{I}(\mathbf{x}, \mathbf{n}, \nu) = -\frac{\sigma(\mathbf{x})\phi(\nu)}{4\pi} \int_0^\infty \phi(\hat{\nu}) \int_{S^2} \mathcal{I}(\mathbf{x}, \hat{\mathbf{n}}, \hat{\nu}) \, d\hat{\omega} \, d\hat{\nu}. \qquad (39)$$

Complete redistribution is also a good approximation within the Doppler profile (28) of a spectral line, and actually provides an excellent first approximation for modelling Lyα line photons in highly redshifted hydrogen clouds (halos).

Another class of problems arises when attention is focused on the angular redistribution of the emitted radiation, and when it is assumed that the scattering is essentially *coherent* (i.e., $\hat{\nu} = \nu$). This situation is of interest when light is scattered by large particles in a planetary atmosphere, for which the earth's atmosphere provides a ready example. Coherence gives

$$R(\hat{\mathbf{n}}, \hat{\nu}; \mathbf{n}, \nu) = g(\hat{\mathbf{n}}, \mathbf{n})\phi(\hat{\nu})\delta(\nu - \hat{\nu}), \qquad (40)$$

where δ is the Dirac function and g is an *angular phase function* normalised such that

$$\frac{1}{4\pi} \int_{S^2} g(\hat{\mathbf{n}}, \mathbf{n}) \, d\hat{\omega} = 1. \qquad (41)$$

For our test calculations further down, we assume isotropic scattering

$$g(\hat{\mathbf{n}}, \mathbf{n}) \equiv 1, \qquad (42)$$

but we would like to emphasise, that our algorithm is also capable of treating the more general case of anisotropic scattering. In astrophysical applications the dipole phase function

$$g(\hat{\mathbf{n}}, \mathbf{n}) = \frac{3}{4}\left(1 + \cos^2(\hat{\mathbf{n}} \cdot \mathbf{n})\right), \qquad (43)$$

is of great importance, which applies to Thomson and Rayleigh scattering by small particles and electrons. Other phase functions arise for scattering by large particles, which are of a size comparable to the wavelength of the light, but these phase functions are often extremely complicated and undergo large, rapid changes when the angle is varied (see [BH83, Hul57]).

For *coherent scattering* the scattering operator in (27) gives

$$\mathcal{S}^{\text{coh}}\mathcal{I}(\mathbf{x}, \mathbf{n}, \nu) = -\frac{\sigma(\mathbf{x}, \nu)}{4\pi} \int_{S^2} g(\hat{\mathbf{n}}, \mathbf{n}) \mathcal{I}(\mathbf{x}, \hat{\mathbf{n}}, \nu) \, d\hat{\omega}. \qquad (44)$$

For a spectral line, coherent scattering occurs only for an infinitely sharp lower state of the line in conjunction with an upper state which is non-perturbed during emission, while the scattering atoms remain at rest in the frame of the observer. But this is usually not the case, however, and line scattering is therefore much better described by complete redistribution.

Finally, the problem of accounting for the effects of angle-averaged *partial redistribution* in spectral line formation has to be addressed. Again, we shall confine the derivation to a two-level atom, while seeking to application the results to Lyα line transfer. For simplicity it is assumed that transitions between bound and unbound states, as with (re-) ionisation processes, will not occur. Let us examine the formation of a resonance line connecting a perfectly sharp lower level to an upper state that is broadened into a distribution of substates by both radiation damping and collisions. Denote the population of the lower level by n_1, the total population of the upper level (summed over all substates) by n_2. The distribution of the atoms over the upper state is specified in terms of the observer's frame *emission profile* $\psi(\nu)$ (cf. (37)), defined as the fraction of all atoms in the upper state that emit photons of frequency ν as seen in the laboratory frame, if the states decay radiatively. For the cases discussed so far, $R(\hat{\nu}, \nu) = R(\nu, \hat{\nu})$, and therefore $\psi(\nu) = \phi(\nu)$ is assumed. This is not true in general, since $\psi(\nu)$ is a unique but also complicated one-to-one mapping of the distribution of atoms over their own rest-frame frequencies and velocities. This transformation is specified by the redistribution function $\psi(\nu)$ which is normalised, so that $n_2 = n_2 \int \psi(\nu) d\nu$. The substate occupation number $n_2(\nu) = n_2 \psi(\nu)$ or, equivalently, $\psi(\nu)$ is specified by a rate equation, which accounts for the losses and gains due to radiative and collisional (de-) excitation. The solution of the transfer equation is now quite complicated because the emission profile $\psi(\nu)$ is *not known a priori*, but needs to be derived from the statistical equilibrium equations (for details see [Mih78]). Unlike the case of complete redistribution, where only the ratio (n_2/n_1) is required to specify the source function and hence only *one* statistical equilibrium equation is needed, we must now *compute* $\psi(\nu)$, and this introduces a fixed-point iteration where the same number of equilibrium equations and radiative transfer equations needs to be solved to define this function to the desired precision. The algorithm presented in this paper is capable of dealing with partial redistribution. The mean intensity $J(\mathbf{x}, \nu)$ is available for all frequencies, since the intensities are stored for all spatial points $\mathbf{x}$, directions $\mathbf{n}$ and frequencies ν. Thus, even angle-dependent redistribution functions may be included.

4.2 Boundary Conditions

Formally, the frequency derivative in the Doppler term of (27) can be viewed as being similar to the time derivative for non-stationary radiative transfer or heat transfer problems, so that a parabolic system of initial value problems is obtained. Unfortunately, the Doppler term also contains a function $w(\mathbf{x}, \mathbf{n})$, which is essentially the gradient of the macroscopic velocity field, multiplied by the photon propagation direction $\mathbf{n}$ (cf. (31)). Depending on the complexity of the velocity field, this function $w(\mathbf{x}, \mathbf{n})$ may change its sign at different points $\mathbf{x}$ and for different directions $\mathbf{n}$ when setting up the initial value either on the lower or upper frequency boundary. In this paper we take this into account by defining an additional frequency boundary value problem, which involves a "frequency inflow boundary" $\Sigma^- = \Omega \times S^2 \times \partial \Lambda$ depending on the sign of $w(\mathbf{x}, \mathbf{n})$. In order to solve (27) correctly, boundary conditions of the form

$$\mathcal{I}(\mathbf{x}, \mathbf{n}, \nu) = p(\mathbf{x}, \mathbf{n}, \nu) \quad \text{on } \Gamma^- \times \Lambda \quad \text{and} \tag{45}$$

$$\mathcal{I}(\mathbf{x}, \mathbf{n}, \nu) = q(\mathbf{x}, \mathbf{n}, \nu) \quad \text{on } \Sigma^- \tag{46}$$

must be imposed. The "spatial inflow boundary" is

$$\Gamma^- \times \Lambda = \left\{ (\mathbf{x}, \mathbf{n}, \nu) \in \Gamma \;\middle|\; \mathbf{n}_\Gamma \cdot \mathbf{n} < 0 \right\}, \tag{47}$$

where $\mathbf{n}_\Gamma$ is the unit vector perpendicular to the boundary surface Γ of the spatial domain Ω. The sign of the product $\mathbf{n}_\Gamma \cdot \mathbf{n}$ describes the "flow direction" of the photons across the boundary of the spatial domain. For our modelling of the spectral Lyα line, we assume that there is no continuum emission "outside of the line" and that there are no light sources "outside the modelled domain". In this case the two boundary conditions for the solution of the transfer equation (27) in moving media are

$$\mathcal{I}(\mathbf{x}, \mathbf{n}, \nu) = 0 \quad \text{on } \Sigma^-, \tag{48}$$

$$\mathcal{I}(\mathbf{x}, \mathbf{n}, \nu) = 0 \quad \text{on } \Gamma^- \times \Lambda. \tag{49}$$

4.3 Finite Element Discretisation

The simultaneous Galerkin discretisation in space-frequency is, from a theoretical point of view, an extremely appealing approach. Considering the memory requirement, the solution of the complete algebraic system resulting from this discretisation is by far too "expensive". With a view to reducing the memory demand, a discretisation first in space and then in frequency is performed. This approach is called method of lines and finally results

122   Erik Meinköhn

in a system of ordinary differential equations. Contrary to the continuous Galerkin discretisation in space, the discretisation in frequency is a discontinuous Galerkin (DG) method. It is important to note that the above discretisation in space and in frequency can be interpreted as a simultaneous Galerkin discretisation in the space-frequency domain. The frequency DG method is performed by using piecewise polynomials of degree zero, which corresponds to an implicit Euler method for N equidistantly distributed frequency points $\nu_i \in \{\nu_1, \nu_2, ..., \nu_N\} \subset \Lambda$ (see [Boe96]). DG methods are not only used for discretising ordinary differential equations, especially initial value problems, but also for the time discretisation and mesh control of partial differential equations as described in [EJ87, EJ91] and [EJT85].

The natural solution space for the frequency-dependent radiative transfer problem involving moving media (cf. (27)) is

$$W^{\text{poly}} = \left\{ \mathcal{I} \in L^2(\Omega \times S^2 \times \Lambda) \mid \mathcal{HI} \in L^2(\Omega \times S^2 \times \Lambda) \right\}. \tag{50}$$

$\Lambda \subset [0, \infty[$ is the frequency space and $\mathcal{H} = \mathcal{T} + \mathcal{D}$ is a combination of the transport operator with the Doppler operator. The space W^{poly} contains all the solutions $\mathcal{I}$ and their derivatives with respect to $\mathbf{x}$ and ν. If we impose homogeneous boundary conditions on both the spatial and frequency inflow boundary, while excluding "external" photon sources (for details see previous section), the solution space is

$$W_0^{\text{poly}} = \left\{ \mathcal{I} \in W^{\text{poly}} \mid \mathcal{I} = 0 \text{ on } \Gamma^- \times \Lambda \text{ and } \mathcal{I} = 0 \text{ on } \Sigma^- \right\}. \tag{51}$$

As with the monochromatic case we have to stabilise this finite element discretisation via the streamline diffusion modification:

$$\begin{aligned} d_\delta(\mathcal{I}, v) &:= \left(\mathcal{TI}, v + \delta \mathcal{T} v \right)_{\Omega \times S^2 \times \Lambda} + \left(\mathcal{DI}, v + \delta \mathcal{T} v \right)_{\Omega \times S^2 \times \Lambda} \\ &\quad + \left(\mathcal{SI}, v + \delta \mathcal{T} v \right)_{\Omega \times S^2 \times \Lambda} \\ &\quad + \left(\chi \mathcal{I}, v + \delta \mathcal{T} v \right)_{\Omega \times S^2 \times \Lambda} \\ &= \left(f, v + \delta \mathcal{T} v \right)_{\Omega \times S^2 \times \Lambda} \qquad \forall v \in W_0^{\text{poly}}. \end{aligned} \tag{52}$$

The boundary condition and the initial value are not included in this weak formulation. They are, rather, implemented in a strong manner, as presented in the previous section. We are going to apply standard Galerkin finite elements to solve (52), and therefore we choose a finite dimensional subspace $W_h^{\text{poly}} \subset W_0^{\text{poly}}$ consisting of functions that are piecewise polynomial with respect to a triangulation $T_h^{\text{poly}} := \mathbb{T}^\Omega \otimes \mathbb{T}^{S^2} \otimes \mathbb{T}^\Lambda$ of $\Omega \times S^2 \times \Lambda$:

$$W_h^{\text{poly}} = W_h^{\Omega} \otimes W_h^{S^2} \otimes W_h^{\Lambda}$$
$$= \Big\{ v \in W^{\text{poly}} \,\Big|\, v(\cdot, \mathbf{n}, \nu)|_{K^{\Omega}} \in Q^1(K^{\Omega}) \forall K^{\Omega} \in \mathbb{T}^{\Omega},\, v|_{K^{\Omega}} \in C^0(\Omega),$$
$$v(\mathbf{x}, \cdot, \nu)|_{K^{S^2}} \in P^0(K^{S^2})\, \forall K^{S^2} \in \mathbb{T}^{S^2},$$
$$v(\mathbf{x}, \mathbf{n}, \cdot)|_{K^{\Lambda}} \in P^0(K^{\Lambda})\, \forall K^{\Lambda} \in \mathbb{T}^{\Lambda} \Big\}. \tag{53}$$

The triangulation of $\mathbb{T}^{\Lambda}$ is realized via a subdivision of the frequency space Λ in finite intervals, or cells, K_i^{Λ} ($i = 1, \ldots, N < \infty$), so that the function space over Λ is given as follows

$$W_h^{\Lambda} := \Big\{ v \in L^2(\Lambda) \,\Big|\, v(\mathbf{x}, \mathbf{n}, \cdot)|_{K^{\Lambda}} \in P^0(K^{\Lambda})\, \forall K^{\Lambda} \in \mathbb{T}^{\Lambda} \Big\}. \tag{54}$$

The discretisation of Λ is analogous to the discretisation of the unit sphere S^2: first we start with a regular subdivision, and then we choose constant basis functions $v(\mathbf{x}, \mathbf{n}, \cdot)|_{K^{\Lambda}} = const$, which approach can be identified with a discontinuous Galerkin method of zeroth degree (DG(0)).

The discrete analog of (52) consists in finding $\mathcal{I}_h \in W_h^{\text{poly}}$, such that

$$d_\delta(\mathcal{I}_h, v_h) := \big(\mathcal{T}\mathcal{I}_h, v_h + \delta\mathcal{T} v_h\big)_{\Omega \times S^2 \times \Lambda} + \big(\mathcal{D}\mathcal{I}_h, v_h + \delta\mathcal{T} v_h\big)_{\Omega \times S^2 \times \Lambda}$$
$$+ \big(\mathcal{S}\mathcal{I}_h, v_h + \delta\mathcal{T} v_h\big)_{\Omega \times S^2 \times \Lambda} \tag{55}$$
$$+ \big(\chi\mathcal{I}_h, v_h + \delta\mathcal{T} v_h\big)_{\Omega \times S^2 \times \Lambda}$$
$$= \big(f, v_h + \delta\mathcal{T} v_h\big)_{\Omega \times S^2 \times \Lambda} \qquad \forall v \in W_h^{\text{poly}}.$$

The cell-wise constant-parameter function δ depends on the local mesh width h and the coefficients κ and σ. As mentioned before in Sect. 3.2, the streamline diffusion modification term will be omitted in the interest of conciseness of the written formulation, but it is assumed to be always present.

Discretisation for Coherent Scattering and Resulting Linear System of Equations

If the operator form of the transport matrix for coherent scattering is introduced as

$$\mathcal{A}^{\text{coh}} = \mathcal{T} + \mathcal{S}^{\text{coh}} + \chi(\mathbf{x}, \nu), \tag{56}$$

the weak formulation of (27) can be written as follows

$$\big(\mathcal{A}^{\text{coh}}\mathcal{I}, \varphi\big)_{\Omega \times S^2 \times \Lambda} + \big(\mathcal{D}\mathcal{I}, \varphi\big)_{\Omega \times S^2 \times \Lambda} = \big(f, \varphi\big)_{\Omega \times S^2 \times \Lambda} \qquad \forall \varphi \in W_0^{\text{poly}}. \tag{57}$$

Since the function $w(\mathbf{x}, \mathbf{n})$ may change its sign, this simple difference scheme for the Doppler term $\mathcal{D}$ reads

$$w(\mathbf{x},\mathbf{n})\nu\frac{\partial \mathcal{I}}{\partial \nu} \longrightarrow w\nu_i\frac{\mathcal{I}_i - \mathcal{I}_{i-1}}{\Delta\nu} \qquad (w_i > 0) \qquad (58)$$

and

$$w(\mathbf{x},\mathbf{n})\nu\frac{\partial \mathcal{I}}{\partial \nu} \longrightarrow w\nu_i\frac{\mathcal{I}_{i+1} - \mathcal{I}_i}{\Delta\nu} \qquad (w_i < 0), \qquad (59)$$

where $\Delta\nu$ is the constant frequency step size. All quantities referring to the discrete frequency point ν_i are denoted by an index "i". Employing the Euler method, we get a semi-discrete representation of the transport operator for coherent scattering

$$\tilde{\mathcal{A}}_i^{\mathrm{coh}} = \mathcal{A}_i^{\mathrm{coh}} + \frac{|w|\nu_i}{\Delta\nu} \qquad (60)$$

and of the source term on the right hand side

$$\tilde{f}_i = f_i + \frac{|w|\nu_i}{\Delta\nu}\begin{cases}\mathcal{I}_{i-1} & (w > 0) \\ \mathcal{I}_{i+1} & (w < 0)\end{cases}. \qquad (61)$$

In the expression (60) for the transport operator $\tilde{\mathcal{A}}_i^{\mathrm{coh}}$, the additional term beyond $\mathcal{A}_i^{\mathrm{coh}}$ causes a slight alteration which can be interpreted as an artificial opacity, which is beneficial when solving the corresponding linear system of equations. The additional term in (61) acts like an artificial source term. The semi-discrete weak formulation for each frequency point ν_i therefore gives

$$(\tilde{\mathcal{A}}_i^{\mathrm{coh}}\mathcal{I}_i, \varphi)_{\Omega\times S^2\times\Lambda} = (\tilde{f}_i, \varphi)_{\Omega\times S^2\times\Lambda} \qquad \forall \varphi \in W_0^{\mathrm{poly}}. \qquad (62)$$

If there are L degrees of freedom in Ω, M directions on the unit sphere S^2 and N frequency points, the overall discrete system has the matrix form

$$\mathbf{A}_{\mathrm{coh}}^{\mathrm{poly}}\mathbf{u} = \mathbf{f}, \qquad (63)$$

where the vector $\mathbf{u}$ contains the discrete intensities and the vector $\mathbf{f}$ contains the values of the source term for all frequency points ν_i. Both vectors have $(L \cdot M \cdot N)$ components and $\mathbf{A}_{\mathrm{coh}}^{\mathrm{poly}}$ therefore is a $(L \cdot M \cdot N) \times (L \cdot M \cdot N)$ matrix. The fact that the function $w(\mathbf{x},\mathbf{n})$ may change its sign, results in a block-tridiagonal structure of $\mathbf{A}_{\mathrm{coh}}^{\mathrm{poly}}$, which gives

$$\begin{pmatrix} \tilde{\mathbf{A}}_1^{\mathrm{coh}} & \mathbf{R}_1 & 0 & \cdots & 0 \\ \mathbf{B}_2 & \tilde{\mathbf{A}}_2^{\mathrm{coh}} & \mathbf{R}_2 & \ddots & \vdots \\ 0 & \ddots & \ddots & \ddots & 0 \\ \vdots & \ddots & \ddots & \ddots & \mathbf{R}_{N-1} \\ 0 & \cdots & 0 & \mathbf{B}_N & \tilde{\mathbf{A}}_N^{\mathrm{coh}} \end{pmatrix}\begin{pmatrix} \mathbf{u}_1 \\ \mathbf{u}_2 \\ \vdots \\ \vdots \\ \mathbf{u}_N \end{pmatrix} = \begin{pmatrix} \mathbf{f}_1 \\ \mathbf{f}_2 \\ \vdots \\ \vdots \\ \mathbf{f}_N \end{pmatrix}. \qquad (64)$$

The block matrices $\mathbf{R}_i$ and $\mathbf{B}_i$ are given in terms of $w(\mathbf{x},\mathbf{n})\nu_i/\Delta\nu$. Depending on the sign of $w(\mathbf{x},\mathbf{n})$, either a redshift or a blueshift is obtained for the photons in the medium, respectively. For a reasonable resolution, the resulting linear system of equations of the total system is too large to be solved directly. Hence, we are treating N quasi-monochromatic radiative transfer problems

$$\tilde{\mathbf{A}}_i^{\mathrm{coh}} \mathbf{u}_i = \tilde{\mathbf{f}}_i, \qquad (65)$$

with a slightly modified right hand side

$$\tilde{\mathbf{f}}_i = \mathbf{f}_i + \mathbf{R}_i \mathbf{u}_{i+1} + \mathbf{B}_i \mathbf{u}_{i-1}. \qquad (66)$$

A detailed description and discussion of the structure and solution strategy of these transfer problems can be found in Sect. 3.2.

When radiation fields are modelled in static media, the resulting linear system decouples into N independent quasi-monochromatic transfer problems, which have to be solved successively only once per frequency point ν_i. When admitting only simple macroscopic velocity fields in moving media, e.g. inflow or outflow motions as described in Sect. 5.2, the sign of the function $w(\mathbf{x}, \mathbf{n})$ is either positive or negative at every point and for all directions, and all the photons experience either a redshift *or* a blueshift. For this case the structure of the resulting linear system is block-bidiagonal, which is of great advantage for our solution effort. The i-th quasi-monochromatic transport problem depends only on quantities of the i-th frequency point and on the solution of the previous frequency step (see (65) and (62)). The computational effort for solving this block-bidiagonal system is essentially the same as for static media. If more general and therefore more complex velocity fields are included, e.g. rotational movements as described in Sect. 5.2, the resulting block-tridiagonal system is solved via a fixed-point frequency iteration loop. Then (65) is solved several times for each frequency point ν_i, depending on the choice of the velocity field. During this fixed-point iteration, changes of the emergent line profile are monitored carefully in a single selected direction. The outer frequency loop is stopped if changes in the emergent line profile are below a given tolerance.

Discretisation for Complete Redistribution and Resulting Linear System of Equations

In analogy to the discretisation approach for coherent scattering, we now introduce the operator form of the transport matrix for complete redistribution

$$\mathcal{A}^{\mathrm{crd}} = \mathcal{T} + \mathcal{S}^{\mathrm{crd}} + \chi(\mathbf{x}, \nu) \qquad (67)$$

so that (27) can be written in the weak formulation as

$$\left(\mathcal{A}^{\mathrm{crd}} \mathcal{I}, \varphi\right)_{\Omega \times S^2 \times \Lambda} + \left(\mathcal{D}\mathcal{I}, \varphi\right)_{\Omega \times S^2 \times \Lambda} = \left(f, \varphi\right)_{\Omega \times S^2 \times \Lambda} \quad \forall \varphi \in W_0^{\mathrm{poly}}. \qquad (68)$$

If complete redistribution is assumed, an implicit Euler scheme suffices to discretise the Doppler term as described above, along with a simple quadrature method for the frequency integral in the scattering operator $\mathcal{S}^{\mathrm{crd}}$ in (39). Starting from N equidistantly distributed frequency points $\nu_i \in \{\nu_1, \nu_2, ..., \nu_N\} \subset \Lambda$ and N weights $q_1, q_2, ..., q_N$, we define a quadrature method

$$Q(\nu_i) := \sum_{j=1}^{N} q_j \xi(\nu_j) \qquad (69)$$

for integrals $\int_\Lambda \xi(\nu')d\nu'$. In the case of complete redistribution the kernel is

$$\xi(\nu_j) = \frac{\phi(\nu_j)}{4\pi} \int_{S^2} \mathcal{I}(\mathbf{x}, \hat{\mathbf{n}}, \nu_j) d\hat{\omega}. \qquad (70)$$

If the terms with the unknown intensities $\mathcal{I}_i$ are separated from the known quantities $\mathcal{I}_j$, the discretised scattering integral (39) is found to give

$$\frac{\sigma_i}{4\pi} \phi_i q_i \int_{S^2} \mathcal{I}_i d\hat{\omega} + \frac{\sigma_i}{4\pi} \sum_{j \neq i}^{N} \phi_j q_j \int_{S^2} \mathcal{I}(\mathbf{x}, \hat{\mathbf{n}}, \nu_j) d\hat{\omega}. \qquad (71)$$

Employing the Euler method, we get a semi-discrete representation of the transport operator for complete redistribution

$$\tilde{\mathcal{A}}_i^{\mathrm{crd}} = \mathcal{T} + \phi_i q_i \mathcal{S}^{\mathrm{coh}} + \chi_i + \frac{|w|\nu_i}{\Delta \nu} \qquad (72)$$

and of the source term on the right hand side of (68)

$$\hat{f}_i = f_i + \frac{\sigma_i}{4\pi} \sum_{j \neq i}^{N} \phi_j q_j \int_{S^2} \mathcal{I}(\mathbf{x}, \hat{\mathbf{n}}, \nu_j) d\hat{\omega} + \frac{|w|\nu_i}{\Delta \nu} \begin{cases} \mathcal{I}_{i-1} & (w > 0) \\ \mathcal{I}_{i+1} & (w < 0) \end{cases}. \qquad (73)$$

As discussed in the previous section, the additional term in (72), which is caused by the macroscopic velocity field, acts like an artificial opacity. In contradistinction to the case of coherent scattering, the semi-discrete righthand side $\hat{f}_i$ in (73) includes two artificial source terms, causing a much stronger coupling between the solutions $\mathcal{I}_j$ ($j = 1, \ldots, N$). Under these circumstances the semi-discrete weak formulation for each frequency point ν_i gives

$$(\tilde{\mathcal{A}}_i^{\mathrm{crd}} \mathcal{I}_i, \varphi)_{\Omega \times S^2 \times \Lambda} = (\hat{f}_i, \varphi)_{\Omega \times S^2 \times \Lambda} \qquad \forall \varphi \in W_0^{\mathrm{poly}}. \qquad (74)$$

Given a discretisation with L degrees of freedom in Ω, M directions on the unit sphere S^2 and N frequency points, the overall discrete system has the matrix form

$$\mathbf{A}_{\mathrm{crd}}^{\mathrm{poly}} \mathbf{u} = \mathbf{f}, \qquad (75)$$

with the vector $\mathbf{u}$ containing the discrete intensities, while the vector $\mathbf{f}$ comprises the values of the source term for all frequency points ν_i. Both vectors are $(L \cdot M \cdot N)$-dimensional and $\mathbf{A}_{\mathrm{coh}}^{\mathrm{poly}}$ therefore represents a $(L \cdot M \cdot N) \times (L \cdot M \cdot N)$

matrix. Unfortunately, global frequency-coupling via the scattering integral (39) results in a full block matrix $\mathbf{A}_{\text{crd}}^{\text{poly}}$ and we get

$$\begin{pmatrix} \tilde{\mathbf{A}}_1^{\text{crd}} & \mathbf{R}_1 + \mathbf{Q}_2 & \mathbf{Q}_3 & \cdots & \mathbf{Q}_N \\ \mathbf{B}_2 + \mathbf{Q}_1 & \tilde{\mathbf{A}}_2^{\text{crd}} & \mathbf{R}_2 + \mathbf{Q}_3 & \ddots & \vdots \\ \mathbf{Q}_1 & \ddots & \ddots & \ddots & \vdots \\ \vdots & \ddots & \ddots & \ddots & \vdots \\ \mathbf{Q}_1 & \cdots & \cdots & \cdots & \tilde{\mathbf{A}}_N^{\text{crd}} \end{pmatrix} \begin{pmatrix} \mathbf{u}_1 \\ \mathbf{u}_2 \\ \vdots \\ \vdots \\ \mathbf{u}_N \end{pmatrix} = \begin{pmatrix} \mathbf{f}_1 \\ \mathbf{f}_2 \\ \vdots \\ \vdots \\ \mathbf{f}_N \end{pmatrix}. \tag{76}$$

The block matrices $\mathbf{B}_i$ and $\mathbf{R}_i$ contain the contribution of the Doppler factor $w(\mathbf{x}, \mathbf{n})\nu_i/\Delta\nu$ which cause a redshift or a blueshift of the photons, depending on the sign of w. The other off-diagonal block matrices $\mathbf{Q}_j$ comprise the terms which originate from the quadrature scheme. As already explained for the case of coherent scattering, we do not solve the complete system, but rather N quasi-monochromatic radiative transfer problems

$$\tilde{\mathbf{A}}_i^{\text{crd}} \mathbf{u}_i = \hat{\mathbf{f}}_i, \tag{77}$$

successively, with a modified right hand side

$$\hat{\mathbf{f}}_i = \mathbf{f}_i + \mathbf{R}_i \mathbf{u}_{i+1} + \mathbf{B}_i \mathbf{u}_{i-1} + \sum_{j \neq i} \mathbf{Q}_j \mathbf{u}_j. \tag{78}$$

More or less independent of the motion of the medium, complete redistribution causes a strong coupling between the solutions $\mathcal{I}_j$, which is incorporated via the artificial source term $\sum_{j \neq i} \mathbf{Q}_j \mathbf{u}_j$ in (78). Then, (77) is solved several times for each frequency point ν_i, depending essentially on the strength of the scattering process, or, to be more precise, on the scattering coefficient and on the total optical depth. During this fixed-point iteration, we monitor the changes of the emergent line profile in a single selected direction. The outer frequency loop is stopped if the changes in the emergent line profile are below a given tolerance.

4.4 Full Solution Algorithm

Equation (62) and (74) are of the same form as the monochromatic radiative transfer equation, cf. (6), for which we proposed a solution method based on a finite element technique in Sect. 3. Our method employs unstructured grids which are adaptively refined by means of an a-posteriori error estimation strategy. Now, we apply this method to (62) or (74). In brief, the full solution algorithm is as follows:

1. Start with $\mathcal{I} = 0$ for all frequencies.
2. Solve (65) or (77) for $i = 1, .., N$.

3. Repeat step 2 until convergence is reached.
4. Refine the grid and repeat step 2 and 3.

We start with a relatively coarse grid, where only the most important structures are pre-refined, and assure that the frequency interval $[\nu_1, \nu_N]$ is wide enough to cover the total line profile. Then, we solve (62) or (74) for each frequency several times depending on the choice of the redistribution function and the velocity field. During this fixed-point iteration, we monitor the changes of the resulting line profile in a single selected direction $\mathbf{n}_{out}$. When there are no more changes, we go over to step 4 and refine the spatial grid. Again, we apply the fixed-fraction grid refinement strategy: The cells are ordered according to the size of the local refinement indicator $\eta_K = max(\eta_K(\nu_i))|_{\nu_i}$, and a fixed portion of the cells, consisting of those with largest η_K, is refined. $\eta_K(\nu_i)$ is an indicator for the error of the solution in cell K at frequency ν_i.

5 Test Calculations

5.1 Monochromatic 3D Transfer Problems

Test I: Searchlight Beam Test

The searchlight beam test is a standard test for the transport operator: A narrow beam of radiation is directed into the computational grid at a certain angle. The beam traverses vacuum ($\kappa = 0, \sigma = 0, f = 0$) without dispersion.

For this test we started with a two-dimensional, globally pre-refined grid with 16×16 cells which covers the domain $\{x = [-1, 1], y = [-1, 1]\}$. This grid consists of 17×17 vertices and has a resolution of $\Delta x = \Delta y = 0.125$. The beam of radiation with intensity $I_0 = 1.0$ impinges at $x = -1$ between $y = [-0.875, -0.750]$ at an angle of approximately $45°$. We calculated the radiation field for the basic grid and 8 successively refined grids, where in each step 25% of the cells with the worst L^2-indicator η_{L^2} (see Sect. 3.3) were refined.

Figure 1 shows the intensity distribution and the grid structure for step 2, 4 and 8. The minimal size of the grid cells at step 2 corresponds to an equidistant grid with 64×64 cells. The dispersion of the beam at this step is similar to what was obtained with the short characteristic method described in [KA88]. As is most clearly apparent near where the beam impinges on the grid, the first refinement steps resolve the beam itself, whereas later steps then resolve its edges. At step 8 the intensity distribution shows a straight beam with negligible dispersion. This result was obtained with about 10^5 vertices, but it should be recognised that an equidistant grid with the same resolution would have required a hundredfold increase in the number of the vertices (see last two rows of Table 1).

Finite Elements for 3D Radiative Transfer Problems   129

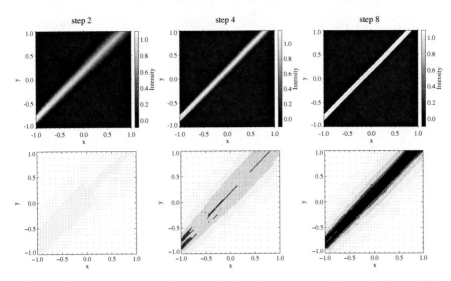

Fig. 1. Test I: Intensity distribution (top) and structure of the grid (bottom) for refinement steps 2, 4 and 8. Each dot marks the location of a vertex. The incoming beam impinges at $x = -1.0$ between $y = [-0.875, -0.750]$

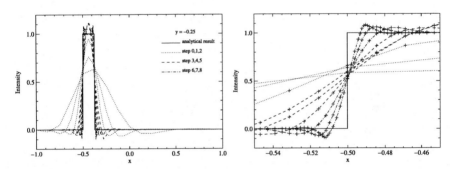

Fig. 2. Test I: Intensity distribution at $y = -0.25$ for different adaptive steps as indicated. The right hand side gives a close-up for the range $x = [-0.55, -0.45]$. The location of the vertices is marked by the crosses

The working of the code is demonstrated in detail in Fig. 2. Here the intensity distribution at $y = -0.25$ is shown for all refinement steps. The right hand side of this figure gives a close-up of the left-hand edge of the beam. The locations of the vertices are indicated by crosses. As was to be expected from the grid structure, the initial steps resolve the beam itself, whereas the following steps lead to a better resolution of the intensity distribution at the edges of the beam, which become increasingly sharp. Since the streamline diffusion discretisation is a second-order method, problems are bound to arise, as negative intensities and overshooting will occur. This is shown in greater detail in

Table 1. Test I which gives for each refinement step the minimum intensity $I_{\min}$, the relative number of vertices with negative intensity values n_{low}, the maximum intensity $I_{\max}$, the relative number of vertices with intensity values greater than unity n_{high}, the total number of vertices N, and the total number of required vertices N_{uni} for a grid with uniformly distributed cells of the same resolution

step	$I_{\min}$	n_{low} in %	$I_{\max}$	n_{high} in %	N	N_{uni}
0	−0.0457	35	1.000	0.37	289	289
1	−0.0805	39	1.076	0.32	631	1 089
2	−0.0833	38	1.104	1.7	1 395	4 225
3	−0.0963	34	1.109	7.6	2 995	16 641
4	−0.1018	36	1.171	14	6 312	66 049
5	−0.0833	36	1.106	24	13 084	263 169
6	−0.0833	34	1.080	24	28 005	1 050 625
7	−0.0929	32	1.095	25	57 606	4 198 401
8	−0.0934	29	1.093	23	118 179	16 785 409

Table 1. For all steps, the minimum intensity $I_{\min}$ and the maximum intensity $I_{\max}$ deviate by about 10% of I_0 from their values as derived from the analytical solution. With increasing resolution the fraction of vertices with negative intensities n_{low} is slightly declining from 40% to 30%, whereas the fraction of vertices with intensity values greater than unity n_{high} is continuously increasing to a fraction of about 25%. The minimum intensities are somewhat better than those obtained with the parabolic upwind interpolation scheme presented in [KA88], who find $I_{\min} \simeq -0.15$ for equally-spaced grids as well as for logarithmic grids. Unsurprisingly, their relative number of cells with negative intensities $n_{\text{low}} \simeq 20\%$ is smaller, because the finite element code strongly favours the generation of finer cells in regions of discontinuities where negative intensity values occur. Negative intensity values cause problems in cases where even the mean intensity J turns out to be negative, so that important physical quantities cannot be determined. A satisfactory treatment of these cases would do away with simply adopting $I = 0$ as a lower limit. Instead, it would require expensive code development involving the tracing of discontinuities in order to switch between second order and first order methods, as with *shock capturing* methods described in the literature (e.g. [EJ93]).

Test II: Radiation Field of a Plane-Parallel Layer

To test the scattering operator we calculate the radiation field of an infinite plane-parallel layer. In this case, we compare our results with those of the separable representation (SR) method, which is one of the analytical methods, and with those of a finite difference (FD) code. Both methods are briefly described in the appendix.

The test starts with a homogeneous plane-parallel layer of thickness Δ. We assume zero boundary conditions and a constant source term $f = \chi(1-\gamma)$, with extinction coefficient $\chi = \kappa + \sigma$ and albedo $\gamma = \sigma/\chi$. We calculate the radiation field for a selection of optical depths $\tau = \chi\Delta = 0.4, 2, 6, 20$ and albedo $\gamma = 0.02, 0.8, 0.98$, and investigate the angular dependence of the specific intensity emanating from the plane-parallel layer.

When using the FE code we approximate the infinitely extended layer by a three-dimensional anisotropic grid with initially 4^3 cells which have an aspect ratio of 100:100:1. For each optical depth and albedo we perform up to a maximum of seven refinement steps and probe the outgoing intensity at the middle of the surface of the layer. We use local refinement criteria η_K taken from (21) for a functional $\mathcal{M}(e) = e(\mathbf{x}_0, \mathbf{n}_0)$, which depends only on the intensity issuing from the domain Ω at *one* point $\mathbf{x}_0$ in *one* particular direction $\mathbf{n}_0$. An aspect ratio of 100:100:1 implies that we get reliable results for $\mu = \geq 0.02$, where μ is the angle-cosine of the ray direction $\mathbf{n}$ relative to the vertical.

The FD code is also three-dimensional (see Appendix B). Here, we use a numerical grid with 32^3 cells with the same aspect ratio of 100:100:1. The iterative procedure for the mean intensity J is carried out until the relative change of J is smaller than 10^{-5}.

The SR method (see Appendix A) gives the solution in terms of an infinite series, so that an approximation by a finite sum is possible. For a wide range of parameters, an excellent agreement between the approximated function and the exact result can already be obtained with $N = 5$, where N is the number of terms of the finite sum. For small μ the exact and the approximated function differ in behaviour, which is taken as an indication of strong errors in the approximation for this particular range of μ-values. But in proportion as N is increased, the results for small values of μ improve. For our test runs, the specific intensity was calculated with setting $N = 10$ for all cases.

Figure 3 shows the angular distribution of the specific intensity escaping from a plane-parallel layer, which was obtained by employing three methods. In general, the three codes yield the same results. Excellent agreement is obtained between the SR method and the FE method. The FE code needed 2 to 3 refinement steps for low optical depth and 5 refinement steps for $\tau = 20.0$ to match the solution of the SR method with excellent precision. The resolution of refinement step 5 corresponds to a globally refined grid with 128^3 cells. The FD code is not able to reproduce the results of the other methods for high optical depth and large albedo. In these cases the results of the FD code are too low. The results can be improved by using a grid with higher spatial resolution, but even a grid with 128^3 cells fails to produce results which are comparable in quality to those obtained with the FE code. Since the FD code only uses a first order discretisation, its excellent results must be attributed to the use of a second-order FE streamline diffusion discretisation. The results of the SR method depend to a lesser extent on the accuracy of J, and the number of ordinates M plays a minor role.

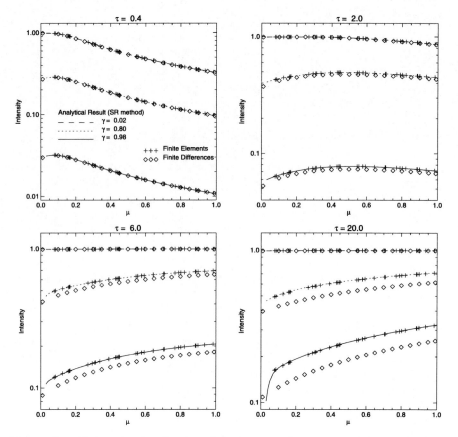

Fig. 3. Test II: Angular distribution of the intensity for various total optical depths τ and albedo γ

The general form of the intensity distributions displayed in Fig. 3 becomes apparent when the change of I with the optical depth $d\tau = \chi ds$ is considered

$$\mu \frac{dI}{d\tau} = -[I - \gamma J - (1-\gamma)]. \tag{79}$$

The solution of this equation for a given J and zero boundary condition is

$$I(\tau,\mu) = \exp(-\tau/\mu) \int_0^{\tau/\mu} (\gamma J + (1-\gamma)) \exp(\tau') d\tau'. \tag{80}$$

For $\gamma \sim 0$ the intensity distribution is

$$I(\tau,\mu) = 1 - \exp(-\tau/\mu), \tag{81}$$

which expresses the decrease of the intensity with increasing μ in the optical thin case and the flat distribution $I(\tau,\mu) \approx 1$ for larger values of the optical

Table 2. Test II: Investigation of the solution of the specific intensity in a particular direction $\mu = 0.705$ for a variety of adaptively refined grids (L2, DUAL) and a structured grid (GLOBAL). The optical depth and the albedo is $\tau = 20$ and $\gamma = 0.80$, respectively

	FE					FD	
DUAL		L2		GLOBAL		GLOBAL	
vertices	value	vertices	value	vertices	value	vertices	value
125	0.4274	125	0.4274	125	0.4274	35 937	0.5760
501	0.5960	579	0.5250	729	0.5912	274 625	0.6162
2 213	0.6609	2 474	0.6397	4 913	0.6420	2 146 689	0.6417
8 103	0.6723	10 935	0.6616	35 937	0.6612		
14 953	0.6740	86 903	0.6687	274 625	0.6686		

depth. For $\gamma \sim 1$ the intensity distribution is

$$I(\tau,\mu) = \exp(-\tau/\mu) \int_0^{\tau/\mu} J \exp(\tau')d\tau', \qquad (82)$$

which strongly depends on the spatial distribution of J within the layer. For a layer which is optically thin, a nearly homogeneous distribution for J is obtained. Hence, $I(\tau,\mu)$ shows the same behaviour for low and high albedo. At large optical depths, J is stronger at the centre of the layer and falls off towards the edges. Since the contributions to the intensity along the path are only significant until $\tau' \sim 1$, the intensity has its maximum at $\mu = 1$.

We use this test problem to show the advantage of employing a refinement process based on adaptive grids as described in Sect. 3.3, by making a comparison with what is obtained when structured grids are used, which is standard practice. To do this, we calculate the specific intensity in a particular direction $\mu = 0.705$ on a variety of grids. The optical depth τ and albedo γ are 20 and 0.80, respectively. Table 2 lists the number of vertices and the computed value of the specific intensity escaping the layer at the point $\mathbf{x}_0$. The FE method uses grids which are generated by using local refinement criteria as described in Sect. 3.3 (DUAL, L2) and a global refinement criterion where all cells are refined in every step (GLOBAL). L2 indicates the control of the L^2-error and DUAL employs the functional for a point value as described at the beginning of this section. The FD code is restricted to the use of structured grids only. First, we discuss the quality of the L2 and the DUAL grid and make a comparison with the quality of structured grids used by the FE code. The quality of a grid is related to how economic it is, which in its turn depends on the number of vertices needed to produce an accurate solution. Table 2 shows that the L2 refinement strategy already improves the quality of the generated grid by a factor of 3 to 4. A dramatic enhancement is obtained by applying

Table 3. Test II: Comparison of the memory and CPU requirements of the three-dimensional codes for modelling an optical thick and an optical thin plane-parallel layer. The albedo is $\gamma = 0.80$ in both cases. The FE method employs the DUAL grid refinement strategy. The final grid has 670 and 4263 vertices for the optical thin and thick case, respectively. The FD code uses a structured grid (GLOBAL) with 65^3 vertices for $\tau = 0.4$ and 129^3 vertices for $\tau = 20$. In the latter case, the resolution of the spatial grid is too low to obtain a solution as accurate as the one from the FE code (see Table 2)

	$\tau = 0.4$						$\tau = 20$				
	FE					FD	FE					FD
processors	1	2	4	8	16	1	1	2	4	8	16	1
memory [MB]	32	17	9	5	3	13	196	100	52	29	15	99
time [s]	847	372	121	56	48	1275	3167	1364	429	207	136	9879

the DUAL refinement process, since the number of vertices needed to produce a solution similar in quality to the one of the GLOBAL case is reduced approximately by a factor of 20. The DUAL local refinement criteria η_K taken from (21) generate a grid that is much more adapted to our test problem. Due to the local weights $\|\nabla^2 z\|_K$ the indicators η_K concentrate refinement on the cells close to the boundary point where the outgoing intensity is measured. Finally, we compare the solution of the FE code (for the GLOBAL grid) with the solution of the FD method in Table 2. To compute a solution of comparable quality, the FD code requires 129^3 vertices, whereas for the FE method 17^3 are already sufficient. Again, this reduction is caused by the higher order of the FE streamline diffusion discretisation.

When modelling three-dimensional radiative transfer problems, accuracy is not the only criterion for the selection of a particular code. It is at least equally important that memory and CPU requirements should be reasonable. Table 2 displays that using a higher order discretisation along with efficient grid refinement strategies reduces the computational costs. A more efficient solver for the linear system of equations and the application of parallelisation techniques are additional means for saving resources. In Table 3 we compare the memory and CPU requirements of the three-dimensional codes for two extreme cases: a layer which is optically thin with $\tau = 0.4$ and a layer which is optically thick with $\tau = 20$, while the albedo is $\gamma = 0.80$ in both cases. The FD code uses a structured grid (GLOBAL) with 65^3 vertices for $\tau = 0.4$ and 129^3 vertices for $\tau = 20$. The FE code starts with a pre-refined grid with 5^3 vertices and stops the refinement process when the result is almost identical to the analytical result. The computations were performed on the Pentium III 650 MHz cluster of the Interdisciplinary Centre for Scientific Computing (IWR) at Heidelberg University. The cluster consists of 16 single nodes with 1 GB memory for each processor. If a single processor is used the FE code roughly needs twice as much memory as the FD code. However, for both cases of τ the execution time of the FE code is shorter. When employing parallelisation, the memory requirements of the FE algorithm can be matched to the requirements of the FD method by employing two processors. This halves the CPU time of the FE code. If we use all nodes of the parallel cluster, the computational costs can be reduced even further. In particular, the execution time of the FE algorithm can be reduced by approximately two orders of magnitude for the optical thick case. The difference in execution time between the two codes is even more dramatic, since a finer grid with at least 257^3 vertices is needed for the FD method to obtain results of the same accuracy as that achieved with the FE code (see Table 2). If the monochromatic FE code is extended to deal with frequency- or time-dependent transfer problems, a reduction in the execution time becomes a necessity. When comparing the memory and CPU requirements of the two codes, it should be recalled that the FE method makes use of all the modern numerical tools described above.

Test III: Propagation of Radiation in a Scattering Halo

The setup for the last test problem is as follows: The 3D computational domain ($x = [-1, 1]$, $y = [-1, 1]$, $z = [-1, 1]$) contains a spherically symmetric cloud with radius $r_c = 0.9$ which is illuminated by an extended central emission region with radius $r_s = 0.125$. The source term is given by

$$f(r) = \begin{cases} 1 & \text{for} \quad r \leq r_s \\ 0 & \text{for} \quad r > r_s \end{cases} \quad . \tag{83}$$

We chose a Lorentz profile for the radial distribution of the extinction coefficient

$$\chi(r) = \begin{cases} \chi_0/(1 + 100 r^2) & \text{for} \quad r \leq r_c \\ \chi(r_c)/100 & \text{for} \quad r > r_c \end{cases} \quad , \tag{84}$$

where the constant χ_0 is calculated from a given optical depth $\tau = \int_0^{r_c} \chi(r) dr$. In this way, $\chi(r)$ varies by almost two orders of magnitude between r_s and r_c.

We designed this test problem so as to have a very simple model for Lyman-α halos associated with high-redshift galaxies. Lyman-α halos are diffuse and clumpy emission-line regions which are more than ten times larger than the corresponding regions from where the Lyman-α line photons originate (cf. [ORC96, ORM97, SAS00]). The diffuse emission-line appearance of the halo is the result of a resonance line scattering process by "cold" neutral hydrogen gas. Observed line profiles reveal the complexity of the velocity field of the intergalactic medium. Hereafter, we intend to model Lyman-α halos with the line transfer version of the finite-element code. This test problem will help us to determine the demands on spatial resolution and the required number of ordinates.

Since the Lyman-α line is a resonance line, we assume $\chi = \sigma$ everywhere. In this case, the zeroth moment equation of the radiative transfer equation reduces to

$$\nabla \cdot \mathbf{F} = \frac{1}{r^2}\frac{d}{dr}r^2 F_r = 4\pi f, \tag{85}$$

where F_r is the radial component of the radiative flux

$$\mathbf{F}(\mathbf{x}) = \int_{S^2} \mathbf{n}\, I(\mathbf{x}, \mathbf{n})\, d\omega.$$

The solution of (85) is

$$F_r(r) = \begin{cases} \frac{4}{3}\pi r & \text{for } r \leq r_s \\ \frac{4}{3}\pi \frac{r_s^3}{r^2} & \text{for } r > r_s \end{cases} \tag{86}$$

The analytical result for the flux from the computational box is $F_r(1.0) = 8.18 \times 10^{-3}$ in every direction, independent of τ.

The numerical calculations start with a pre-refined grid with 16^3 cells. Thus, cell size at step zero is equal to r_s. The grid refinement process is based on the global L^2-indicators η_{L^2} taken from (23). We perform the calculations for $\tau = 0.1, 1, 10$ and $M = 20, 80, 320, 1280$. Due to limited memory resources, the number of feasible refinement steps critically depends on M. In Table 4 we show the values for $F_r(1.0)$, averaged over all directions, for all calculations. In comparison to the analytical result, step zero overestimates and step one underestimates the radiative flux. With step two in the refinement process, the emission region is resolved sufficiently well and the relative error of the numerical value for the radiative flux is of order 10^{-3} for all τ and M. This result proves the global flux-conserving property of the FE implementation. In contradistinction to the FE code, the FD code is not flux-conserving. If the FD code is employed, and if a global spatial resolution comparable to step 2 is selected, the radiative flux for the same problem is about 6% smaller than that obtained from the analytical solution.

Table 4 also shows, that refinement step 2 is not sufficient for calculations with $\tau \leq 1.0$ and $M \geq 320$. In these cases, the error estimator prefers to refine cells in the outer regions of the cloud. Since we now know the minimal requirements as to the resolution of the emission region, this effect can be avoided with a proper local pre-refinement of the central region at step 0.

Table 4. Test III: Flux, in 10^{-3} of a scattering dominated halo, for a range of optical depths τ, number of ordinates M and refinement steps

τ	0.1				1.0				10.0			
M	20	80	320	1280	20	80	320	1280	20	80	320	1280
step 0	9.48	9.48	9.48	9.48	9.48	9.48	9.48	9.48	9.48	9.48	9.48	9.48
step 1	8.04	8.04	8.04	8.04	8.04	8.04	8.04	8.04	8.04	8.04	8.04	8.04
step 2	8.19	8.19	8.04	–	8.19	8.22	8.04	–	8.19	8.19	8.19	–
step 3	8.19	8.19	–	–	8.19	8.19	–	–	8.20	8.20	–	–
step 4	8.20	–	–	–	8.20	–	–	–	8.18	8.20	–	–

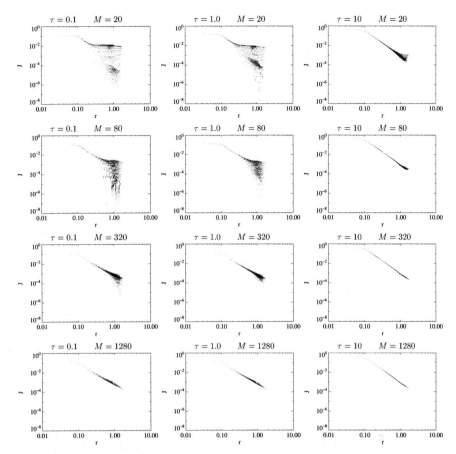

Fig. 4. Test III: Dependence of the mean intensity J on the distance from the centre r for different τ and M

For the spherically symmetric test case, it is sufficient to use only 20 ordinates to determine the radiative flux with good precision. But a completely different requirement results as to the number of ordinates, if the interest is in local quantities, for example the radiative flux from isolated regions, when making a comparison with spatially resolved spectroscopic observations or with the mean intensity in order to determine a temperature distribution. In Fig. 4 we show the distribution of the mean intensity $J(r)$ at all vertices for step 2, except for $M = 1280$, where we show step 1. Instead of a smooth distribution we find a large scatter of about three orders of magnitudes for small τ and $M = 20$. The scatter is typical for discrete-ordinates methods and decreases with higher optical depth and with increasing number of ordinates. At high optical depth it seems sufficient to use $M = 80$, but at small optical depth one has to use at least $M = 320$.

Table 5. Parameters used for all calculations. Distances are given in units of the computational unit cube

r_h	r_c	α	r_s	v_D	v_0	r_0	R_0
1.0	0.2	10^3	0.2	$10^{-3}c$	$-10^{-3}c$	0.2	1.0

5.2 Polychromatic 3D Line Transfer Problems

The test calculations are performed using a spherically symmetric 3D distribution of the extinction coefficient $\chi(\mathbf{x}) = \chi(x, y, z)$ of the form

$$\chi(\mathbf{x}) = \begin{cases} \chi_0/(1 + \alpha r_c^2) & \text{for } r \leq r_c \\ \chi_0/(1 + \alpha r^2) & \text{for } r_c < r \leq r_h \\ \chi_0/(1 + \alpha r_h^2)/10^3 & \text{for } r > r_h \end{cases} \quad (87)$$

where $r^2 = x^2 + y^2 + z^2$. It should be noted, that the value of the extinction coefficient is constant within the centre core of radius r_c and outside the halo of radius r_h. χ_0 is determined from the optical depth at the line centre

$$\tau = \int_{r_c}^{r_h} \chi(\mathbf{x}) \phi(\nu_0) \, \mathbf{n} \, d\mathbf{x} \quad (88)$$

between r_c and r_h along the photon direction $\mathbf{n}$. The spatial distribution of χ is determined by a total of three parameters: the radii r_c and r_h, the dimensionless parameter α, and the optical depth τ. For r_c, r_h and α we use the values given in Table 5.

Since we are predominantly interested in the radiation transfer in resonance lines like Lyα, we assume $\sigma(\mathbf{x}) = \chi(\mathbf{x})$ and $\kappa(\mathbf{x}) = 0$ for all our calculations, which implies a restriction to purely non-thermal source functions. In particular, we treat a spatially confined source region with radius r_s centred at the position $\mathbf{x}_i = 0$:

$$f(\mathbf{x}, \nu) = \begin{cases} \phi(\nu) & \text{for } |\mathbf{x} - \mathbf{x}_i| \leq r_s \\ 0 & \text{for } |\mathbf{x} - \mathbf{x}_i| > r_s \end{cases} \quad (89)$$

The function $\phi(\nu)$ is the Doppler profile defined in (28).

In general, the finite-element code is capable of dealing with arbitrary velocity fields. In the case of velocity fields which are defined on a discrete grid, e.g. those which result from hydrodynamical simulations, the velocity gradient in direction $\mathbf{n}$ must be derived numerically. Presently, for the sake of a simple illustration, two simple velocity fields are assumed, so that the function w can be obtained in closed form.

The first velocity field corresponds to a spherically symmetric inflow ($v_0 < 0$) or outflow ($v_0 > 0$) and is of the form

$$\mathbf{v}_{io} = v_0 \left(\frac{r_0}{r}\right)^l \frac{\mathbf{x}}{r}, \qquad (90)$$

where $r = |\mathbf{x}|$ and v_0 is the scalar velocity at radius r_0. For this choice of the velocity field, the function w is given as follows

$$w(\mathbf{x}, \mathbf{n}) = v_0 \left(\frac{r_0}{r}\right)^l \left(\frac{1}{r} - (l+1)\frac{|\mathbf{nx}|}{r^3}\right). \qquad (91)$$

For the second velocity field, we assume a rotational flow around the z-axis

$$\mathbf{v}_{rot} = v_0 \left(\frac{R_0}{R}\right)^l R^{-1} \begin{pmatrix} y \\ -x \\ 0 \end{pmatrix}, \qquad (92)$$

where $R^2 = x^2 + y^2$ is the distance from the axis of rotational, and v_0 is the scalar velocity at distance R_0. If $\mathbf{n} = (n_x, n_y, n_z)$, the w function is

$$w = v_0 \left(\frac{R_0}{R}\right)^l (l+1) \left(\frac{xy(n_y^2 - n_x^2) + n_x n_y(x^2 - y^2)}{R^3}\right). \qquad (93)$$

Fig. 5 shows the results of the finite-element code for different optical depths, velocity fields and redistribution functions. We used 41 frequencies equally spaced in the interval $(\nu - \nu_0)/\Delta\nu_D = [-4, 6]$ and 80 directions. Starting with a grid of 4^3 cells and a pre-refined source region, 3–5 spatial refinement steps were needed.

The simplest case is a static model with coherent isotropic scattering. Figure 5a displays the emergent line profiles for different values of τ. As expected, the Doppler profile is preserved and the flux F_ν is independent of τ. The analytical solution, which is marked by crosses, deviates only slightly from the numerical results. The line profiles for $\tau = 0.1$ and $\tau = 1$ are identical even for the enlargement as displayed by the insert of Fig. 5a. For $\tau = 100$, conservation of total flux is still better than 99%. This result demonstrates that the frequency-dependent version of our finite element code works correctly.

Next, we consider an inflow halo with coherent scattering and show the effects of frequency coupling due to the Doppler term. The emergent line profiles in Fig. 5b are plotted for different exponents l of the velocity field $\mathbf{v}_{io}$ defined in (90). The line profiles are redshifted for $\tau = 1$ (thin lines). Most of the photons traverse the halo straightforwardly, while moving away from the observer. Since the Doppler term is proportional to the gradient of the velocity field, the redshift becomes larger as the exponent l is increased. For $\tau = 10$ (thick lines), the line profiles are blueshifted. Before photons escape from the optically thick halo in front of the source, they are backscattered and blueshifted in the approaching halo behind the source. The blueshift is less pronounced for the accelerated inflow with $l = 2$, because the strong gradient of the velocity field leads to a slight redshift in the very

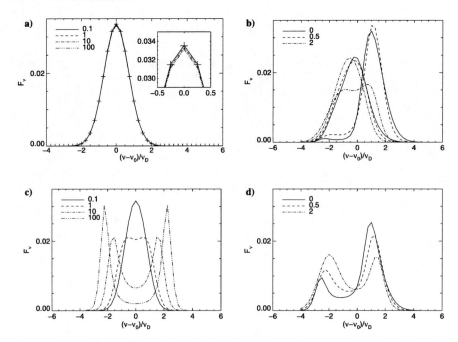

Fig. 5. Lyα line profiles calculated with the finite element code for a spherically symmetric model configuration: a) a static halo with coherent scattering, b) an inflow halo with coherent scattering, c) a static halo with complete redistribution, and d) an inflow halo with complete redistribution. For the static cases a) and c) the line styles refer to calculations with different optical depths τ as indicated. The insert in a) gives a close-up of the vicinity at the peak of the curve. The crosses mark the results of the analytical solution. For the moving halos we show in b) the results for $\tau = 1$ (thin lines) and $\tau = 10$ (thick lines) and in d) only for $\tau = 10$. In these cases, the line styles refer to the exponent l used for the velocity fields

inner parts of the halo. In this region, the total optical depth is still small. Further out, where the total optical depth increases, the line profile becomes blueshifted.

Complete redistribution is another method of frequency coupling, which leads to a stronger coupling than the Doppler effect (see Sect. 3). The line profiles obtained for a static model with complete redistribution are displayed in Fig. 5c for different values of τ. An increase in optical depth leads to a marked increase in the number of those photons which escape through the line wings. For $\tau \geq 1$, we get a doubly-peaked line profile with an absorption trough at the line centre. The greater τ the larger the distance between the peaks, and the deeper the absorption trough. Since our frequency resolution is too poor for the pointed wings, flux conservation is only 96% for $\tau = 100$.

Figure 5d shows the results for an inflow halo with complete redistribution for $\tau = 10$ and different exponents l. For $l = 0$ and $l = 0.5$ the inflow motion of the halo enhances the blue wing of the doubly-peaked line profile. Similarly, an outflow halo would enhance the red peak. In the particular case of $l = 2$, and for an inflow halo, the red peak is slightly enhanced due to the strong velocity gradient, as outlined above. This example affords an insight into how resonance lines form and gives evidence for the necessity of a treatment which must in essence be multi-dimensional.

6 Applications

All calculations to be discussed in this section were performed with complete redistribution. We used 49 equidistant frequencies covering the interval $(\nu - \nu_0)/\Delta\nu_D = [-6, 6]$, and 80 directions. We started from a grid with 4^3 cells and pre-refined source regions, which resulted in several 10^3 cells for the initial mesh. 3–7 refinement steps were performed and lead to approximately 10^5 cells for the finest grid.

6.1 Elliptical Halos

As a first step towards a fully three-dimensional problem without any symmetries, we investigated an axially symmetric, disk-like model configuration ($a : b : c = 3 : 3 : 1$) with a single source located at $\mathbf{x} = 0$, for which we selected several values of $\tau = \tau(\mathbf{n}_{\text{thick}})$ and several different velocity fields. The particular direction $\mathbf{n}_{\text{thick}}$ which indicates where the optical thickness is largest, is in the equatorial plane of the disk. The direction perpendicular to the equatorial plane is given by the z-axis which by tradition is called rotational axis even for cases without rotational symmetry.

The static case is used to demonstrate what kind of information can be extracted from the application of our FE code. The emergent line profiles provide global information on the underlying system. They are displayed in Fig. 6a for different optical depths and viewing angles, with the viewing angle is defined to be the angle between the rotational axis and the direction towards the observer. For instance, a viewing angle of $90°$ implies that the disk is viewed edge-on. Again, we obtain the characteristic doubly-peaked line profile for $\tau \geq 1$. The optical thickness decreases for smaller viewing angles. For a viewing angle of $0°$ the effective optical depth is only $\tau(\mathbf{n}_{\text{thick}})/3$. Consequently, the distance between the peaks in the line profile as well as the depth of the absorption feature are increasing in proportion to increases in the viewing angle. It should be noted that the total relative flux $F(\mathbf{n} = 0°)/F(\mathbf{n} = 90°)$ escaping along the z-axis increases in proportion to any increase in $\tau(\mathbf{n}_{\text{thick}})$. An explanation of Figure 6b will be given further down.

Far more information is contained in the spatial distribution of the line intensity projected into the s-t-plane which is the plane perpendicular to the

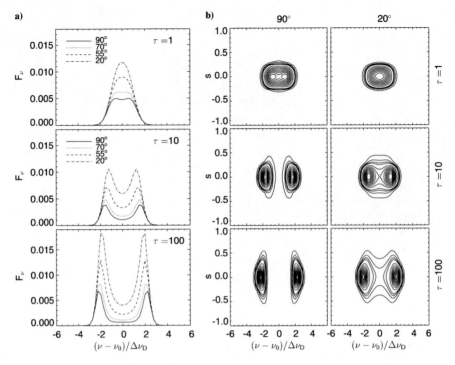

Fig. 6. Results obtained for a static, disk-like model configuration: a) Emergent line profiles for different viewing angles and optical depths $\tau = \tau(\mathbf{n}_{\text{thick}})$. b) Two-dimensional spectra for an edge-on view (90°) and a nearly plan view (20°) and different optical depths. The position and width of the slit is indicated in Fig. 7a. The contours are given for 2.5%, 5%, 7.5%, 10%, 20%, ..., 90% of the maximum value

viewing direction. In Fig. 7a the frequency-integrated intensity distribution is shown for $\tau = 10$ and different viewing angles. At 90° the Lyα image shows an elliptical emission line region. The diffuse halo of scattered radiation is clearly visible. With decreasing viewing angle the diffuse halo becomes less pronounced. For a nearly plan view (20°), the halo is optically thinner and the intensity distribution appears spherically symmetric.

Figure 7b displays the spatial distribution of the Lyα intensity for different viewing angles (columns) at selected frequencies (rows). For the edge-on view, the intensity distribution strongly depends on the frequency. In the outer line wings at $(\nu - \nu_0)/\Delta\nu_D = \pm 2$ the contours are elliptical and reproduce the disk-like form of the source region. The peaks of the line profile are at $(\nu - \nu_0)/\Delta\nu_D \sim \pm 1$. Here, the major axis of the elliptical intensity distribution in the inner parts of the model is parallel to the rotational axis, whereas the major axis of the elliptical contour lines in the outer parts remains parallel to the equatorial plane of the disk. At

Finite Elements for 3D Radiative Transfer Problems 143

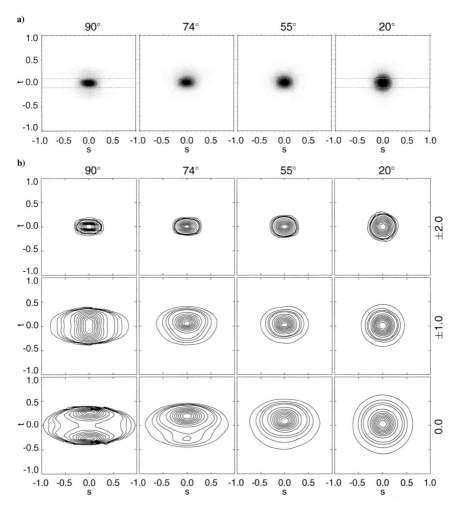

Fig. 7. Spatial intensity distribution for a static disk-like model configuration with $\tau(\mathbf{n}_{\text{thick}}) = 10$ for different viewing angles: a) Frequency integrated intensity distribution. b) Intensity distribution for specific frequencies. The frequency is given in Doppler units relative to the line centre $(\nu - \nu_0)/\Delta\nu_D$ at the right hand side

the very centre of the line $((\nu - \nu_0)/\Delta\nu_D = 0)$, the direct view on the source region is blocked by material which is optically thick. Two knots of Lyα emission appear directly above and below the disk, which are surrounded by an extended elliptical halo. The Lyα knot below the disk vanishes with decreasing viewing angle. At 20° the intensity distribution is spherically symmetric for all frequencies. At the frequency of the line centre, the observable emission originates from a region of maximal expansion.

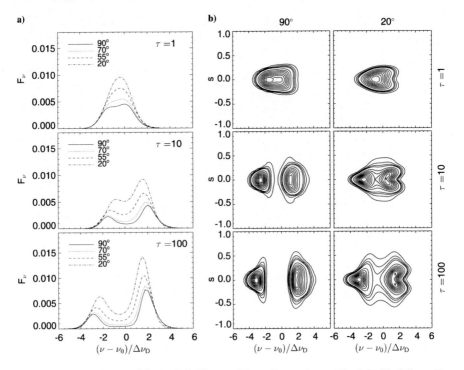

Fig. 8. Results obtained for a disk-like model configuration with global infall motion ($l = 0.5$): a) Emergent line profiles for different viewing angles and optical depths $\tau = \tau(\mathbf{n}_{\text{thick}})$. b) Two-dimensional spectra for an edge-on view (90°) and a nearly plan view (20°) and different optical depths. The position and width of the slit is indicated in Fig. 7a. The contours are given for 2.5%, 5%, 7.5%, 10%, 20%, ..., 90% of the maximum value

Two-dimensional spectra from high-resolution spectroscopy provide frequency-dependent data for only one spatial direction. To be able to compare our results with these observations we calculated two-dimensional spectra using the data within the slits which are marked in Fig. 7a for the edge-on and nearly face-on view. The results are shown in Fig. 6b for different optical depths. We find that the form of the two-dimensional spectra depends only weakly on the width of the slit. For $\tau = 1$ a single emission region is visible. But already at 90°, the highest contour lines reproduce the faint absorption trough of the corresponding line profile (see Fig. 6a). The higher the optical depth the wider the gap and the spatial extent of the two peaks. It should be noted that the spatial extent of the outer contour line only depends on $\tau(\mathbf{n}_{\text{thick}})$ and not on the viewing direction.

The changes that arise when imposing a macroscopic velocity field are shown in the following two figures. First, we consider an inflow velocity field with $l = 0.5$ which is to model a gravitational collapse. Figure 8a displays the

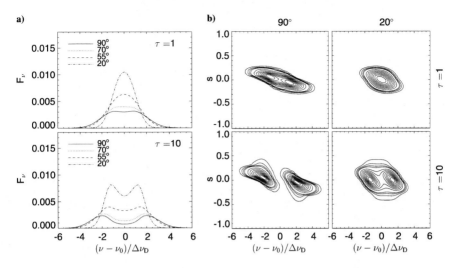

Fig. 9. Results obtained for a disk-like model configuration with Keplerian rotation ($l = 0.5$) around the z-axis: a) Emergent line profiles for different viewing angles and optical depths $\tau = \tau(\mathbf{n}_{\text{thick}})$. b) Two-dimensional spectra for an edge-on view (90°) and a nearly plan view (20°) and different optical depths. The position and width of the slit is indicated in Fig. 7a. The contours are given for 2.5%, 5%, 7.5%, 10%, 20%, ..., 90% of the maximum value

calculated line profiles for different values of τ and various viewing angles. As expected, the blue peak of the line is enhanced. The higher τ the stronger the blue peak. Figure 8b shows the corresponding two-dimensional spectra, obtained with the same width and the same position of the slit as in the static case. For low optical depths, the contour lines form a triangle. Photons changing frequency while trying to escape via the blue wing are also scattered in space, this leading to a greater spatial spread of the blue wing. While the gap between the two peaks is growing in proportion to an increase in optical depth, the general shape remains triangular.

Next, we investigated the elliptical model configuration with Keplerian rotation ($l = 0.5$) about the z-axis as axis of rotation. The results are plotted in Fig. 9 for $\tau = 1$ and $\tau = 10$. The emergent line profiles are symmetric with respect to the line centre and show the same behaviour as in the static case when the optical depth is increased. However, the spread of the line wings towards higher and lower frequencies increases strongly with increasing viewing angle because of the growing effect of the velocity field. Rotation is clearly visible in the two-dimensional spectra (see Fig. 9b) where the sheared contour lines indicate rotational motion. For an edge-on view at $\tau = 10$, Keplerian rotation produces two banana-shaped emission regions.

In spite of the relatively moderate optical depths of the simple model configurations, our results reflect the form of line profiles and the patterns in two-dimensional spectra, as observed for many high-redshift galaxies. For example, the two-dimensional spectra of the Lyα blobs published in [SAS00] (see Fig. 8) are comparable to the results obtained for inflow (Fig. 8b) and rotating (Fig. 9b) halos. In their collection of characteristic specimens, selected from among the presently known high-redshift galaxies, [ORM97] included many radio galaxies which have singly-peaked and doubly-peaked Lyα profiles. The corresponding two-dimensional spectra show asymmetrical emission regions which are more or less extended in space. The statistical study of emission lines from highly redshifted radio galaxies in [BRM00] indicates that a triangular shape of the Lyα emission is a characteristic pattern in the two-dimensional spectra of high-redshift radio galaxies. Since the emission of the blue peak of the line profile is predominately less pronounced, most of the associated halos should be in the state of expansion.

6.2 Multiple Sources

Highly redshifted radio galaxies are found in the centre region of proto clusters. In such an environment, the Lyα emission originating from several different galaxies can be scattered so as to form a common halo. To investigate this scenario, we started with a spherically symmetric distribution for the extinction coefficient and with three source regions located at $x_1 = [0.5, 0.25, 0]$, $x_2 = [-0.5, 0.25, 0]$ and $x_3 = [0, -0.25, 0]$ forming a triangle in the x-y-plane. The following figures give results for $\tau = 10$.

We begin with the static model. Figure 10a shows Lyα images for four selected viewing directions. The angle as specified in the figure corresponds to the viewing angle between the viewing direction and the x-y-plane. It should be noted that source locations in the plane have a different orientation for each image. Viewing the configuration from a direction almost perpendicular to the x-y-plane (70°) renders all three source regions visible, because the source regions are situated in the optically thinner outer parts of the halo. It should be recalled that most of the scattering matter is located in and around the centre of the system. For other angles, some of the source regions are located behind the centre and therefore their images are less discernible. Figure 10b shows images at different frequencies. In the line wings at $(\nu - \nu_0)/\Delta\nu_D = \pm 2$, the number and position of the source regions can be determined for all viewing angles. However, for frequencies form around the line centre at $(\nu - \nu_0)/\Delta\nu_D = \pm 1$ and 0, the number of visible sources depends on the viewing angle. Some of the images show only one, other two or three sources.

The corresponding line profiles and two-dimensional spectra are displayed in Fig. 11 for the static case as well as for a halo with global inflow and Keplerian rotation. Width and position of the slits are depicted in Fig. 10a. We get doubly-peaked line profiles for almost all cases, except for the rotating halo, where the line profiles are very broad for viewing angles lower than 70°. Additional features, e.g., dips or shoulders, are visible in the red or blue

Finite Elements for 3D Radiative Transfer Problems 147

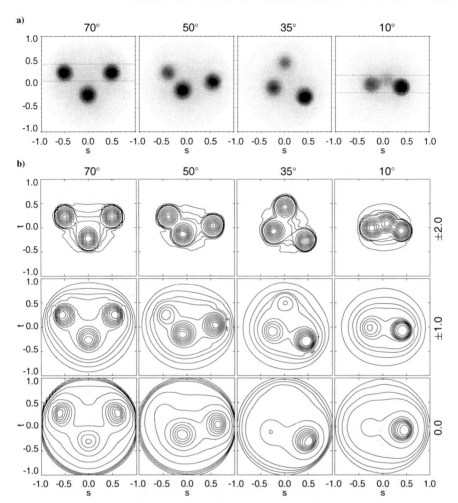

Fig. 10. Spatial intensity distribution for a static spherically symmetric model configuration with $\tau = 10$ and three source regions for different viewing angles: a) Frequency integrated intensity distribution. b) Intensity distribution for specific frequencies. The frequency is given in Doppler units relative to the line centre $(\nu - \nu_0)/\Delta\nu_D$ at the right hand side

wing. They arise because the three sources have significantly different velocity components with respect to the observer.

The slit for a viewing angle of 70° contains two sources. They show up as four emission regions in the two-dimensional spectra (Fig. 11b). The pattern is very symmetric, even for the moving halos. For a viewing angle of 10°, the slit covers all source regions. Nevertheless, the two-dimensional spectra are dominated by two pairs of emission regions resulting from the sources located

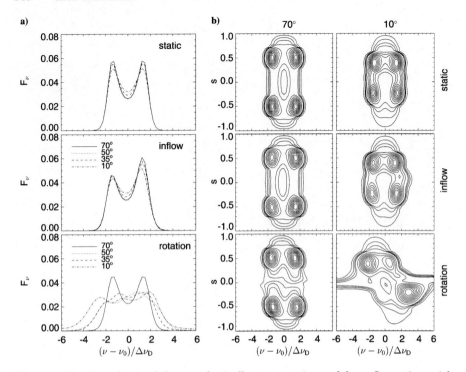

Fig. 11. Results obtained for a spherically symmetric model configuration with $\tau = 10$ and three source regions for the static case and two different velocity fields (inflow and rotation): a) Emergent line profiles for different viewing angles. b) Two-dimensional spectra for a nearly face-on view (70°) and a nearly edge-on view (10°). The position and width of the slit is indicated in Fig. 10a. The contours are given for 2.5%, 5%, 7.5%, 10%, 20%, ..., 90% of the maximum value

closer to the observer. In the case of global inflow, the third source region only shows up as a faint emission in the blue part. In the case of rotation, emission regions from a third source are present, but the overall pattern is very irregular and prevents their clear identification.

This example clearly demonstrates the complexity of three dimensional problems. In a clumpy medium, the determination of the number and position of Lyα sources would be extremely difficult if there were only frequency-integrated images available. Two-dimensional spectra may help, but could prove to be too complicated also. An approach based on images obtained from different parts of the line profile or on information from other emission lines of OIII or Hα in a manner proposed in [KPR02] is obviously more promising.

7 A Multi-Model Preconditioning Scheme

It is often necessary to solve radiation transfer problems in composite domains with mixed transport regimes, so that in some parts of the domain there is diffusion control, whereas radiative transfer dominates elsewhere. The character of the transfer changes strongly, giving a change in the type of equation, if diffusion control changes to radiation control. In case of such a mixed regime, therefore, the application of standard solvers and of standard preconditioning methods for the discrete problem is usually marred by an extremely slow rate of convergence. In this section, we propose a solution method combining the radiative transfer model that turns out to be equivalent to a linear Boltzmann model with its diffusion approximation. Our approach is designed to yield good convergence rates for both types of the equation. For various reasons, the diffusion approximation alone is not sufficient, and the full radiation transfer equation must be solved. First, we do not assume a lower bound for the scattering cross section and, therefore, the approximation may become inaccurate in parts of the domain. Furthermore, the question arises as to what boundary conditions for the diffusion problem are physically acceptable. These problems can be fully resolved only by solving the Boltzmann equation for radiation transfer. Since the spectrum of the discrete operator becomes degenerate in the limit when scattering dominates, the solution of the corresponding discrete linear system is particularly challenging.

When the parameters of the equation become homogeneous, efficient solution techniques can be found easily. If scattering is small, a simple Gauss-Seidel process yields good convergence rates (see [JP86]). In the limit of high scattering, the solution is isotropic in the interior of the domain yielding an elliptic equation of second order, which can be solved efficiently by using multilevel preconditioning. The scheme proposed in this section combines these two preconditioners in a way which is related to two-level multigrid algorithms, by exploiting the good convergence properties of both methods in all the regimes of the equation. Similar schemes have been proposed under the names of quasi diffusion and diffusion-synthetic acceleration method (for details see [Lar91, AL02] and the literature cited therein).

7.1 Discretisation and Diffusion Approximation

Let us consider radiative transfer problems which are monochromatic, and rewrite the partial integro-differential equation (2) for a multidimensional spatial domain $\Omega \subset \mathrm{R}^d$ ($d = 2, 3$) as follows

$$\mathbf{n} \cdot \nabla_x I(\mathbf{x}, \mathbf{n}) = -\chi(\mathbf{x}) \Big(I(\mathbf{x}, \mathbf{n}) - \frac{1-\epsilon}{2(d-1)\pi} \int_{S^{d-1}} I(\mathbf{x}, \mathbf{n}') \, d\omega' - \epsilon B(\mathbf{x}) \Big), \quad (94)$$

where $\epsilon = \kappa/\chi$ is the thermalization factor and $1 - \epsilon = \sigma/\chi$ is the albedo. For the sake of simplicity, isotropic scattering (i.e., a phase function $P(\mathbf{n}', \mathbf{n}) \equiv 1$), a constant monochromatic Planck function $B(\mathbf{x}) = 1$ and – unless stated otherwise – zero boundary conditions

$$I(\mathbf{x}, \mathbf{n}) = 0 \qquad (95)$$

are adopted on the "inflow boundary"

$$\Gamma_- = \{(\mathbf{x}, \mathbf{n}) \in \Gamma \,|\, \mathbf{n}_\Gamma \cdot \mathbf{n} < 0\}. \qquad (96)$$

$\mathbf{n}_\Gamma$ is the unit normal of the surface Γ of Ω and is taken to point outward, so that the sign of the product $\mathbf{n}_\Gamma \cdot \mathbf{n}$ indicates the direction of the photon flux across the boundary. The test calculations which follow are exclusively performed for 2D configurations (i.e., $d = 2$). For the discretisation of the integral operator in (94), a collocation method is used, again as described in Sect. 3.2, using a set of equally spaced points on the unit circle. Then, the transport equation (94) gives for an arbitrary fixed direction $\mathbf{n}_i$

$$\mathbf{n}_i \cdot \nabla_x I(\mathbf{x}, \mathbf{n}_i) + \chi(\mathbf{x})\Big(I(\mathbf{x}, \mathbf{n}_i) - \frac{1-\epsilon}{2\pi} \sum_{k=1}^{k=M} \omega_{ik} I(\mathbf{x}, \mathbf{n}_k)\Big) = \chi(\mathbf{x})\epsilon B(\mathbf{x}), \quad (97)$$

where discretisation is by the discontinuous Galerkin finite-element method (DGFEM). This particular approach has been proposed previously in [LR74] for neutron transport without scattering, and also in [KM04] for the full radiative transfer problem.

Let $\mathbb{T}_h$ be a mesh of quadrilaterals (hexahedron) T covering the domain Ω exactly. For any point $\mathbf{x}$ at the boundary of a mesh cell $T \in \mathbb{T}_h$, we assume that $\mathbf{x}$ is either at the boundary of Ω or belongs to the boundary of a second mesh cell T'. Cell boundaries do not have to match and in particular the case of hanging nodes is allowed. The set of edges (edge will always mean surface in three dimensions) is denoted by $\mathbb{E}$. In the case of non-matching cells, the edges of both cells are divided such that an edge $E \in \mathbb{E}$ either has two neighbouring grid cells or is on the boundary. Accordingly, the set of edges $\mathbb{E}$ is decomposed into the set of interior edges $\mathbb{E}^i$ and edges on the boundary $\mathbb{E}^\Gamma$. For a mesh cell T, we denote its diameter by h_T. Furthermore, we introduce the function $h(\mathbf{x}) = h_T$ if $\mathbf{x} \in T$.

Multiplying by a test function ϕ, the weak form of the DG method on a single grid cell T reads

$$\int_T \Big(n_i \cdot \nabla I(\mathbf{x}, \mathbf{n}_i) + \chi(\mathbf{x})\big[I(\mathbf{x}, \mathbf{n}_i) - \frac{1-\epsilon}{2\pi} \sum_{k=1}^{k=M} \omega_{ik} I(\mathbf{x}, \mathbf{n}_k)\big]\Big)\phi(\mathbf{x})\, d\mathbf{x}$$

$$+ \int_{\delta_- T} \big(I(\mathbf{x}, \mathbf{n}_i) - I_-(\mathbf{x}, \mathbf{n}_i)\big)\phi(\mathbf{x})ds = \int_T \chi(\mathbf{x})\epsilon B(\mathbf{x})\phi(\mathbf{x})\, d\mathbf{x},$$

$$i = 1, \ldots, M. \quad (98)$$

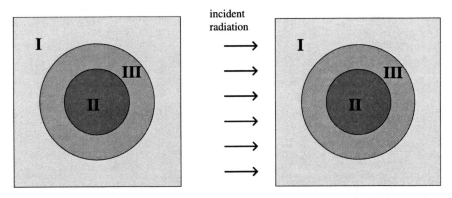

Fig. 12. On the left-hand side, different shadings designate the three radiative transfer problems (I)–(II), for a 2D configuration with an inhomogeneous density distribution ($\chi \sim \exp(5 - 5/(1 - (x + y)))$). Region (I) corresponds to the transport dominated, (II) to the scattering dominated, optically thick, and (III) to the intermediate transfer problem. The right-hand side is essentially the same configuration with the exception that we now allow for incident radiation on the boundary of the computational domain. This situation is typical for the illumination of a molecular cloud or a nebula by a distant star

Here, $\delta_- T$ denotes the inflow boundary (see (96)) of T and $I_-(\mathbf{x}, \mathbf{n}_i)$ is the exterior trace of I at the boundary point $\mathbf{x}$. If $\delta_- T \subset \Gamma_-$, this is the boundary value 0, otherwise it is the function I on the neighbouring cell.

Now, we must restrict the variational equation (98) to a finite dimensional function space V_h in order to achieve a discretisation. Since we do not require continuity of the finite element functions, this space can be generated by concatenating local spaces $\mathcal{P}$ on each mesh cell,

$$V_h = \{ v \in L^2(\mathbb{T}_h) \,|\, v|_T \in \mathcal{P}_T,\ \forall T \in \mathbb{T}_h \}. \tag{99}$$

$\mathcal{P}_T$ is either the space P_k of polynomials of degree k on T or the space Q_k of isoparametric tensor-product polynomials of degree k on T. Usually, we use Q_k, since the solutions are more accurate, even taking the larger number of degrees of freedom into account. We assume $k \geq 1$, which gives a discretisation of at least second order.

A crude analysis of (94) leads to the definition of three different radiative transfer problems: (I) transport dominated problems ($\epsilon \approx 1$) where scattering is negligible; (II) scattering dominated problems in optically thick media ($\chi \gg 1$ and $\kappa \ll 1 \longrightarrow \epsilon \ll 1$), where photon transport may be approximately treated as a diffusion process; (III) the intermediate problem, where the conditions of both extreme cases (I) and (II) hold jointly. The problems (I)-(III) are illustrated in Fig. 12 for an inhomogeneous density distribution ($\chi \sim \exp(5 - 5/(1 - (x + y)))$).

For transport dominated problems (I), discretisation of the first order differential operator $\mathbf{n} \cdot \nabla_x$ gives a condition number of $O(h^{-1})$. Such h-

dependence is unfavourable since it leads to extremely bad convergence on fine spatial grids, but this can fortunately be set right by choosing an appropriate preconditioning scheme. In astrophysics, it is standard practice to solve (94) by the so-called Λ-iteration. This is a Richardson method with nearly block-Jacobi-preconditioning, when taking a broad view of the whole discrete system in its overall structure. In order to achieve a somewhat better convergence for transport-dominated problems of type (I), a full block-Jacobi preconditioner can be used as described in [Tur93]. Furthermore, the degrees of freedom of each block are renumbered according to their transport direction, which then gives a triangular matrix structure which is highly desirable from a computational point of view. These front-solving algorithms converge very fast in the case of transport-dominated problems, because the transport operator can be inverted explicitly (cf. [JP86]). In Tables 6, 7 and 8 it is clearly apparent that for transport-dominated problems (I) this inversion is indeed very cheap since only a few matrix-vector multiplications have to be performed to achieve excellent accuracy. In the case of a purely absorbing medium without scattering, the front-solving block-Jacobi preconditioning scheme is already exact and therefore only one matrix-vector multiplication is needed. On the other hand, if scattering becomes increasingly prominent, i.e., for transfer problems of type (III) or in the limit of $\epsilon \to 0$ and $\chi \gg 1$ for problems of type (II), the front-solving block-Jacobi preconditioning scheme is not very efficient. The condition number is of order $O(\epsilon^{-1})$ and the lowest eigenvalues are in the range of ϵ, whereas the largest eigenvalue is 1 or at least bounded away from zero. Stationary iteration techniques for solving the corresponding linear system of equations, like Jacobi- or Gauss-Seidel-iterations (SOR), are marred by bad convergence or tend to break down for $\epsilon \to 0$ and $\chi \gg 1$ (for details see [Tur93]). Because of the well-known drawbacks with these stationary techniques, it is nearly impossible to model astrophysical configurations for scattering-dominated transport problems in optically thick media of higher spatial dimensions. Better convergence for case (II) and (III) problems can be achieved if an adequate solver is used, which performs better when some eigenvalue are separated by a finite distance, and if better preconditioning schemes are employed. Krylov-space methods like GMRES or bi-CGSTAB (see e.g. [SF93] and references therein) are particularly suited to those cases where a finite distance occurs between some of the eigenvalues. Our approach is based on a multi-model preconditioner (MMP), which combines the acceleration effect of a front-solving algorithm for transport dominated problems with an efficient preconditioner for the scattering dominated problem in optically thick media. Additional numerical tests will then demonstrate the special advantages of the MMP algorithm.

At large optical depths ($\chi \gg 1$) radiative transfer in scattering dominated media ($\epsilon \to 0$) becomes simple because all length scales are larger than the length of the photon mean free path ($\sim 1/\chi(\mathbf{x})$). At small scales, such a medium is homogeneous, because the photons are trapped locally, even though many random scattering events occur. Consequently, the radiation field is isotropic. Furthermore, the density is sufficiently large so that collisional photon destruction far outweighs photon scattering. Such a medium is therefore approaching thermodynamic equilibrium (TE), so that local thermodynamic equilibrium (LTE) a valid assumption (cf. [Mih78]). Simple approximations to the transport problem (II) lead in a diffusion equation (cf. [CZ67]), if ad-hoc physical assumptions such as Fick's law are made (cf. [Lam65, DL00]) or if a polynomial approximation is adopted, which assumes that the flux has an expansion in terms of the first two Legendre Polynomials. To derive the diffusion limit for the transport problem (II), a point $\mathbf{x}$ must be selected which is far from the boundary at an optical depth $\tau \gg 1$, and also far from those regions of the domain Ω where the coefficients κ and σ vary strongly. In the 1970s and 1980s, the mathematical relationship between transport and diffusion theory has been clarified in a large body of work, which showed that diffusion theory corresponds to some asymptotic limit of transport theory. A discussion of these aspects is given in [LK74, HM75, Pap75, LPB83, LMM87].

When modelling celestial objects, density gradients of extraordinary steepness, open boundaries and deviations from LTE are often encountered. At least some of the severe restrictions of the asymptotic (diffusion) limit can be overcome by using the Eddington diffusion approximation, which is based on the so-called *first Eddington* or *Eddington-Milne approximation* (cf. [Edd26, Kou52, Kou63])

$$K_0(\mathbf{x}) \approx \frac{1}{d} J(\mathbf{x}), \tag{100}$$

where $K_0(\mathbf{x}) = (1/d)\,\mathrm{Tr}[K(\mathbf{x})]$ is the trace of the tensor $K(\mathbf{x})$, divided by the dimension d of the spatial domain. $K(\mathbf{x})$ is one of the first three moments of the specific intensity with respect to $\mathbf{n}$:

$$\begin{aligned}
J(\mathbf{x}) &= \frac{1}{2\pi(d-1)} \int_{S^{d-1}} I(\mathbf{x}, \mathbf{n}) d\omega, \\
H(\mathbf{x}) &= \frac{1}{2\pi(d-1)} \int_{S^{d-1}} I(\mathbf{x}, \mathbf{n}) \mathbf{n} d\omega, \\
K(\mathbf{x}) &= \frac{1}{2\pi(d-1)} \int_{S^{d-1}} I(\mathbf{x}, \mathbf{n}) \mathbf{n}\mathbf{n} d\omega.
\end{aligned} \tag{101}$$

The mean intensity $J(\mathbf{x})$ represents the total energy of the photons contained in $d\mathbf{x}$. $H(\mathbf{x})$ is called the Eddington flux and is almost identical to the astrophysical photon flux. $K(\mathbf{x})$ is related to the radiation pressure and represents a tensor of second rank, because $\mathbf{n}\mathbf{n}$ is a dyadic product.

The first Eddington approximation may be derived from approximations of $J(\mathbf{x})$ and $K(\mathbf{x})$ for large optical depths, but it is also obtained by simply assuming that the radiation field is isotropic, i. e. $J(\mathbf{x}) \approx I(\mathbf{x}, \mathbf{n})$. Then, the equation for the second momentum $K(\mathbf{x})$ gives

$$K_0(\mathbf{x}) \equiv \frac{1}{d}\operatorname{Tr}[K(\mathbf{x})] = \frac{1}{d}\frac{1}{2\pi(d-1)}\operatorname{Tr}\left[\int_{S^{d-1}} I(\mathbf{x},\mathbf{n})\mathbf{nn}d\omega\right].$$

$$\approx \frac{1}{d}\frac{1}{2\pi(d-1)}J(\mathbf{x})\operatorname{Tr}\left[\int_{S^{d-1}} \mathbf{nn}d\omega\right] \quad (102)$$

$$\approx \frac{1}{d}J(\mathbf{x}).$$

The off-diagonal elements of **nn** vanish due to the assumption of radiation isotropy, and the trace of $K(\mathbf{x})$ is the average of its diagonal components, so that $K_0(\mathbf{x})$ represents a *mean second moment* which is derived from the *mean radiation pressure*. It is important to bear in mind that $K(\mathbf{x})$ is not necessarily isotropic, and therefore does not reduce to a simple scalar hydrostatic pressure. The anisotropy of $K(\mathbf{x})$ reflects the anisotropy in the distribution of $I(\mathbf{x},\mathbf{n})$, which is induced by an efficient photon exchange between regions with significantly different physical properties, particularly in the presence of strong gradients or an open boundary. It is also evident that the radiation field approaches isotropy in the asymptotic (diffusion) limit of the transfer problem (II). But the main advantage is that the Eddington approximation does not require LTE. Besides being valid for optical depths $\tau \gg 1$, the Eddington approximation also tends to hold quite well in layers with $\tau \approx 1$, and it may even be useful for $\tau < 1$.

To derive a reduced equation modelling photon transport in regions where the Eddington approximation is applicable, a second-order transport equation is introduced, which is given in terms of the moments of $I(\mathbf{x},\mathbf{n})$ defined by (101). This equation is as follows:

$$\nabla_x \cdot \left(\frac{1}{\chi}\nabla_x K(\mathbf{x})\right) = \kappa\left(J(\mathbf{x}) - B(\mathbf{x})\right). \quad (103)$$

Equation (103) is essentially obtained by multiplying (94) with **n**, followed subsequently by angular averaging. Only one single assumption needs to be made, namely that the source terms in (94) must be isotropic (for details see [Edd26, Kou52, Kou63, Mih78]). It should be noted that this assumption does not necessarily require the radiation field to be isotropic. Only when substituting the Eddington approximation (100) into the moment equation (103), the assumption of isotropic radiation is needed, yielding the final diffusion equation that is used for our test calculations

$$-\nabla_x \cdot \left(\frac{1}{d\chi}\nabla_x J(\mathbf{x})\right) + \kappa J(\mathbf{x}) = \kappa B(\mathbf{x}), \quad (104)$$

where a fixed boundary condition $J(\mathbf{x}) = 0$ is to be imposed on the spatial boundary surface Γ. The advantage of the approach presented here is the simple assumptions to be made, allowing for a basic validation of the derived diffusion equation. In the following the diffusion equation is used in a preconditioning scheme for the Boltzmann equation. Since it is not intended to substitute the full radiative transfer problem by a much simpler diffusion process, we do not have to determine the accuracy of this approximation and we avoid the following questions. What are estimates of the error? What are appropriate boundary conditions? What about the validity of the diffusion equation, if open boundaries and strong gradients occur? Finally, how can we improve on the results of diffusion theory?

When a family of solutions I_δ to the radiative transfer equation (94) with parameters $\chi \sim 1/\delta$ and $\epsilon \sim \delta^2$ for $\delta \to 0$ is considered, the difference to the solution to (104) admits the estimate

$$\mathcal{I}_\delta(\mathbf{x}, \mathbf{n}) - J(\mathbf{x}) = \mathcal{O}(\delta), \tag{105}$$

in the interior of the domain Ω (see e.g. [DL00]). It is this property that guided us in designing the experiments in Section 7.3.

While (105) suggests that (104) gives a good approximation to the radiative transfer problem in regions with dominant scattering (see Fig. 13), this is not true in the advection dominated case. Moreover, applications we consider may contain parts where $\chi(\mathbf{x}) = 0$. It is quite obvious, that this involves a division by zero, so that the diffusion approximation cannot be applied straightforwardly. Division by zero can be circumvented by defining a so-called "cut-off" $\widehat{\chi}$ and replace $\chi(\mathbf{x})$ by $\max\{\widehat{\chi}, \chi(\mathbf{x})\}$. Such a modification of $\chi(\mathbf{x})$ is still consistent, since it is to be used only in the diffusion equation, but not in the radiative transfer equation. The results in Section 7.3 show that this is a feasible approach. Having this modification in mind, will still use χ in the following in order not to confuse the notation.

We remark that by multiplying equation (104) by $\delta = \max_{\mathbf{x}} \chi$, the diffusion equation becomes independent of the parameter δ. Therefore, the solutions to (104) tend to the solution of a generalised Poisson equation.

In order to maintain compatible data structures, we choose to discretise the diffusion equation (104) by a discontinuous Galerkin scheme, namely the interior penalty method [Arn82]. On a single grid cell T, the weak form of this method reads

$$\int_T \left(\frac{1}{d\chi(\mathbf{x})} \nabla J(\mathbf{x}) \nabla \phi(\mathbf{x}) + \chi(\mathbf{x}) J(\mathbf{x}) \phi(\mathbf{x}) \right) d\mathbf{x}$$

$$+ \int_{\partial T} \left(\frac{\alpha}{h_T} J(\mathbf{x}) [\![\phi(\mathbf{x})]\!] - \partial_\mathbf{n} J(\mathbf{x}) \{\!\{ \phi(\mathbf{x}) \}\!\} - \{\!\{ J(\mathbf{x}) \}\!\} \partial_\mathbf{n} \phi(\mathbf{x}) \right) ds$$

$$= \int_T \chi(\mathbf{x}) B(\mathbf{x}) \phi(\mathbf{x}) \, d\mathbf{x}. \tag{106}$$

Here, we used the customary abbreviations ($\tilde{u}$ is the value on the other cell adjacent to a particular edge)

156 Erik Meinköhn

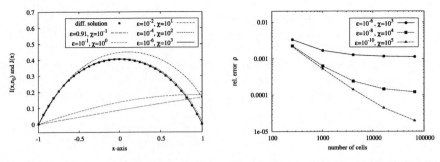

Fig. 13. Illustration of the process by which the diffusion limit is obtained from the full radiative transfer problem, which gives a reduced transport problem describing a diffusion process. The left-hand part shows the specific intensity $I(\mathbf{x}, \mathbf{n}_0)$ (various line types) and the solution $J(\mathbf{x})$ (filled squares) of the diffusion problem, respectively, along the x-axis ($y = 0$) of the 2D-domain $\Omega = [-1,1]^2$. The photon propagation direction $\mathbf{n}_0$ points from $x = -1$ to $x = +1$, which is parallel to the x-axis. The density structure is homogeneous, which results in the constancy of the coefficients ϵ and χ, with values as specified in the inserts. The right-hand part shows the relative error $\rho = (J(\mathbf{x}_0) - I(\mathbf{x}_0, \mathbf{n}_0))/J(\mathbf{x}_0)$, which measures the relative deviation from the assumption $J(\mathbf{x}) \approx I(\mathbf{x}, \mathbf{n})$ at a particular point $\mathbf{x}_0 = (0,0)$ far from the surface boundary, and in a particular direction $\mathbf{n}_0$, against the number of cells for various structured spatial grids

$$\{\!\{u(\mathbf{x})\}\!\} = \frac{u(\mathbf{x}) + \tilde{u}(\mathbf{x})}{2} \qquad [\![u(\mathbf{x})]\!] = \begin{cases} u(\mathbf{x}) - \tilde{u}(\mathbf{x}) & \mathbf{x} \in \Omega \\ u(\mathbf{x}) & \mathbf{x} \in \Gamma \end{cases}$$

on interior edges and for $\mathbf{x} \in \partial\Omega$

$$\{\!\{u(\mathbf{x})\}\!\} = u(\mathbf{x}) \qquad [\![u(\mathbf{x})]\!] = u(\mathbf{x}) - \int_{S^{d-1}} g(\mathbf{x}, \mathbf{n}) \, d\omega.$$

When polynomials of degree $k \geq 1$ are employed, this gives a discretisation of second order at least. The linear system of equations resulting from (106) has the block structure

$$\mathbf{A}_{\text{diff}} \mathbf{J} = \mathbf{f} \tag{107}$$

where $\mathbf{J}$ is the vector of discrete mean intensities and $\mathbf{f}$ represents the data (thermal radiation and boundary data). The $(n \times n)$-matrix $\mathbf{A}_{\text{diff}}$ arises from the DGFEM discretisation of the left-hand side of (104).

Figure 13 illustrates the diffusion limit for the full radiative transfer problem as given by (94). In this limit, a reduced transport problem is obtained which describes a diffusion process as modelled by the diffusion equation (104). The left-hand part of Fig. 13 shows the specific intensity $I(\mathbf{x}, \mathbf{n}_0)$ and the solution $J(\mathbf{x})$ of the diffusion problem, respectively, along the x-axis ($y = 0$) which traverses the 2D-domain $\Omega = [-1,1]^2$. The direction $\mathbf{n}_0$ of photon propagation points from $x = -1$ to $x = +1$ and is therefore parallel to the x-axis. The density is homogeneous, so that coefficients ϵ and χ have constant values, as specified in the inserts of Fig. 13. The solution $J(\mathbf{x})$ of the diffusion problem (104) is denoted by filled squares. It was obtained with a grid of 1024 cells, and the following values were assigned to the coefficients: $\epsilon = 10^{-8}$ and $\chi = 10^4$. The shape of $J(\mathbf{x})$ is essentially independent of the spatial resolution and also – at least for $\chi \geq 1$ – independent of the values which were selected for the coefficients. In Figure 13, the influence of the coefficients ϵ and χ on the specific intensity $I(\mathbf{x}, \mathbf{n}_0)$ is shown, where the various line styles represent different value assignments for the coefficients. In contradistinction to the solution of the diffusion problem, the specific intensity varies significantly for all assignments to the coefficients with $\chi \leq 100$. This discrepancy is primarily caused by the decreasing validity of the Eddington approximation, due to which $J(\mathbf{x})$ deviates increasingly from $I(\mathbf{x}, \mathbf{n}_0)$. Furthermore, in the case of a medium which is optically thin, the effect of the fundamentally different boundary conditions becomes increasingly noticeable. For the diffusion problem (104), the condition $J(\mathbf{x}) = 0$ is imposed at all locations $\mathbf{x}$ of the bounding surface Γ, so that a symmetric distribution of $J(\mathbf{x})$ results, as shown in Fig. 13. In great contrast with this, the boundary condition for the full radiative transfer problem is only imposed on the "inflow boundary" $\Gamma_- = \{(\mathbf{x}, \mathbf{n}_0) \in \Gamma | \mathbf{n}_\Gamma \cdot \mathbf{n}_0 < 0\}$, where $\mathbf{n}_\Gamma$ is the unit vector perpendicular to the bounding surface Γ. The value of $I(\mathbf{x}, \mathbf{n}_0)$ is therefore only fixed at $x = -1$, whereas it is unrestricted at $x = +1$ where the bounding surface remains open. As a consequence, for optically thin and moderately thick problems at least, a non-symmetric distribution of the specific intensity $I(\mathbf{x}, \mathbf{n})$ can arise, as shown in Fig. 13. If the asymptotic (diffusion) limit, i.e. $\epsilon \ll 1$ and $\chi \gg 1$, is approached, both parts of Fig. 13 show that $I(\mathbf{x}, \mathbf{n})$ approaches $J(\mathbf{x})$ with increasing precision, which attests to the fact that in this limit the assumption $J(\mathbf{x}) \approx I(\mathbf{x}, \mathbf{n})$ is definitively justified. The right-hand part of Fig. 13 plots the relative error $\rho = (J(\mathbf{x}_0) - I(\mathbf{x}_0, \mathbf{n}_0))/J(\mathbf{x}_0)$, which gives a measure of the relative deviation from the assumption $J(\mathbf{x}) \approx I(\mathbf{x}, \mathbf{n})$ at a particular point $\mathbf{x}_0 = (0, 0)$ far from the surface boundary and in a particular direction $\mathbf{n}_0$, against the number of cells for various structured spatial grids for three different assignments for the values of the coefficients ϵ and χ. Fig. 13 gives a good illustration of the approach to the diffusion limit, which is characterised by a $O(h^{-1})$ dependence of the relative error ρ for the best selection of coefficient values $\epsilon = 10^{-10}$ and $\chi = 10^5$.

7.2 Preconditioning

The linear system of equations resulting from the DGFEM discretisation of (94) has the following block structure

$$\left(\begin{pmatrix} A_1 & & \\ & \ddots & \\ & & A_M \end{pmatrix} - \begin{pmatrix} \omega_{11}X & \cdots & \omega_{1M}X \\ \vdots & & \vdots \\ \omega_{M1}X & \cdots & \omega_{MM}X \end{pmatrix} \right) \begin{pmatrix} u_1 \\ \vdots \\ u_M \end{pmatrix} = \begin{pmatrix} f_1 \\ \vdots \\ f_M \end{pmatrix}. \quad (108)$$

Here, u_i is the vector of discrete intensities and f_i is the vector which contains the data, i.e. thermal radiation and boundary data, for any fixed direction $\mathbf{n}_i$ of (98). Each $(n \times n)$-matrix A_i corresponds to the DGFEM discretisation of the transport and absorption part of (98) for a single radiation direction $\mathbf{n}_i$, and X discretises the scattering part of (98), which is the multiplication with $(1-\epsilon)\chi$. Equation (108) can be written in the following operator form

$$\mathbf{A}_{\mathrm{rte}}\,\mathbf{u}=\mathbf{f}, \tag{109}$$

where $\mathbf{A}_{\mathrm{rte}} = \mathbf{T} - \mathbf{S}$ is the difference between the transport operator $\mathbf{T}$ and the scattering operator $\mathbf{S}$.

For the system of linear equations (109) and for some iterate $\mathbf{u}^i$, the *iteration error* is defined as

$$\mathbf{e}^i := \mathbf{u} - \mathbf{u}^i \tag{110}$$

so that the error is determined by a defect equation which is as follows

$$\mathbf{A}_{\mathrm{rte}}\,\mathbf{e}^i = \mathbf{A}_{\mathrm{rte}}\left(\mathbf{u}-\mathbf{u}^i\right) = \mathbf{f} - \mathbf{A}_{\mathrm{rte}}\,\mathbf{u}^i = \mathbf{d}^i. \tag{111}$$

An approximation $\mathbf{g}(\mathbf{x},\mathbf{n})$ for $\mathbf{e}^i$ can be computed by solving

$$\mathbf{M}\mathbf{g} = \mathbf{d}^i \quad \Longrightarrow \quad \mathbf{g} = \mathbf{M}^{-1}\left(\mathbf{f} - \mathbf{A}_{\mathrm{rte}}\,\mathbf{u}^i\right). \tag{112}$$

$\mathbf{M}$ should be an approximation of $\mathbf{A}_{\mathrm{rte}}$ that is easy to invert, such as:

$\mathbf{M}$ identity matrix	Richardson
$\mathbf{M}$ diagonal of $\mathbf{A}_{\mathrm{rte}}$	Jacobi
$\mathbf{M}$ lower triangle of $\mathbf{A}_{\mathrm{rte}}$	Gauss-Seidel.

A new iterate is then obtained by setting

$$\mathbf{u}^{i+1} = \mathbf{u}^i + \mathbf{g}. \tag{113}$$

After the degrees of freedom according to the photon transport direction $\mathbf{n}_i$ have been renumbered, the matrices A_i in (108) can be inverted easily by a front-solving algorithm (see [JP86]). A simple preconditioner for the system (108) is therefore given by the matrix

$$\mathbf{\Lambda} = \begin{pmatrix} A_1^{-1} & & \\ & \ddots & \\ & & A_M^{-1} \end{pmatrix}. \tag{114}$$

This particular preconditioner is also used in the well-known Λ- or Source Iteration method. When setting $\mathbf{M} \equiv \mathbf{\Lambda}$, the new iterate in (113) is obtained

$$\mathbf{u}^{i+1} = \mathbf{\Lambda}\left(\mathbf{f} + \begin{pmatrix} \omega_{11}X & \cdots & \omega_{1M}X \\ \vdots & & \vdots \\ \omega_{M1}X & \cdots & \omega_{MM}X \end{pmatrix} \mathbf{u}^i \right). \tag{115}$$

Table 6. Comparison of the two algorithms, with and without multi-model preconditioning, which solve the full radiative transfer problem for constant coefficients on several structured grids. The details are given in the text below

ϵ	0.91	0.1	10^{-2}	10^{-4}	10^{-6}	10^{-8}
χ_0	0.1	1	10	10^2	10^3	10^4
cells	\multicolumn{6}{c}{constant coefficients without MMP}					
256	2	5	15	88	423	1424
1024	2	5	16	108	1427	4929
4096	2	5	16	122	2871	13759
16384	2	5	17	131	4718	—
t [s]	0.5	1.3	4.8	103	3319	16457
cells	\multicolumn{6}{c}{constant coefficients with MMP}					
256	2	4	7	12	24	26
1024	2	4	7	10	21	24
4096	2	4	7	9	16	23
16384	2	4	7	8	14	24
t [s]	2.8	4.6	8.2	10.0	14.5	24.8

This preconditioner performs well only if the extinction χ is small or if the albedo $1 - \epsilon$ is bounded away from 1. Preconditioning can be enhanced by replacing the blocks A_i in Λ by $A_i - \omega_{ii}X$ (cf. [Kan96, Tur93]). Nevertheless, this version still suffers from the same limitations as the Λ-preconditioner in the case when scattering is dominant. In fact, if acceleration for the Λ- or Source Iteration is brought about by an application of methods using matrix algebra, as is the case with Jacobi and Gauss-Seidel preconditioning, the results have been unsatisfactory. As it turned out, acceleration methods based on some direct manipulation of the transport equation itself have a better performance, as described in great detail in [Lar91, AL02].

We are now going to formulate a preconditioning scheme for the Boltzmann problem (109), which uses the solution of the diffusion problem (107) to improve the extremely poor convergence for configurations which are optically thick and dominated by scattering. Transport-dominated problems are also solved by the preconditioning front-solving algorithm described above, which performs efficiently. Our scheme combines these two preconditioners, so as to exploit the good convergence of both methods in the different regimes, and our method is therefore similar to a two-level multigrid algorithm (see [Hac85]). As described in [AL02], a two-level multigrid algorithm places much of the earlier work on transport acceleration in a common unified framework (for details see [Lar91]). The basic idea is to work with two grids, one of which has a "fine mesh" on which the original problem (109) is to be solved, whereas the other one has a "coarse mesh" on which "corrections" (107) to the iterative transport solution are derived. In general, a truly multigrid algorithm uses a whole sequence of increasingly coarse spatial meshes. However, for the particular type of our transport problems where space and angle variables occur, the method we propose solves the transport problem on a single coarse grid, independent of the photon direction. In addition to this coarse grid a single fine grid is used, so that our method qualifies as a two-level multigrid algorithm, for which we propose the acronym MMP, which stands for Multi-Model Preconditioner.

In comparison to standard Λ- or Krylov-type iteration methods with acceleration by Jacobi or Gauss-Seidel preconditioning, our MMP algorithm is designed to give less of an iteration error $\mathbf{e}^i$ as defined by (110), while requiring a significantly smaller number of iteration steps. Like any other two-level multigrid method, our MMP algorithm combines a smoothing operation (115) with a "coarse grid" correction, which is obtained from solving the diffusion problem (107). When this is done, the vector $\mathbf{d}^i(\mathbf{x}, \mathbf{n})$ can be composed, which is the defect iterate of the discretised Boltzmann problem (cf. (109)–(112)). The MMP method then employs the following succession of steps:

1. "fine grid" pre-smoothing by transport preconditioner Λ

$$\mathbf{g}_1(\mathbf{x}, \mathbf{n}) = \Lambda \, \mathbf{d}^i(\mathbf{x}, \mathbf{n}), \tag{116}$$

2. projection of the "fine grid" residual to the "coarse grid" (restriction)

$$\bar{\mathbf{g}}_2(\mathbf{x}) = \frac{1}{2\pi(d-1)} \int_{S^{d-1}} \left(\mathbf{d}^i(\mathbf{x}, \mathbf{n}) - \mathbf{A}_{\text{rte}} \, \mathbf{g}_1(\mathbf{x}, \mathbf{n}) \right) d\omega \tag{117}$$

3. "coarse grid" solution of diffusion problem (cf. (107))

$$\mathbf{A}_{\text{diff}} \, \bar{\mathbf{g}}_3(\mathbf{x}) = \bar{\mathbf{g}}_2(\mathbf{x}) \tag{118}$$

4. "coarse grid" correction (prolongation)

$$\mathbf{g}_4(\mathbf{x}, \mathbf{n}_j) = \mathbf{g}_1(\mathbf{x}, \mathbf{n}_j) + \bar{\mathbf{g}}_3(\mathbf{x}) \qquad \forall \text{ directions } j = 1, \ldots, M \tag{119}$$

5. "fine grid" post-smoothing by transport preconditioner Λ

$$\mathbf{g}_5(\mathbf{x}, \mathbf{n}) = \mathbf{g}_4(\mathbf{x}, \mathbf{n}) + \Lambda \left(\mathbf{d}^i(\mathbf{x}, \mathbf{n}) - \mathbf{A}_{\text{rte}} \, \mathbf{g}_4(\mathbf{x}, \mathbf{n}) \right). \tag{120}$$

The accuracy of the high-frequency eigenvectors improves due to error reduction by what is known as the smoothing property of the iteration (for details see [Hac85]). These high-frequency eigenvectors are determined by the local part of the transfer operator $\mathbf{A}_{\text{rte}}$, i.e. they arise from pure transport and absorption. The idea of the MMP scheme is to construct an iteration that is complementary to the smoother, so that a reduction in the error for the low-frequency eigenvectors is obtained. These errors are determined by the non-local part of the transfer operator, i.e., they arise from the scattering contribution to $\mathbf{A}_{\text{rte}}$. A reduction of the low-frequency errors can be obtained from using a "coarser grid", so that less computational effort is needed.

Table 7. Comparison of the two algorithms, with and without the multi-model preconditioning, solving the full radiative transfer problem for non-constant coefficients on several structured grids. For details see the text below

ϵ	0.91	0.1	10^{-2}	10^{-4}	10^{-6}	10^{-8}
χ_0	0.1	1	10	10^2	10^3	10^4
cells	\multicolumn{6}{c}{inhomogeneous coefficients without MMP}
256	2	4	8	37	221	27274
1024	2	4	8	41	1429	52138
4096	2	4	8	41	1298	55738
16384	2	4	8	37	1367	—
t [s]	0.5	1	2.4	18.9	1664	71219
cells	\multicolumn{6}{c}{inhomogeneous coefficients with MMP}
256	2	3	5	6	18	46
1024	2	3	5	6	14	40
4096	2	4	4	6	10	34
16384	2	4	4	5	8	29
t [s]	8.5	18.0	18.3	28.9	48.8	179

7.3 Numerical Results

The performance of the MMP algorithm is tested by running three model problems (A), (B) and (C) on the unit square $\Omega = [-1,1]^2$. For all three models, the coefficient ϵ is kept constant on the whole domain Ω. In the case of model (A), $\chi(\mathbf{x})$ is also kept constant on Ω, with $\chi(\mathbf{x}) \equiv \chi_0$. In model (B), on the other hand, $\chi(\mathbf{x})$ is given by an inhomogeneous distribution which is defined in terms of χ_0 as follows

$$\chi(\mathbf{x}) = \chi_0 \begin{cases} \exp(5 - \frac{5}{1-(x^2+y^2)}) & \text{for } (x^2+y^2)^{1/2} < 1 \\ 0 & \text{for } (x^2+y^2)^{1/2} \geq 1 \end{cases} . \quad (121)$$

The values for χ_0 and ϵ as used in the three models are given in the Tables 6–8.

It is quite obvious, that use of (121) in (104) involves a division by zero, so that the diffusion approximation cannot be applied straightforwardly. Division by zero can be circumvented by fixing a minimum value for $\chi(\mathbf{x})$ that is finite, a so-called "cut-off". Such a modification of $\chi(\mathbf{x})$ is still consistent, since it is to be used only in the diffusion equation, but not in the radiative transfer equation. For all test runs the opacity cut-off is at a value $\chi_{cut}(\mathbf{x}) = \chi_0/50$. While zero boundary conditions are imposed in (A) and (B), model (C) describes an externally illuminated, inhomogeneous gas cloud with the density distribution (121). A boundary condition is imposed only for the inflow bounding surface $\Gamma_- = \{(\mathbf{x}, \mathbf{n}_0) \in \Gamma | \mathbf{n}_\Gamma \cdot \mathbf{n}_0 < 0\}$, where $\mathbf{n}_\Gamma$ is the unit normal of the bounding surface Γ. In particular, test run (C) is carried out while fixing the value $I(\mathbf{x}, \mathbf{n}_0) = 16$ for the boundary point $\mathbf{x} = (-1, 0)$. The photon propagation direction $\mathbf{n}_0$ is to point from $x = -1$ to $x = +1$, which is parallel to the x-axis. Test runs with Model (C), with its non-zero

Table 8. Comparison of the two algorithms, with and without the multi-model preconditioning, solving the full radiative transfer problem on several structured grids for an externally illuminated gas cloud with an inhomogeneous density distribution $(\chi(\mathbf{x}) \sim \chi_0 \exp(5 - 5/(1 - (x^2 + y^2))))$. For details see the text below

ϵ	0.91	0.1	10^{-2}	10^{-4}	10^{-6}	10^{-8}	0
χ_0	0.1	1	10	10^2	10^3	10^4	10^4
cells	inhomogeneous coefficients without MMP							
256	2	4	10	56	652	4343	5898
1024	2	4	10	56	692	9800	13429
4096	2	4	9	54	681	12019	11536
16384	2	3	9	54	609	—	—
t [s]	0.5	1.0	2.5	22.4	459	10554	10548
cells	inhomogeneous coefficients with MMP							
256	2	4	5	6	14	32	32
1024	2	4	5	5	10	24	24
4096	2	3	5	5	9	19	19
16384	2	3	5	5	8	17	17
t [s]	9.4	14.1	23.0	22.8	39.4	80.2	81.3

boundary conditions, are designed to demonstrate the robustness of the MMP algorithm. The Eddington approximation fails in these cases, because there are no defined boundary conditions in terms of incident radiation for the variable $J(\mathbf{x})$ in the diffusion equation (104). To compute the radiation field correctly, the full transfer problem (94) with appropriate boundary conditions for the specific intensity $I(\mathbf{x}, \mathbf{n}_0)$ must be solved. For non-zero boundary conditions, particularly, the limiting case of a purely scattering medium with $\epsilon = 0$ is assumed. The setting of model (B) and (C) is depicted in Fig. 12.

All MMP examples use the GMRES method which includes the so-called right preconditioning method. Convergence is therefore measured in terms of the norm of the original residual, rather than the preconditioned one. The test runs for all three models are performed on 4 differently-refined structured grids, and for a selection of values for the scale-of-extinction coefficient χ_0 and the thermalization parameter ϵ, to demonstrate the efficiency of the MMP algorithm for all transport problems (I), (II), (III) defined above. The unit circle S^1 is discretised by of N_COMP= 16 chords. Tables 6–8 give the number of GMRES iteration steps which were needed to solve the full radiative transfer problem (94) for models (A), (B) and (C). Solution is reached when the Euclidean norm of the initial residual of (94) is reduced by a factor of 10^{-5}. Since a single smoothing step is sufficient for the MMP method, only post-smoothing as given in (120) is applied at each GMRES iteration step. The computations without MMP use the transport preconditioner Λ (cf. (114)), so that an operation results which is identical to the MMP pre-smoothing step in (116). The results show clearly, that the number of iteration steps depends only moderately on the parameters, which was what we intended to achieve. In particular, these numbers are considerably smaller than the

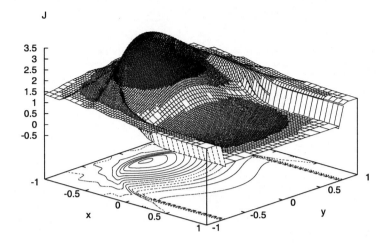

Fig. 14. A surface plot along with a contour plot of the mean intensity $J(\mathbf{x})$ for an externally illuminated gas cloud with a non-constant density distribution and $\chi_{\max} = 10^4$. In addition, the surface plot also shows the structure of the adaptively refined 2D grid

number of steps reported for a preconditioning Λ-iteration which required the GMRES method to be restarted every 150 steps. We chose a large basis to ensure a fair comparison. Since the numerical effort for the MMP algorithm is evidently larger than for the case without MMP, the CPU time t in seconds is also given. The measurements were only performed for the grid with 4096 cells, since our choice of a large basis required a bigger machine, with different runtime characteristics, when using the Λ-preconditioner on the finer grid.

In comparison with computations without MMP, test runs with MMP for all three models (A), (B) and (C) show a remarkable reduction of the number of iteration steps for configurations with $\chi \geq 10^2$. This reduction is accompanied by a reduction of the CPU time t, which demonstrates the efficiency of the MMP acceleration technique not only for transfer problems in purely diffusive media, but also for configurations with varying coefficients, where the diffusion approximation alone is not applicable. When the MMP method is applied, the results clearly show that the total number of iteration steps depends only moderately on the parameters and on the configuration, which was what we intended to achieve. On the other hand, for test runs with models (A), (B) or (C), the runtime t may vary by a factor < 10. This is caused by convergence problems which arise from the use of the non-preconditioned conjugate gradient method. This method gives an exact solution of the diffusion equation in heterogeneous media, with an increase by two orders of magnitude in accuracy in comparison with the final outer residual of the Boltzmann problem (109). Using this conjugate gradient method in combination with the multi-level variable V-cycle preconditioner studied in [GK03] gives a sizeable increase in efficiency, since only 10–20 inner iteration steps are required to solve the diffusion problem of (118) exactly. Nevertheless, it seems to be advisable to study also the particular preconditioner which results if the exact solution is replaced by just a single-variable V-cycle step. The results in [KM04] show, that this simplification causes only a moderate increase in the number of iteration steps, whereas the runtime is reduced by a factor of 2–3.

Figure 14 displays a surface plot along with a contour plot of the mean intensity $J(\mathbf{x})$ for an externally illuminated gas cloud with a non-constant density distribution, as used for model (C). The coefficients are assigned the following values: $\epsilon = 0$ and $\chi_0 = 10^4$. The surface plot also shows the structure of the adaptively refined 2D grid, where adaptation is effected via an a-posteriori error indicator similar to (23).

8 Summary

A finite element code for solving the resonance line transfer problem in moving media is presented. Simple velocity fields and complete redistribution are considered. The code is applicable to any three-dimensional model configuration with optical depths up to 10^3–10^4. The solution approach for the frequency-dependent line transfer problem uses an iterative procedure, which solves successively a set of elementary monochromatic radiative transfer problems. Thus, a fast and accurate solution strategy for these monochromatic transfer problems is crucial if the extended frequency-dependent model is to be simulated efficiently. The use of adaptively refined grids and ordinate parallelisation facilitates the computation of radiation fields in complex three-dimensional configurations, where steep gradients of both the solution and the coefficients are allowed to occur.

The searchlight beam test revealed that the monochromatic code is capable of resolving discontinuities in nothing more than a few adaptive steps. To achieve the same resolution with an equidistant grid, on the other hand, would require a hundredfold increase in the number of cells. Negative intensities and overshooting in the vicinity of discontinuities would also occur because the streamline discretisation is of second order.

The monochromatic radiation field emanating from an infinite plane-parallel layer was used to test the scattering operator. The angular distribution of the intensity was compared with the results of an analytical method and with those of a simple finite difference technique for various values of optical depth and albedo. The results of the finite element code turned out to be in close agreement with the analytical results. No matter how many ordinates were considered, the finite element code needed 2–5 adaptive steps to reach convergence. The finite difference code also gave fairly good results for small optical depths. But for large optical depths and for a large albedo the analytical results could only be reproduced by resorting to grid resolutions which had to be extraordinarily high. This drawback is obviously caused by the lower order of the finite difference discretisation. Due to efficient grid adaptation in combination with an application of parallelisation techniques, the finite element code can produce results of greater accuracy about 20–70 times faster than the finite difference code. The total memory requirement of the two codes is of the same order of magnitude. The good performance of the monochromatic finite element code is of crucial importance for the overall performance of an efficient 3D frequency-dependent radiative transfer algorithm.

The final test runs were applied to monochromatic transfer problems, where the propagation of radiation was examined which emanated from an extended object embedded in a scattering, spherically symmetric halo. This three-dimensional problem gives the basic model setup by which line transfer calculations for Lyα halos associated with high-redshift galaxies were performed. The results showed that the total flux is already determined after a few adaptive steps, independent of the number of ordinates. This is of importance because a high number of ordinates is required in order to obtain reliable results for the spatial distribution of any quantity, of which the mean intensity is a fairly typical example.

The application of the computational procedure to the hydrogen Lyα line for model configurations of slight optical thicknesses ($\tau \leq 10^2$) was demonstrated and the resulting line profiles were discussed in detail. The paper proceeded from very simple to more complex models, which gave the following results:

- An optical depth of $\tau \geq 1$ causes the appearance of the characteristic doubly-peaked line profile with a central absorption trough, which is as expected from analytical studies (e.g. [Neu90]). This particular profile results from scattering in space and frequency. Photons escape via the wings of the line where the optical depth is much lower.
- Global velocity fields destroy the symmetry of the line profile. Generally, the blue peak of the profile is enhanced for models with inflow motion and the red peak is enhanced for models with outflow motion. But there are certain velocity fields (e.g. those with steep gradients) and spatial distributions of the extinction coefficient, where the formation of a prominent peak is suppressed.
- Doubly-peaked line profiles show up as two emission regions in the two-dimensional spectra. Global inflow or outflow leads to an emission region which is of overall triangular shape. Rotation produces a shear pattern resulting in banana-shaped emission regions for optical depths ≥ 10.
- For non-symmetrical model configurations, the optical depth varies with the line of sight. Thus, the total flux, the depth of the absorption trough and the pattern in the two-dimensional spectra strongly depend on the viewing direction.

The applications clearly demonstrate the capacity of the finite element code and show that the three-dimensional structure, the kinematics and the total optical depth of the model configurations are very important. The latter point is crucial, since the convergence of the algorithm is extremely poor for optical depths $\geq 10^3$–10^4. These large depths are related to photons at the line centre. Photons in the line wings or those emerging close to the boundary of the Lyα halo hardly interact with the cold hydrogen gas and will escape almost freely towards the observer. These objects are especially challenging from a numerical point of view. With the objective of modelling multidimensional radiation fields of Lyα halos with optical depths $\geq 10^4$, a multi-model preconditioning (MMP) scheme is presented in this paper. This method combines two distinct preconditioners that are designed to accelerate the solution of advection dominated and highly scattering radiative transfer problems in optically thick media. The development of the MMP method was stimulated by the theory of two-level multigrid algorithms which exploit the good convergence properties of both advection and scattering acceleration methods in different spatial regimes. The monochromatic model problems clearly demonstrate the considerable improvement in run-time efficiency of the MMP scheme. In particular, the paper presents a preconditioner which is instrumental in generating a fast solution of radiative transfer problems where coefficients are variable, so that the diffusion approximation itself is not applicable.

The MMP method can easily be extended to non-isotropic scattering, in which case the scalar factor $1/d\chi$ in (104) needs to be replaced by an appropriate tensor. If the model is extended to deal with frequency-dependent problems, use of the multi-model preconditioner will lead to great improvements in run-time efficiency, because of the large number of quasi-monochromatic radiative transfer problems which must be solved (see [MR02, Mei02]). Even though only two-dimensional results for low-order polynomials were presented in the paper, the method is obviously applicable in spaces of dimensions higher than two, with approximation spaces of any higher order.

Acknowledgements. The author would like to thank Sabine Richling, Rainer Wehrse, Guido Kanschat and Rolf Rannacher for their collaboration and many fruitful discussions. This work is supported by the Deutsche Forschungsgemeinschaft (DFG) within the SFB 359 "Reactive Flows, Diffusion and Transport".

A Separable Representation Method

The separable representation method presented in [EWW95, EWW97]) allows to obtain an analytical solution of the plane-parallel radiative transfer equation

$$\mu \frac{dI}{d\tau} = -I + \frac{\gamma}{2} \int_{-1}^{1} I \, d\mu' + (1 - \gamma), \tag{122}$$

in the following form

$$I_{\text{out}}(\mu) = \frac{1-\gamma}{\gamma} \frac{2}{\sqrt{\mu}} \frac{1}{w^{(0)}(\mu,\mu') - w^{(2)}(\mu,\mu')} \frac{E(\mu'^2)}{\sqrt{\mu'}}, \qquad (123)$$

where the integration over μ' from 0 to 1 is implied, and the operators $w^{(0)}(\mu,\mu')$, $w^{(2)}(\mu,\mu')$ and the function $E(\mu^2)$ are given by

$$E(\mu^2) = \frac{4\gamma}{\tau} \sum_{m=0}^{\infty} \frac{1}{C_m} \cdot \frac{\mu^2}{1+y_m^2 \mu^2},$$

$$w^{(0)}(\mu,\mu') = \left(1 + \tanh\left(\frac{\tau}{2\mu}\right)\right) \delta(\mu - \mu'),$$

$$w^{(2)}(\mu,\mu') = \frac{4\gamma}{\tau} \sum_{m=0}^{\infty} \frac{y_m^2}{C_m} \cdot \frac{\mu^{3/2}}{1+y_m^2\mu^2} \cdot \frac{\mu'^{3/2}}{1+y_m^2\mu'^2},$$

where

$$y_m = \frac{2\pi}{\tau}\left(m + \frac{1}{2}\right), \qquad C_m = 1 - \gamma \frac{\arctan(y_m)}{y_m}.$$

γ and τ are albedo and the total optical thickness of the slab, respectively.

Because of the very slow convergence of the sums (several thousands terms must be taken in order to achieve the required accuracy), this solution can hardly be applied. Appropriate approximations of $E(\mu^2)$ and $w^{(2)}(\mu,\mu')$ are necessary. The Stieltjes-Markov theory of orthogonal functions (see [Per57]) allows to represent the infinite sums by means of finite ones in the following way:

$$E(\mu^2) \approx E_N(\mu^2) = \sum_{n=1}^{N} \frac{a_n \mu^2}{1 + A_n \mu^2},$$

where a_n and A_n depend on τ and γ. The operator $w^{(2)}(\mu,\mu')$ can be expressed through the function $E(\mu^2)$ as

$$w^{(2)}(\mu,\mu') = \mu^{3/2} \cdot \frac{\frac{E(\mu^2)}{\mu^2} - \frac{E(\mu'^2)}{\mu'^2}}{\mu'^2 - \mu^2} \cdot \mu'^{3/2},$$

that leads to the separable representation of $w_N^{(2)}(\mu,\mu')$:

$$w_N^{(2)}(\mu,\mu') \approx \sum_{n=1}^{N} \frac{\mu^{3/2}}{1 + A_n \mu^2} \cdot a_n A_n \cdot \frac{\mu'^{3/2}}{1 + A_n \mu'^2}.$$

Thus, the final solution becomes

$$I_{\text{out}}(\mu) = \frac{1-\gamma}{\gamma}\left(1 + e^{-\tau/\mu}\right) \times$$

$$\times \left(\frac{1}{\mu} E_N(\mu^2) + \sum_{nn'}^{N} \frac{\mu}{(1+A_n\mu^2)} S_{nn'} K_{n'}\right), \qquad (124)$$

where $S_{nn'}$ is a matrix and $K_{n'}$ a vector which are calculated only once for given parameters τ and γ.

B Finite Difference Method

The finite difference code is based on an implicit discretisation of the transfer equation (2) in Cartesian coordinates. The resulting linear system of equations is solved by a combination of recursion and iteration (see [SSW91]):

When $\mathbf{n} = (n_x, n_y, n_z)$ defines a specific direction and Δx, Δy and Δz are the spatial resolutions, the radiative transfer equation is discretized in a first order upwind scheme

$$\frac{n_x}{\Delta x}(I_{i,j,k} - I_{i-1,j,k}) + \frac{n_y}{\Delta y}(I_{i,j,k} - I_{i,j-1,k})$$
$$+\frac{n_z}{\Delta z}(I_{i,j,k} - I_{i,j,k-1}) = -\chi_{i,j,k}(I_{i,j,k} - S_{i,j,k}) \qquad (125)$$

and recursively solved for

$$I_{i,j,k} = \left(\frac{n_x}{\Delta x}I_{i-1,j,k} + \frac{n_y}{\Delta y}I_{i,j-1,k} + \frac{n_z}{\Delta z}I_{i,j,k-1}\right.$$
$$\left. +\chi_{i,j,k}S_{i,j,k}\right) / \left(\frac{n_x}{\Delta x} + \frac{n_y}{\Delta y} + \frac{n_z}{\Delta z} + \chi_{i,j,k}\right). \qquad (126)$$

With extinction coefficient $\chi = \kappa + \sigma$ and albedo $\gamma = \sigma/\chi$ the source function $S(\mathbf{x})$ is given by

$$S(\mathbf{x}) = \frac{f(\mathbf{x})}{\chi(\mathbf{x})} + \gamma(\mathbf{x}) \int_{S^2} P(\mathbf{n}', \mathbf{n}) I(\mathbf{x}, \mathbf{n}') d\omega', \qquad (127)$$

and determined in an iterative procedure using a simultaneous overrelaxation (SOR) method. The iteration is continued until $max\left|(J^n - J^{n-1})/J^n\right| < \epsilon$, where J^n is the mean intensity of the nth iteration step and ϵ a very small number. This relatively simple solution method is stable for all optical depths and guarantees positive intensities everywhere. But it gets into difficulties with steep gradients and shows slow convergence in case of high albedo.

References

[Ada72] Adams, T.F.: The escape of resonance-line radiation from extremely opaque media. Astrophys. J., **174**, 439 (1972)

[AL02] Adams, M.L., Larsen, E.W.: Fast iterative methods for discrete-ordinates particle transport calculations. Prog. Nucl. Energy, **40**, No. 1, 3–159 (2002)

[ALL01] Ahn, S.-H., Lee, H.-W., Lee, H.M.: Lyα line formation in starburst galaxies. I. Moderately thick, dustless and static H I media. Astrophys. J., **554**, 604–614 (2001)

[ALL02] Ahn, S.-H., Lee, H.-W., Lee, H.M.: Lyα line formation in starburst galaxies. II. Extremely thick, dustless and static H I media. Astrophys. J., **567**, 922–930 (2002)

[Arn82] Arnold, D.N.: An interior penalty finite element method with discontinuous elements. SIAM J. Numer. Anal., **19**, 742–760 (1982)
[AM69] Auer, L.H., Mihalas, D.: Non-LTE model atmospheres. III. a complete-linearization method. Astrophys. J., **158**, 641 (1969)
[Bal01] Balsara, D.: Fast and accurate discrete ordinates methods for multidimensional radiative transfer. Part I, basic methods. J. Quant. Spectrosc. Radiat. Transfer, **69**, 671–707 (2001)
[BR01] Becker, R., Rannacher, R.: An optimal control approach to a posteriori error estimation in finite element methods. Acta Numerica 2001, Cambridge University Press (2001), http://www.iwr.uni-heidelberg.de/sfb359/Preprints2001.html
[Boe96] Böttcher K.: Adaptive Schrittweitenkontrolle beim Unstetigen Galerkin-Verfahren für Gewöhnliche Differentialgleichungen. Diplomarbeit, Universität Heidelberg (1996)
[BH83] Bohren, C.F., Huffman, D.R.: Absorption and Scattering of Light by Small Particles. Wiley, New York (1983)
[BL00] Brantley, P.S., Larsen, E.W.: The simplified P_3 approximation. Nucl. Sci. Eng., **134**, Part 1, 1–21 (2000)
[CZ67] Case, K.M., Zweiffel, P.F.: Linear Transport Theory. Addison-Wesley Publishing Company, Reading, Massachusetts (1967)
[Cha50] Chandrasekhar, S.: Radiative Transfer, Clarendon Press, Oxford (1950)
[Cha60] Chandrasekhar, S.: Radiative Transfer, Dover, New York (1960)
[DL00] Dautray, R., Lions, J.-L.: Mathematical Analysis and Numerical Methods for Science and Technology. Vol. **6**. Springer, Berlin Heidelberg New York (2000)
[BRM00] De Breuck, C., Röttgering, H., Miley, G., van Breugel, W., Best, P.: A statistical study of emission lines from high redshift radio galaxies. Astron. Astrophys., **362**, 519–543 (2000)
[DT00] Dullemond, C.P., Turolla, R.: An efficient algorithm for two-dimensional radiative transfer in axisymmetric circumstellar envelopes and disks. Astron. Astrophys., **360**, 1187–1202 (2000)
[Edd26] Eddington, A.S.: The Internal Constitution of the Stars. Cambridge Univ. Press, Cambridge (1926)
[EWW95] Efimov, G.V., Waldenfels, W. von, Wehrse, R.: Analytic solution of the non-discretized radiative transfer equation for a slab of finite optical depth. J. Quant. Spectrosc. Radiat. Transfer, **53**, 59–74 (1995)
[EWW97] Efimov, G.V., Waldenfels, W. von, Wehrse, R. Mathematical aspects of the plane-parallel transfer equation. J. Quant. Spectrosc. Radiat. Transfer, **58**, 355 (1997)
[EJ93] Eriksson, K., Johnson, C.: Adaptive streamline diffusion finite element methods for stationary convection-diffusion problems. Math. Comput., **60**, No. 201, 167–188 (1993)
[EJT85] Eriksson, K., Johnson, C., Thomée, C.: Time discretisation of parabolic problems by the discontinuous Galerkin method. RAIRO, Modélisation Math. Anal. Numér., **19**, 611–643 (1985)
[EJ87] Eriksson, K., Johnson, C.: Error estimates and automatic time step control for nonlinear parabolic problems. I. SIAM J. Numer. Anal., **24**, 12–23 (1987)

[EJ91] Eriksson, K., Johnson, C.: Adaptive finite element methods for parabolic problems. I: A linear model problem. SIAM J. Numer. Anal., **28**, 43–77 (1991)

[EEH95] Eriksson, K., Estep, D., Hansbo, P., Johnson, C.: Introduction to adaptive methods for differential equations. Acta Numerica 1995, Cambridge University Press (1995)

[FK97] Führer, C., Kanschat, G.: A posteriori error control in radiative transfer. Computing, **58**, No. 4, 317–334 (1997)

[FTM00] Fynbo, J.U., Thomsen, B., Møller, P.: Lyα emission from a Lyman limit absorber at z=3.036. Astron. Astrophys., **353**, 457–464 (2000)

[FMT01] Fynbo, J.U., Møller, P., Thomsen, B.: Probing the faint end of the galaxy luminosity function at z= 3 with Lyα emission. Astron. Astrophys., **374**, 443–453 (2001)

[GK03] Gopalakrishnan, J., Kanschat, G.: A multilevel discontinuous Galerkin method, Numer. Math., **95**, Vol. 3, 527–550 (2003)

[HM91] Haardt, F., Maraschi, L.: A two-phase model for the X-ray emission from Seyfert galaxies. Astrophys. J., **380**, L51–L54 (1991)

[HM75] Habetler, G.J., Matkowsky, B.J.: Uniform asymptotic expansions in transport theory with small mean free paths, and the diffusion approximation. J. Math. Phys., **16**, 846–854 (1975)

[Hac85] Hackbusch, W.: Multi-Grid Methods and Applications. Springer, Heidelberg (1985)

[Hac93] Hackbusch, W.: Iterative Lösung Grosser Schwachbesetzter Gleichungssysteme. Teubner, Stuttgart (1993)

[HM93] Hippelein, H., Meisenheimer, K.: Imaging of a Lyman-α absorption cloud in front of the radio galaxy 4C41.17. Nature, **362**, 224–226 (1993)

[Hog98] Hogerheijde, M.: The Molecular Environment of Low-Mass Protostars, Ph.D. Thesis, Rijks Universiteit Leiden (1998)

[HCM98] Hu, E.M., Cowie, L.L., McMahon, R.G.: The density of Ly-α emitters at very high redshift. Astrophys. J., **502**, L99–L104 (1998)

[Hub03] Hubeny, I.: Accelerated lambda iteration: an overview. In: Hubeny I., Mihalas D., Werner K. (eds.) Stellar Atmosphere Modeling. ASP Conf. Ser. Vol. **288**, San Francisco: Astronomical Society of the Pacific, 17–30 (2003)

[Hul57] Hulst, H.C. van de: Light Scattering by Small Particles. Wiley, New York (1957)

[HK80] Hummer, D.G., Kunasz, P.B.: Energy loss by resonance line photons in an absorbing medium. Astrophys. J., **236**, 609–618 (1980)

[JP86] Johnson, C., Pitkäranta, J.: An analysis of the discontinuous Galerkin method for a scalar hyperbolic equation. Math. Comput., **46**, 1–26 (1986)

[Juv97] Juvela, M.: Non-LTE radiative transfer in clumpy molecular clouds. Astron. Astrophys., **322**, 943–961 (1997)

[Kal87] Kalkofen, W. (ed.): Numerical Radiative Transfer. Cambridge Univ. Press, Cambridge (1987)

[Kan96] Kanschat, G.: Parallel and Adaptive Galerkin Methods for Radiative Transfer Problems. Ph.D. Thesis, University of Heidelberg (1996), http://www.iwr.uni-heidelberg.de/sfb359/Preprints1996.html

[Kan00] Kanschat, G.: Solution of multi-dimensional radiative transfer problems on parallel computers. In: Björstad P. and Luskin M. (eds.) Parallel

Solution of PDE, IMA Vol. in Math. and its Appl., **120**, 85–96, Springer, Berlin Heidelberg New York (2000)

[KM04] Kanschat, G., Meinköhn, E.: Multi-model preconditioning for radiative transfer problems. SFB359 Preprints, University of Heidelberg (2004), http://www.iwr.uni-heidelberg.de/sfb359/Preprints2004.html

[Kou52] Kourganoff, V.: Basic Methods in Transfer Problems. Clarendon Press, Oxford (1952)

[Kou63] Kourganoff, V.: Basic Methods in Transfer Problems. Clarendon Press, Oxford (1963)

[Kro55] Krook, M.: On the solutions of the equations of transfer I. Astrophys. J., **122**, 488–495 (1955)

[KMF00] Kudritzki, R.-P., Mendez, R.H., Feldmeier, J.J., et al., Discovery of nine Lyα emitters at redshift z$\sim$ 3.1 using narrowband imaging and VLT spectroscopy. Astrophys. J., **536**, 19–30 (2000)

[KA88] Kunasz, P., Auer, L.H.: Short characteristic integration of radiative transfer problems - Formal solution in two-dimensional slabs. J. Quant. Spectrosc. Radiat. Transfer, **39**, 67–79 (1988)

[KPR02] Kurk, J.D., Pentericci, L., Röttgering, H.J.A., Miley, G.K.: Observations of radio galaxy MRC 1138-262: Merging galaxies embedded in a giant Lyα halo. In: Henney, W.J., Steffen, W., Raga, A.C., Binette, L. (eds) Emission Lines from Jet Flows. RMxAA (Serie de Conferencias), Vol. **13**, 191–195 (2002)

[Lam65] Lamarsh, J.R.: Introduction to Nuclear Reactor Theory. Addison-Wesley Publishing Company, Reading, Massachusett (1965)

[LK74] Larsen, E.W., Keller, J.B.: Asymptotic solution of neutron transport problems for small scale mean free paths. J. Math. Phys., **15**, 75–81 (1974)

[LPB83] Larsen, E.W., Pomraning, G.C., Badham, V.C.: Asymptotic analysis of radiative transfer problems. J. Quant. Spectrosc. Radiat. Transfer, **29**, 285–310 (1983)

[LMM87] Larsen, E.W., Morel, J.E., Miller, W.F. jun.: Asymptotic solutions of numerical transport problems in optically thick, diffusive regimes. J. Comput. Phys., **69**, 283–324 (1987)

[Lar91] Larsen, E.W.: Transport acceleration methods as two-level multigrid algorithms. Oper. Theory, Adv. Appl., **51**, 34–47 (1991)

[LMM96] Larsen, E.W., Morel, J.E., McGhee, J.M.: Asymptotic derivation of the multigroup P_1 and simplified P_N equations with anisotropic scattering. Nucl. Sci. Eng., **123**, 328 (1996)

[LTK02] Larsen, E.W., Thömmes, G., Klar, A., Seaid, M., Götz, T.: Simplified P_N approximations to the equations of radiative heat transfer and applications. J. Comput. Phys., **183**, 652–675 (2002)

[LR74] LeSaint, P., Raviart, P.-A.: On a finite element method for solving the neutron transport equation. In: de Boor (ed.) Mathematical Aspects of Finite Elements in Partial Differential Equations. Academic Press, New York, 89-123 (1974)

[MR02] Meinköhn, E., Richling, S.: Radiative transfer with finite elements. II. Lyα line transfer in moving media. Astron. Astrophys., **392**, 827–839 (2002)

[Mei02] Meinköhn, E.: Modeling Three-Dimensional Radiation Fields in the Early Universe. Ph.D. Thesis, University of Heidelberg (2002), http://www.iwr.uni-heidelberg.de/sfb359/Preprints2002.html

[Mih78] Mihalas, D.: Stellar Atmospheres. W.H. Freeman and Company, Second Edition, San Francisco (1978)

[MW84] Mihalas, D., Weibel-Mihalas, B.: Foundation of Radiation Hydrodynamics. Oxford University Press, New York (1984)

[MCK94] Murray, S., Castor, J., Klein, R., McKee, C.: Accretion disk coronae in high-luminosity systems. Astrophys. J., **435**, 631–646 (1994)

[Neu90] Neufeld, D.A.: The transfer of resonance-line radiation in static astrophysical media. Astrophys. J., **350**, 216–241 (1990)

[Pap75] Papanicolaou, G.C.: Asymptotic analysis of transport processes. Bull. Am. Math. Soc., **81**, 330–392 (1975)

[PH98] Park, Y.-S., Hong, S.S.: Three-dimensional Non-LTE Radiative Transfer of CS in Clumpy Dense Cores. Astrophys. J., **494**, 605 (1998)

[Per57] Perron, O.: Die Lehre von den Kettenbrüchen, Band II. B.G. Teubner Verlagsgesellschaft, Stuttgart (1957)

[PSS79] Pozdniakov, L.A., Sobol, I.M., Suniaev, R.A.: The profile evolution of X-ray spectral lines due to Comptonization - Monte Carlo computations. Astron. Astrophys., **75**, 214–222 (1979)

[RA02] Rampp, M., Janka, H.-T.: Radiation hydrodynamics with neutrinos. Astron. Astrophys., **396**, 361–392 (2002)

[RMD00] Rhoads, J.E., Malhotra, S., Dey, A.: First results from the large-area Lyman-α survey. Astrophys. J., **545**, L85–L88 (2000)

[RM01] Richling, S., Meinköhn, E., Kryzhevoi, N., Kanschat, G.: Radiative transfer with finite elements. I. Basic method and tests. Astron. Astrophys., **380**, 776–788 (2001)

[Sch05] Schuster, A.: Radiation through a foggy atmosphere. ApJ 21, 1 (1905) Reprod. in: Menzel D.H. (ed.) Selected papers on the transfer of radiation. New York: Dover (1966)

[SF93] Sleijpen, G.L.G., Fokkema, D.R.: Bicgstab(L) for linear equations involving unsymmetric matrices with complex spectrum. Electronic Transactions on Numerical Analysis, Vol. **1**, 11–32 (1993), http://etna.mcs.kent.edu

[SPY95] Sonnhalter, C., Preibisch, T., Yorke, H.: Frequency dependent radiation transfer in protostellar disks. Astron. Astrophys., **299**, 545 (1995)

[Spa96] Spaans, M.: Monte Carlo models of the physical and chemical properties of inhomogeneous interstellar clouds. Astron. Astrophys., **307**, 271 (1996)

[SAS00] Steidel, C.C., Adelberger, K.L., Sharply, A.E., Pettini, M., Dickinson, M., Giavalisco, M.: Lyα Imaging of a Proto-Cluster Region at $\langle z \rangle = 3.09$. Astrophys. J., **532**, 170–182 (2000)

[SSW91] Stenholm, L.G., Störzer, H., Wehrse, R.: An efficient method for the solution of 3-D radiative transfer problems. J. Quant. Spectrosc. Radiat. Transfer, **45**, 47–56 (1991)

[Tur93] Turek, S.: An efficient solution technique for the radiative transfer equation. Imp. Comput. Sci. Eng., **5**, No. 3, 201–214 (1993)

[ORC96] van Ojik, R., Röttgering, H.J.A., Carilli, C.L., Miley, G.K., Bremer, M.N., Macchetto, F.: A powerful radio galaxy at z= 3.6 in a giant rotating Lyman-α halo. Astron. Astrophys., **313**, 25–44 (1996)

[ORM97] van Ojik, R., Röttgering, H.J.A., Miley, G.K., Hunstead, R.W.: The nature of the extreme kinematics in the extended gas of high redshift radio galaxies. Astron. Astrophys., **317**, 358–384 (1997)
[Var00] Varga, R.S.: Matrix Iterative Analysis. Springer, Berlin Heidelberg (2000)
[VBF99] Villar-Martín, M., Binette, L., Fosbury, R.A.E.: The gaseous environments of radio galaxies in the early Universe: kinematics of the Lyman-α emission and spatially resolved H I absorption. Astron. Astrophys., **346**, 7–12 (1999)
[Ver96] Verfürth, R.: A Review of A Posteriori Error Estimation and Adaptive Mesh Refinement Techniques. Wiley-Teubner, New York-Stuttgart (1996)
[WWM04] Warsa, J.S., Wareing, T.A., Morel, J.E.: Krylov iterative methods and the degraded effectiveness of diffusion synthetic acceleration for multidimensional S_n calculations in problems with material discontinuities. Nucl. Sci. Eng., **147**, 218–248 (2004)
[WMK99] Wehrse, R., Meinköhn, E., Kanschat, G.: A review of Heidelberg radiative transfer equation solutions. In: Stee P. (ed.) Radiative transfer and hydrodynamics in astrophysics. EAS Publication Series, Vol. **5**, 13–30 (2002)
[WBW00] Wehrse, R., Baschek, B., von Waldenfels, W.: The diffusion of radiation in moving media: I. Basic assumptions and formulae. Astron. Astrophys., **359**, 780–787 (2000)
[WHS99] Wolf, S., Henning, T., Stecklum, B.: Multidimensional self-consistent radiative transfer simulations based on the Monte-Carlo method. Astron. Astrophys., **349**, 839–850 (1999)
[YBL93] Yorke, H., Bodenheimer, P., Laughlin, G.: The formation of protostellar disks. I - 1 M(solar). Astrophys. J., **411**, 274–284 (1993)
[ZM02] Zheng, Z., Miralda-Escudé, J.: Monte Carlo simulation of Lyα scattering and application to damped Lyα systems. Astrophys. J., **578**, 33–42 (2002)

Radiative Transfer in 4D: The Inclusion of Kinematical Information

Maarten Baes[1,2]

[1] Sterrenkundig Observatorium, Universiteit Gent, Krijgslaan 281 S9, B-9000 Gent, Belgium, maarten.baes@ugent.be
[2] Department of Physics and Astronomy, Cardiff University, 5 The Parade, Cardiff CF24 3YB, Wales, UK, maarten.baes@astro.cf.ac.uk

Summary. Absorption and scattering by dust grains affect all observable properties of galaxies, including the observed kinematics. We demonstrate how kinematical information can be included into the radiative transfer framework, and argue that the Monte Carlo method is a very elegant method to approach such radiative transfer problems. We present the SKIRT code, a new Monte Carlo radiative transfer code that is developed to calculate the observed kinematics of dusty galaxies. As an application we demonstrate that dust absorption and scattering have significant effects on the observed kinematics of elliptical galaxies. In particular, scattering by dust grains can form an alternative explanation for the stellar kinematical evidence of dark matter haloes around elliptical galaxies.

1 Introduction

It is obvious from the appearance of strong dust lanes in edge-on systems that spiral galaxies generally contain a significant amount of interstellar dust. The actual optical depth of spiral galaxies has been the subject of debate for nearly 15 years however. Detailed models of dust lanes in edge-on spiral galaxies indicate a face-on optical depth of order unity in the optical bands (Xilouris et al. 1999). This is more or less in agreement with the observed far-infrared emission from spiral galaxies: the most recent results from the ISO satellite indicate that about 30 percent of the total bolometric luminosity of normal spiral galaxies is absorbed and re-emitted as thermal far-infrared radiation by dust grains (Popescu & Tuffs 2002).

Elliptical galaxies on the other hand were for a long time considered to be virtually devoid of interstellar dust. This traditional viewpoint has gradually changed, however. Signatures from interstellar dust patches have now been detected in the centre of most ellipticals, which infer modest dust masses of the order 10^4–10^5 $M_\odot$ (van Dokkum & Franx 1995; Ferrari et al. 1999). The minority of elliptical galaxies which were detected by IRAS and ISO however, have dust masses estimated from the far-infrared emission which exceed those from the optical by factors of 10 to 100 (Goudfrooij & de Jong 1995; Bregman et al. 1998; Temi et al. 2003). Hopefully, the new generation of far-infrared instrumentation (Spitzer, SOFIA, Herschel, ...) will clarify this issue.

Dust grains strongly interact with optical radiation through the processes of continuum absorption and scattering. The presence of interstellar dust in galaxies hence needs to be taken into account when we want to study their intrinsic photometric properties such as magnitudes, colours and scalelengths. Dust attenuation also affects the observed kinematics of galaxies. The kinematical information on external galaxies is obtained from extracting line-of-sight velocity information from the spectra, through measuring the Doppler shifts of either absorption lines (stars) or emission lines (gas). Often these spectra are taken in the optical regime and therefore they are subject to dust attenuation.

We have started a project to investigate how the observed kinematics of galaxies are affected by dust attenuation. This requires the inclusion of kinematical information into the radiative transfer formalism. Galaxies can be considered as equilibrium dynamical systems, and so what we need is a recipe to include velocity information into the time-independent radiative transfer formalism. In this contribution we demonstrate how kinematical information can be included in the radiative transfer equation (section 2), we describe SKIRT, a Monte Carlo code that we have developed to tackle such radiative transfer problems (section 3), and we shortly describe one application of the SKIRT code (section 4).

2 Radiative Transfer in Dusty Galaxies

2.1 The General Radiative Transfer Equation

The basis for any study of attenuation is the radiative transfer equation (RTE), which statistically describes the interaction between matter and light. The processes we have to take into account in our present situation are the emission of photons by stars, and the absorption and scattering of photons by interstellar dust grains. Dust absorption also accounts for an additional source term, because the energy absorbed by grains is re-emitted at far-infrared wavelengths. As we are primarily interested in the optical regimes, however, this source term can safely be neglected. For a system with these physical processes taken into account, the time-independent RTE can be written as

$$\frac{dI_\lambda}{ds}(\boldsymbol{r}, \boldsymbol{k}) = \ell_\lambda(\boldsymbol{r}) - \kappa_\lambda(\boldsymbol{r}) I_\lambda(\boldsymbol{r}, \boldsymbol{k}) + \omega_\lambda(\boldsymbol{r}) \kappa_\lambda(\boldsymbol{r}) \iint \frac{d\Omega'}{4\pi} I_\lambda(\boldsymbol{r}, \boldsymbol{k}') \Phi_\lambda(\boldsymbol{r}, \boldsymbol{k}, \boldsymbol{k}'), \quad (1)$$

where s is the path length, $I_\lambda(\boldsymbol{r}, \boldsymbol{k})$ the specific intensity of radiation at a position $\boldsymbol{r}$ into a direction $\boldsymbol{k}$, $\ell_\lambda(\boldsymbol{r})$ the stellar emissivity, $\kappa_\lambda(\boldsymbol{r})$ the total (absorption plus scattering) opacity, $\omega_\lambda(r)$ the scattering albedo and $\Phi_\lambda(k,k')$ the scattering phase function. The three terms in the right hand side of the equation represent respectively the stellar emissivity, the fraction of photons removed from the beam by the combination of absorption and scattering, and the fraction of photons scattered into the beam from other directions.

If scattering is neglected, the RTE is an ordinary differential equation, which can be solved by a simple weighted integration along the line of sight. Scattering off dust grains is not a rare phenomenon, however: for typical Milky Way dust grains, the probability for scattering even slightly exceeds the probability for absorption in the optical wavelength range. The inclusion of scattering turns the RTE into a integro-differential equation, far more complicated than the ordinary differential equation if only absorption is taken into account. In particular, the scattering term is responsible for the coupling of the RTE along different paths: due to the integration over the angle, we cannot solve the RTE for a single path, but we have to solve it for all paths at the same time.

The complexity of the RTE does not only depend on which physical processes are taken into account (the right-hand side), but also on the geometry of the system. Indeed, the path length appearing in the left-hand side of the RTE is in general a function of position and direction, i.e. $s = s(r,k)$. As a consequence, the (time-independent) RTE is a partial differential equation with five independent variables. This complexity can be reduced if the system has symmetries. Particularly interesting are the cases of plane-parallel or spherical geometry, where only two of these five independent coordinates need to be considered. Apart from these very simple geometries, however, the transfer equation remains a difficult integro-differential equation with many independent variables, and advanced techniques are indispensible to solve it.

2.2 Including Kinematical Information

As long as scattering is not taken into account, the inclusion of velocity information in the RTE is rather straightforward. Indeed, instead of taking into account the entire stellar emissivity, we can just consider the light from those stars whose velocity component in the direction of the observer equals u. The emissivity of these stars is given by $\ell(r)\,\phi(r,k,u)$, where $\phi(r,k,u)$ is the spatial LOSVD, i.e. the probability for a star at position r to have a velocity component u in the direction k. Solving the RTE with this emissivity, we obtain the LOSVDs observed in the plane of the sky. Hence, if either dust attenuation is completely neglected or only absorption by dust grains is taken into account, the observed LOSVDs can be calculated from the spatial LOSVDs through a single weighted integration along the line of sight. This has been applied by Baes & Dejonghe (2000) to investigate the effects of dust absorption on the observed kinematics of elliptical galaxies.

When scattering is included, however, the inclusion of kinematical information becomes much more complicated, because the velocities of both the

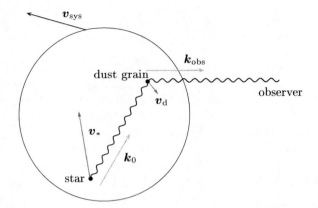

Fig. 1. The inclusion of kinematical information in the radiative transfer equation. This figure shows the trajectory of a photon through the galaxy (undulating line): after being emitted by the star, it is scattered once before it leaves the galaxy. See text for more details

stars and the dust grains that scatter their light need to be taken into account. Consider, for example, figure 1, where a photon is emitted in the direction k_0 by a star moving with a velocity v_*. The line-of-sight velocity information carried by the photon is of course the component $u = v_* \cdot k_0$. When the photon is coherently scattered into the direction k_{obs} by a dust grain with velocity v_{d}, the line-of-sight velocity information carried by the photon is altered and contains a contribution of the scattering dust grain

$$u = v_* \cdot k_0 + v_{\mathrm{d}} \cdot (k_{\mathrm{obs}} - k_0). \tag{2}$$

When more scattering events are involved in the photon's trajectory, the relative velocities of each pair of subsequent dust grains have to be taken into account. If we denote the total number of scattering events as M, the velocities of the dust grains as v_{d_i} and the propagation directions as k_i (for $i = 1 \ldots M$), we obtain

$$u = v_* \cdot k_0 + \sum_{i=1}^{M} v_{\mathrm{d}_i} \cdot (k_i - k_{i-1}). \tag{3}$$

3 Description of the SKIRT Code

From the discussion in the previous section, it might be clear that the inclusion of kinematical information into radiative transfer problems is fairly complex. We constructed a Monte Carlo code to solve our radiative transfer problem. Usually, the main argument against Monte Carlo methods is that they are computationally rather expensive compared to other methods. They have other advantages, however, which make them very competitive, in particular in an era when CPU time is not the most stringent limitation anymore. The ability to include velocity information in an elegant way, for example, is one of these advantages. It might be possible to solve such radiative transfer problem using traditional methods such as ray-tracing techniques. Still, the Monte Carlo technique offers a straightforward and transparent approach and we doubt whether other methods could offer the same advantages.

The basic characteristics of Monte Carlo radiative transfer have been explained at length by various authors (e.g. Witt 1977; Bianchi et al. 1996; Niccolini et al. 2003). In essence, a Monte Carlo radiative transfer code follows the life of a very large number of individual photons. A photon is, at each stage in its existence, characterized by various quantities, such as its position, propagation direction and wavelength. Each photon propagates on straight lines through the interstellar medium until either it interacts with a dust grain or it leaves the galaxy. The various interactions alter the properties of the photon, according to random numbers generated from the appropriate probability functions. When at last the photon escapes from the galaxy, its final properties are recorded. After recording a large numbers of photons in this way, the global observed properties of the system can be calculated.

We will not repeat an in-depth description of the principles and the numerous equations of Monte Carlo radiative transfer here, as they are well described in the articles listed above. We will only shortly describe the basic steps in the Monte Carlo radiative transfer process, and focus on the main aspect which makes the SKIRT code different from the existing ones. More details on the construction of the SKIRT code can be found in Baes & Dejonghe (2002) and Baes et al. (2003).

3.1 The Emission Process

When no kinematical information is included in the radiative transfer calculations, photons are characterized at each moment by a wavelength λ, a position r and a direction k. Initial values for these quantities must be generated randomly from the emitting stellar system, which can be composed of several stellar components. The initial emission direction k_0 can be sampled from the unit sphere, the initial position r_0 is sampled from the spatial distribution of the stellar component, and the initial wavelength λ_0 is sampled from the SED of the stellar component at the position r_0. The SKIRT code contains a library with a set of common stellar components from which it is straightforward to sample a random position. Another library contains a set of SEDs, including both simple analytical SEDs and realistic tabulated stellar and galaxy SEDs.

When the SKIRT code is used to calculate the observed kinematics of a dusty galaxy, photons must also carry with them the kinematical signature of the star that has emitted them. The initial value of the line-of-sight velocity carried by the photon is the component of the star in the direction of the original emission, $u_0 = \boldsymbol{v}_* \cdot \boldsymbol{k}_0$. To complete the initialization of a photon, we therefore need to generate a line-of-sight velocity from the appropriate probability distribution. We approximate the stellar velocity distribution at each position in the galaxy as a local trivariate gaussian distribution. The orientation of the velocity ellipsoid and the values of the mean velocities and velocity dispersions on the three principle axes can vary from position to position, such that this can represent a fairly general dynamical structure. It has the advantage that the spatial LOSVDs can be calculated exactly for any direction $\boldsymbol{k}$: it is straightforward to show that the spatial LOSVD will also be a gaussian univariate distribution, such that the random generation of the photons' initial line-of-sight velocities is straightforward.

3.2 The Dust Iteration

The life of a single photon can be thought of as a loop, whereby at each iteration (representing a physical process) we must update its position $\boldsymbol{r}$, propagation direction $\boldsymbol{k}$, wavelength λ and line-of-sight velocity u. The initial values are determined randomly at the emission phase, and change at every scattering event. As we only consider coherent scattering, and we are not taking into account the re-emission of photons at longer wavelength, the only wavelength change of a photon along its path is due to the varying Doppler shift. These wavelength variations are so tiny that the optical properties of the dust (extinction curve, dust albedo and asymmetry parameter) do not vary. The wavelength of a photon can hence be considered fixed throughout its lifetime. We must therefore only update the three quantities $\boldsymbol{r}$, $\boldsymbol{k}$ and u at each iteration, from their old values ($\boldsymbol{r}_{i-1}\ldots$) to the new ones ($\boldsymbol{r}_i\ldots$). This proceeds in several steps.

The first step in the iteration consists of determining whether an interaction with a dust grain will take place or whether the photon will leave the galaxy. To do this, we sample an optical depth τ_λ from an exponential distribution and compare it to $\tau_{\text{path},\lambda}$, the total optical depth along the path. When $\tau_\lambda > \tau_{\text{path},\lambda}$, the photon will leave the galaxy; in the other case, the photon will interact with a dust grain. This interaction can either be a scattering or an absorption event, which is determined by the scattering albedo. In the former case, the photon will continue its journey through the galaxy; in the latter case, its life cycle is ended.

The second step in the loop, if the photon is scattered, is the determination of the position of the scattering. We have to translate the sampled optical depth τ_λ to a physical path length s. This comes down to solving the equation

$$\int_0^s \kappa_\lambda(\boldsymbol{r}_{i-1} + s'\boldsymbol{k}_{i-1})\,\mathrm{d}s' = \tau_\lambda \qquad (4)$$

for s. This integration through the dust is usually the most time-consuming operation in Monte Carlo simulations. We have implemented a 3D cartesian grid structure in which this integration can efficiently be performed by trilinear interpolation (see Baes et al. 2003 for details). With this path length known, the new position is $r_i = r_{i-1} + sk_{i-1}$.

The third step in the iteration is the determination of the new propagation direction k_i of the photon. It is found by sampling a direction from the probability density $p(k_i) = \Phi_\lambda(k_{i-1}, k_i)$, which represents the scattering phase function. Various phase functions are built into the SKIRT code, and usually the anisotropic Henyey-Greenstein phase function is adopted. The sampling of a direction from this phase functions can be performed analytically.

The final step is to update the photon's line-of-sight velocity information. As shown in section 2.2, the relative orientation of the propagation directions before and after the scattering event cause a change in the line-of-sight velocity information $u_i - u_{i-1} = v_{d_i} \cdot (k_i - k_{i-1})$. In order to update the velocity information, we need to sample a dust grain velocity from the dust velocity distribution at the position r_i of the scattering event. Therefore, the dust components in the SKIRT library must not only be specified by their spatial distribution, but by their distribution function $F_d(r, v)$ in six-dimensional phase space. Analogous to the stellar distribution function, we assume that the dust velocity field can be described by a trivariate gaussian distribution. Similarly as in the stellar case, we don't have to sample a full three-dimensional velocity vector for the scattering dust, but only a component in the direction

$$k'_i = \frac{k_i - k_{i-1}}{||k_i - k_{i-1}||}. \tag{5}$$

3.3 Calculation of the Observed Properties

If a photon leaves the galaxy, we record its final position, propagation direction, wavelength and line-of-sight velocity information. These can be used to calculate the coordinates on the plane of the sky and the observed line-of-sight velocities for any arbitrary observer. The photons are then classified into a four-dimensional histogram (two spatial, one wavelength and one velocity dimension), from which the light profile, the observed SEDs and the observed kinematics can be calculated.

3.4 Optimization Techniques

In the previous subsections we have presented a description of the most elementary Monte Carlo technique. For realistic radiative transfer problems this approach would be extremely slow. Therefore, a number of optimization techniques have been invented to increase the efficiency of Monte Carlo radiative transfer codes. These techniques usually consist of adding a weight to each photon and replacing a number of probabilistic elements by the corresponding deterministic elements. A simple example is the determination of the nature of an interaction between a photon and a dust grain. There are two possibilities: either the photon is absorbed and it disappears from the radiation field, or it is scattered. In the basic Monte Carlo routine, the nature of the interaction would be determined in a probabilistic way, by sampling a random number from a probability function (which is discrete here with only two possible values). A smarter way is to force all interactions to be scattering events and to correct for this manipulated behaviour by changing the weight of the scattered photons. The advantage is that no photons are lost due to absorption, and that a source of Poisson noise is avoided.

We have included many of these techniques into the SKIRT code in order to increase the efficiency, such as the forced first scattering principle, the peel-off technique, and the splitting of the stellar and dust grain contributions in the line-of-sight information carried by the photons. For more details see Baes & Dejonghe (2002) and Baes et al. (2003).

4 Application: Elliptical Galaxy Kinematics

As an application of the SKIRT code, we present some results of our investigation of the effect of dust attenuation on the observed kinematics of elliptical galaxies. A more detailed account of this study can be found in Baes & Dejonghe (2000, 2001, 2002) and Baes, Dejonghe & De Rijcke (2000).

We have constructed a simple synthetic galaxy model, consisting of a stellar component and a diffuse dust component. For the stellar component we use a spherical Hernquist or Plummer model, for which most of the kinematics can be calculated analytically in the optically thin case. The dust has optical properties as the typical Milky Way dust grains and is distributed smoothly over the galaxy. The total optical depth of the dust is a free parameter in the models. Using the SKIRT code, we calculate the stellar kinematics of these dusty galaxy models, taking both absorption and scattering into account. Some results are given in figure 2. As a result of scattering, the projected velocity dispersion profile increases at large radii, and the Gauss-Hermite h_4 parameter, which measures the shape of the line profiles, strongly increases. Both effects become stronger with increasing optical depth.

These results are important concerning the determination of the dark matter content of elliptical galaxies from stellar kinematics. According to the popular CDM cosmological paradigm, all galaxies are embedded in a massive dark matter halo. The easiest way to trace such a dark halo is by studying its dynamical effect on a tracer population. For spiral galaxies, the ubiquitous H I gas can be observed to very large galactocentric radii and forms very strong evidence for the presence of dark matter. Elliptical galaxies lack such a simple ubiquitous tracer. One possibility to search for dark matter in ellipticals is to use the stellar kinematics. The main signature of a dark matter halo on the observed kinematics of elliptical galaxies is an increase of the velocity dispersion at large projected radii. However, such a signature can also be the result of

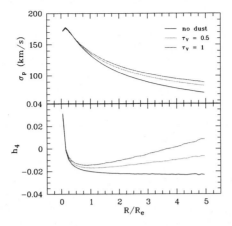

Fig. 2. The effect of dust absorption and scattering on the observed kinematics of an elliptical galaxy. The top panel shows the velocity dispersion profile, the bottom panel shows the Gauss-Hermite h_4 parameter profile. The lay-out for the different models is indicated in the top right corner

a tangentially anisotropic dynamical structure (Gerhard 1993). This degeneracy can be broken by studying the Gauss-Hermite h_4 parameter: a tangential orbital structure leads to negative h_4 values, whereas a dark matter halo gives positive h_4 values. Various authors have recently used stellar kinematical evidence measured out to a few effective radii to demonstrate the presence of dark matter around elliptical galaxies (Rix et al. 1997; Gerhard et al. 1998; Kronawitter et al. 2000).

Our results demonstrate however, that these effects can also be the result of scattering by interstellar dust grains: dust attenuation has the same signature on elliptical galaxy kinematics as a dark matter halo. We examined this further by modelling the dust-affected kinematics of our galaxy models with a dynamical modelling technique where dust is not taken into account (as any modeller would do with real data). The result of this exercise is that the dust-affected kinematics cannot be fitted with a constant mass-to-light model, i.e. a dark matter halo is required to explain the slow decrease of the dispersion and the large h_4 values. This demonstrates that dust can act as an additional or alternative explanation for the presence of dark matter in elliptical galaxies, and strongly advocates for the inclusion of radiative transfer calculations in dynamical modelling codes.

References

Baes M., Dejonghe H., 2000, MNRAS, **313**, 153–164
Baes M., Dejonghe H., De Rijcke S., 2000, MNRAS, **318**, 798–808
Baes M., Dejonghe H., 2001, ApJ, **563**, L19–L22

Baes M., Dejonghe H., 2002, MNRAS, **335**, 441–458
Baes M., Davies J.I., Dejonghe H., Sabatini S., Roberts S., Evans R., Linder S.M., Smith R.M., de Blok W.J.G., 2003, MNRAS, **343**, 1081–1094
Bianchi S., Ferrara A., Giovanardi C., 1996, ApJ, **465**, 127–144
Bregman J.N., Snider B.A., Grego L., Cox C.V., 1998, ApJ, **499**, 670–676
Ferrari F., Pastoriza M.G., Macchetto F., Caon N., 1999, A&AS, **136**, 269–284
Gerhard O.E., 1993, MNRAS, **265**, 213–230
Gerhard O., Jeske G., Saglia R.P., Bender R., 1998, MNRAS, **295**, 197–215
Goudfrooij P., de Jong T., 1995, A&A, **298**, 784–798
Kronawitter A., Saglia R.P., Gerhard O., Bender R., 2000, A&AS, **144**, 53–84
Niccolini G., Woitke P., Lopez B., 2003, A&A, **399**, 703–716
Popescu C.C., Tuffs R.J., 2002, MNRAS, **335**, L41–L44
Rix H., de Zeeuw P.T., Cretton N., van der Marel R.P., Carollo C.M., 1997, ApJ, **488**, 702–719
Temi P., Brighenti F., Mathews W.G., Bregman J.D., 2004, ApJS, in press (astro-ph/0312248)
van Dokkum P.G., Franx M., 1995, AJ, **110**, 2027–2037
Witt A.N., 1977, ApJS, **35**, 1–6
Xilouris E.M., Byun Y.I., Kylafis N.D., Paleologou E.V., Papamastorakis J., 1999, A&A, **344**, 868–878

A Problem-Orientable Numerical Algorithm for Modeling Multi-Dimensional Radiative MHD Flows in Astrophysics – the Hierarchical Solution Scenario

A. Hujeirat

Institute of Applied Mathematics, University of Heidelberg, 69120 Heidelberg, Germany
ahmad.hujeirat@iwr.uni-heidelberg.de

Summary. We present a hierarchical algorithm for the adaptation of numerical solvers in high energy astrophysics.

This approach is based on clustering the entries of the global Jacobian in a hierarchical manner that enables employing a variety of solution procedures ranging from a purely explicit time-stepping up to fully implicit schemes.

A gradual coupling of the radiative MHD equation with the radiative transfer equation in higher dimensions is possible.

Using this approach, it is possible to follow the evolution of strongly time-dependent flows with low/high accuracies and with efficiency comparable to explicit methods, as well as searching quasi-stationary solutions for highly viscous flows.

In particular, it is shown that the hierarchical approach is capable of modeling the formation of jets in active galactic nuclei and reproduce the corresponding spectral energy distribution with a reasonable accuracy.

Key words: Computational methods in fluid dynamics, Radiation transfer, Magnetohydrodynamics, Physical processes
PACS: 02.60, 02.70, 47.70 95.30 95.30.Qd

1 Introduction

Within the last two decades, a tremendous progress has been made in both computational fluid dynamics (CFD) algorithms and the computer hardware technologies. The computing speed and memory capacity of computers have increased exponentially during this period. Similarly is in astrophysical fluid dynamics (AFD), which is a rapidly growing research field, and in which modern numerical methods are extensively used to model the evolution of rather complicated flows. Unlike CFD, in which implicit methods are frequently used, the majority of the methods used in AFD are explicit. Several of them became very popular, e.g., ZEUS [36], NIRVANA+ [41], FLASH [8], VAC [37], THARM [9]. The popularity of explicit methods arises from their being easy to construct, vectorizable, parallelizable and even more efficient as long as dynamical evolutions of compressible flows are concerned. Specifically, for modeling the dynamical evolution of HD-flows in two and three dimensions explicit methods are highly superior to-date. For modeling relativistic flows, Koide and collaborators [21, 22] and [28, 23] have developed pioneering general relativistic MHD solvers. A rather complete review of numerical approaches for relativistic fluid dynamics is given in [27, 7]. A ZEUS-like scheme for general relativistic MHD has also been developed and is described in [40].

These methods, however, are numerically stable as far as the Courant-Friedrich-Levy number is smaller than unity. The corresponding time step size decreases dramatically with the incorporation of real astrophysical effects. Specifically, they may even stagnate if self-gravity, radiative and chemical effects are included. Moreover, explicit methods break down if the flow is weakly or strongly incompressible, and if the domain of calculations is subdivided into a strongly stretched mesh. In an attempt to enhance their robustness, several alternatives have been suggested, such as semi-explicit, semi-implicit or even implicit-explicit methods [20, 37]. Nevertheless, their rather limited range of applications has lead to the fact that most of the interesting astrophysical problems remained, indeed, not really solved. A simple example is the evolution of a steady turbulent accretion disk. It was found by [1] that weak magnetic fields in accretion disks are amplified, generate turbulence, which in turn redistribute the angular momentum in the disk. However, whether this instability leads to the long-sought global steady accretion rate, or is it just a transient phenomenon in which the generation of turbulence is subsequently suppressed by dynamo action are not at all clear. Other notable phenomena are the formation and acceleration of the observed superluminal jets in quasars and in microquasars, the origin of the quasi-periodic oscillation in low mass X-ray binaries or the progenitors of gamma ray burst are still spectacular.

Explicit methods rely on time-extrapolation procedures for advancing the solution in time. However, in order to provide physically consistent solutions, it is necessary that these procedures are numerically stable. The usual approach for examining the stability of numerical methods is to perform the so called von Neumann analysis [see 10, for further details]. This yields the so called Courant-Friedrich-Levy condition (CFL) which is known to limit the range of application and severely affects the robustness of explicit methods. In particular, equations corresponding to physical processes occurring on much shorter time scales than the hydro-time scale (e.g., radiation, self-gravitation and chemical reactions) cannot be followed explicitly. Furthermore, these methods are not suited for searching solutions that correspond to evolutionary phases occurring on time scales much longer than the hydro-time scale. Using high performance computers to perform a large number of explicit time steps may lead to accumulation of round-off errors that can easily distort the propagation of information from the boundaries and cause divergence of the

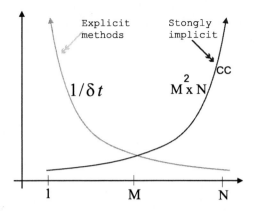

Fig. 1. A schematic description of the time step size and the computational costs versus the band width M of the Jacobian. N is the number of unknowns. Explicit methods correspond to $M = 1$ and large $1/\delta t$. They require minimum computational costs (CC). Large time steps (i.e., small $1/\delta t$) can be achieved using strongly implicit methods. These methods generally rely on the inversion of matrices with large band width, hence computationally expensive, and, in most cases, are inefficient

solution procedure, especially if Neumann type conditions are imposed at the boundaries.

In contrast to explicit methods, implicit methods are based on solving a matrix equation of the form $Ax = b$, where A is the Jacobian matrix corresponding to the system of equations to be solved, b is the right hand side vector of known quantities, and x is the solution vector sought. These methods have two major drawbacks. First, constructing the matrix A is difficult, time consuming, and may considerably influence the robustness of the method (see Fig. 1). Second, the inversion procedure must be stable and efficient. In general, conservative discretization of the MHD equations give rise to sparse matrices, or even to narrow band matrices. Therefore, any efficient matrix inversion procedure must take the advantage of A being sparse. Using Gaussian elimination to invert A directly, and without employing a bandwidth optimizer, requires approximately N^3 algebraic operations, where N is the number of unknowns. If the flow is multi-dimensional and a high spatial resolution is required, the number of operations can be prohibitive even on modern supercomputers. Krylov Sub-Iterative Methods (KSIMs), on the other hand, are most suited for sparse matrices and avoid the fill-in procedure. In the latter case, A is not directly involved in the process, but rather its multiplication with a vector. The convergence rate of KSIMs has been found to depend strongly on the proper choice of the pre-conditioner. For advection-dominated flows, incomplete factorization such as ILU, IC and LQ, approximate factorization, ADI, line Gauss-Seidel are only a small sub-set of possible sequential pre-conditioners [see 34, and the references therein]. Another powerful way of accelerating relaxation techniques is to use the multi-grid method as a direct solver or as a pre-conditioner [3, 38]. For parallel computations, Red-Black ordering in combination with GRMES and Bi-CGSTAB as well as domain decomposition are among the popular pre-conditioners [see 4, for further discussion.]

Towards studying the jet-disk connection around black holes (BHs) and in active galactic nuclei (AGNs) and micro quasi-stellar-objects (μ-QSOs) a series of multi-dimensional calculations have been performed [e.g., 31, 39, 29, 13, 15, 16]. Specifically, these studies revealed that:

(i) Counter-rotating disks with respect to the BH-spin generate jets that propagate approximately twice as fast as in the co-rotating case.

(ii) Jets formed are found to be relatively slow, i.e., the corresponding Γ-factors did not reach the desired large values. This was found in both cases: when the spins of the disk and the BH are parallel and when they are anti-parallel. Moreover, disks surrounding Kerr BHs have been verified to produce jets that are more powerful than in the Schwarzschild case. These jets are driven primarily by strong magnetic fields (MFs) that are created by the frame dragging effect.

(iii) Large Γ-factors[1] are obtainable if the Alfvén speed due to the poloidal magnetic fields (PMF) is equal to or even larger than the local escape velocity [see 29, and the references therein].

(iv) Poliodal magnetic fields may extract rotational energy from the disk plasma, and from a geometrically thin super-Keplerian layer between the disk and the overlying corona. The outflowing plasma in this layer is dissipative, two-temperature, virial-hot, advective and electron-proton dominated. The innermost part of the disk in this model is turbulent-free, sub-Keplerian rotating and advective-dominated. This part ceases to radiate as a standard disk, and most of the accretion energy is converted into magnetic and kinetic energies that go into powering the jet.

Nevertheless, jet-structures, their formation, acceleration, their linkage to the accretion phenomena and the nature of their plasma are still a matter of debate. Furthermore, the flood of observational data makes it even more essential than ever to perform sophisticated numerical calculations to gain a more precise insight of their evolution.

In this paper we focus on the architecture of the global solution procedure rather than on local details, such as order of accuracies, physical consistency, types of advection schemes or fulfilling the solenoidal condition. Specifically, we discuss strategies for enhancing the robustness of solvers through constructing various pre-conditionings to implement a variety of solution methods in arbitrary dimensions. Special attention is given to radiative MHD solvers and their possible coupling with the radiative transfer equation in higher dimensions.

---

[1] Γ-factors are measures that characterize how relativistic the flow is.

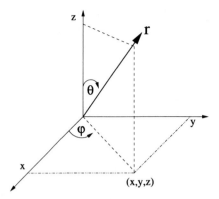

Fig. 2. The radiative MHD equations are solved using Spherical coordinates

2 The Governing Equations

2.1 The 3D Axi-Symmetric Radiative MHD Equations

Spherical geometry is the most appropriate geometry for capturing flow configurations in the vicinity of black holes (Fig. 2). Taking into account the perfect axi-symmetry of black holes, and that their gravitational pull dominates the forces exerting on the surrounding flows, we conclude that axi-symmetry is a reasonable assumption that may characterize accretion flows in their vicinities. Moreover, in applying spherical geometry the transformation $\bar{\theta} = \pi/2 - \theta$ has been used[2]. We note that the dynamical time scale near the event horizon is extremely short, therefor giving rise to multi-component flows, such as electron and ion plasmas.

In the following we describe the set of radiative MHD equations [see 30, 12, and the references therein], and list the scaling variable that may be used for transforming them into non-dimensional form (see Table 1).

– Continuity equation:
$$\frac{\partial \rho}{\partial t} + \nabla \cdot \rho V = 0 \tag{1}$$

– Radial momentum equation:
$$\frac{\partial m}{\partial t} + \nabla \cdot mV = \frac{\partial P}{\partial r} + \rho \frac{(V_\theta^2 + V_\varphi^2)}{r} + \rho \frac{\partial \psi}{\partial r} + \lambda_{\text{FLD}} \frac{\partial E}{\partial r} + F_L^r + Q_{\text{vis}}^r \tag{2}$$

– Vertical momentum equation:
$$\frac{\partial n}{\partial t} + \nabla \cdot nV = \frac{\partial P}{\partial \theta} - \rho V_\varphi^2 \tan\theta + \rho \frac{\partial \psi}{\partial \theta} + \lambda_{\text{FLD}} \frac{\partial E}{\partial \theta} + F_L^\theta + Q_{\text{vis}}^\theta \tag{3}$$

---
[2] This transformation allows simple analogy with and into cylindrical coordinates.

Table 1. An illustrative example of scaling variables that may be used for reformulating the MHD equations in non-dimensions

Scaling variables:

Mass:	$\tilde{\mathcal{M}} = 3 \times 10^8 M_\odot = 5.967 \times 10^{41}$ g
Accretion rate:	$\tilde{\dot{\mathcal{M}}} = 10^{-1} \dot{\mathcal{M}}_{Edd} = 7 \times 10^{25}$ g s^{-1}
Distance:	$\tilde{R} = R_{in} = 3R_S = 2.7 \times 10^{14}$ cm, where $R_S = 2G\tilde{\mathcal{M}}/c^2$
Temperature:	$\tilde{T} = 5 \times 10^7$ K
Velocities:	$\tilde{V} = \tilde{V}_S = [\gamma \mathcal{R}_{gas} \tilde{T}/\mu_i]^{1/2} = 7.506 \times 10^7$ cm s^{-1}, $\mu_i = 1.23$
Density:	$\tilde{\rho} = \tilde{\mathcal{M}}/(\tilde{H}_d \tilde{R}_{out} \tilde{V}_S) = 2.5 \times 10^{-12}$ g cm^{-3}
Magnetic Fields:	$\tilde{B} = \tilde{V}_S/\sqrt{4\pi\tilde{\rho}} = 420.6$ Gauss

- Angular momentum equation:

$$\frac{\partial \ell}{\partial t} + \nabla \cdot \ell V = F_L^\varphi + Q_{vis}^\varphi \tag{4}$$

- Internal equation of the ions:

$$\frac{\partial \mathcal{E}_i^d}{\partial t} + \nabla \cdot \mathcal{E}_i^d V = -(\gamma - 1)\mathcal{E}_i^d \nabla \cdot V + \Phi - \Lambda_{i-e} + \nabla \cdot \kappa_i^{cond} \nabla T_i \tag{5}$$

- Internal equation of the electrons:

$$\frac{\partial \mathcal{E}_e^d}{\partial t} + \nabla \cdot \mathcal{E}_e^d V = -(\gamma-1)\mathcal{E}_e^d \nabla \cdot V + \Lambda_{i-e} - \Lambda_B - \Lambda_C - \Lambda_{Syn} + \nabla \cdot \kappa_e^{cond} \nabla T_e \tag{6}$$

- Equation of the zero moment of the radiation field:

$$\frac{\partial E}{\partial t} + \nabla \cdot EV = \nabla \cdot [\lambda_{FLD} \nabla E] - \Lambda_B + \Lambda_C + \Lambda_{Syn} \tag{7}$$

- The induction equation:

$$\frac{\partial B}{\partial t} = \nabla \times (V \times B + \alpha_{dyn} B - \nu_{mag} \nabla \times B). \tag{8}$$

- Gravitational potential: the Poisson equation:

$$\Delta \psi = 4\pi G \rho, \tag{9}$$

where ψ is the gravitational potential and G is the gravitational constant.

The above set of eleven equations describes the time-evolution of the eleven variables: $(\rho, m, n, \ell, \mathcal{E}_i^d, \mathcal{E}_e^d, E, B_r, B_\theta, B_T, \psi)$, which correspond to density, radial, horizontal and angular momentums, ion and electron internal energies, radiative density, radial, horizontal and toroidal components of the magnetic field and to the gravitational potential, respectively. The equation of state $p = p(\rho, T_{i,e})$ is used to close the system of equations (see Table 2 for further details). In Table (2) we list part of the variables used and their definitions.

Table 2. Variables used and their definitions (see Table 1 for possible scaling variables)

Symbols:

V	$= (V_r, V_\theta, V_\varphi)$	velocity field
B	$= (B_r, B_\theta, B_T) = (B_p, B_T)$	magnetic field
∇	$= (\frac{\partial}{\partial r}, \frac{1}{r}\frac{\partial}{\partial \theta})$	gradient in spherical coordinates
$\nabla \cdot$	$= \frac{1}{r^2}\frac{\partial}{\partial r}r^2 + \frac{1}{r\cos\theta}\frac{\partial}{\partial \theta}\cos\theta$	divergence in spherical coordinates
$T^{e,i}$	=	electron and ion temperatures
$P^{e,i}$	$= \mathcal{R}_{gas}\rho(T_i/\mu_i + T_e/\mu_e))$	electron and ion pressure
$\mathcal{E}^{e,i}$	$= P^{e,i}/(\gamma - 1)$,	electron and ion internal energies
$\kappa^{e,i}$	$= 7.8 \times T_e^{3/2}, 3.2 \times T_i^{3/2}$	electron and ion conductivities (erg cm s^{-1} K$^{-3/2}$)
(m, n, ℓ)	$= \rho(V_r, r V_\theta, r\cos\theta V_\varphi)$	momentum
$\nu(=\eta/\rho), \nu_{mag}$		turbulent and magnetic diffusivities (cm^2 s^{-1})
Φ	$= \Phi_{HD} + \Phi_{MHD}$	HD and MHD turbulent dissipation (J s^{-1}).

Further, the subscripts "i" and "e" correspond to ion and electron plasmas, where $\gamma = 5/3$, $\mu_i = 1.23$ and $\mu_e = 1.14$ are used. α_{dyn}, η_{mag} correspond to the α-dynamo and the magnetic diffusivity, respectively. The radiative diffusion coefficient λ_{FLD} is a radiative flux limiter which forces the radiative flux to adopt the correct form in optically thin and thick regions, i.e.,

$$\nabla \cdot \lambda_{FLD} \nabla E = \begin{cases} \nabla \cdot \frac{1}{3\chi} \nabla E & \text{if} \quad \tau \gg 1 \\ \nabla \cdot nE & \text{if} \quad \tau \ll 1, \end{cases} \quad (10)$$

and provides a smooth matching in the transition regions. Here $\chi = \rho(\kappa_{abs} + \sigma)$ and $n = \nabla E/|\nabla E|$, where κ_{abs} and σ are the absorption and scattering coefficients. Λ_B, Λ_{i-e}, Λ_C, Λ_{syn} correspond to Bremsstrahlung cooling, Coulomb coupling between the ions and electrons, Compton and synchrotron coolings, respectively [33]. These processes read:

$$\Phi = \nu_{mag}|\nabla \times B|^2/\mathcal{N}$$

$$\Lambda_{i-e} = 5.94 \times 10^{-3} n_i n_e c k \frac{(T_i - T_e)}{T_e^{3/2}}/\mathcal{N}$$

$$\Lambda_B = 4ac\kappa_{abs}\rho(T^4 - E)/\mathcal{N}, \quad (11)$$

$$\Lambda_C = 4\sigma n_e c(\frac{k}{m_e c^2})(T_e - T_{rad})E/\mathcal{N},$$

where $\mathcal{N} = [(\gamma - 1)/\gamma](\tilde{V}^2 \tilde{V}_\varphi/\tilde{R})$ is a normalization quantity. n_e, n_i are the electron- and ion-number densities. E is the density of the radiative energy, i.e., the zero-moment of the radiative field. The radiative temperature is defined as $T_{rad} = E^{1/4}$. The Lorenz forces acting on charged plasma in the MHD approximation read:

$$F_L^r = \frac{B_\theta}{r}\frac{\partial B_r}{\partial \theta} - \frac{1}{r}\frac{\partial}{\partial r}r(B_\theta^2 + B_T^2) + \frac{1}{2}\frac{\partial}{\partial r}(B_\theta^2 + B_T^2)$$
$$F_L^\theta = B_r\frac{\partial}{\partial r}rB_\theta - \frac{1}{2}\frac{\partial}{\partial \theta}B_r^2 - [\frac{1}{\cos\theta}\frac{\partial}{\partial \theta}\cos\theta\, B_T^2 - \frac{1}{2}\frac{\partial}{\partial \theta}B_T^2] \quad (12)$$
$$F_L^\varphi = B_p \cdot \nabla \bar{B} = B_r\frac{\partial}{\partial r}(r\cos\theta B_T) + \frac{B_\theta}{r}\frac{\partial}{\partial \theta}(r\cos\theta B_T).$$

The turbulent-diffusive terms read:

$$Q_{vis}^r = \frac{1}{r^2}\frac{\partial}{\partial r}(r^2 T_{rr}) + \frac{1}{r\cos\theta}\frac{\partial}{\partial \theta}(\cos\theta T_{r\theta}) + \frac{T_{rr}}{r}$$
$$Q_{vis}^\theta = \frac{1}{r^2}\frac{\partial}{\partial r}(r^2 T_{r\theta}) + \frac{1}{r\cos\theta}\frac{\partial}{\partial \theta}(\cos\theta T_{\theta\theta}) + T_{\phi\phi}\tan\theta \quad (13)$$
$$Q_{vis}^\phi = \frac{1}{r^2}\frac{\partial}{\partial r}(r^2 T_{r\phi}) + \frac{1}{r\cos\theta}\frac{\partial}{\partial \theta}(r\cos^2\theta T_{\theta\varphi}),$$

where

$$T_{rr} = 2\eta(\frac{\partial V_r}{\partial r} - \frac{1}{3}(\frac{1}{r^2}\frac{\partial r^2 V_r}{\partial r} + \frac{1}{r\cos\theta}\frac{\partial}{\partial \theta}(\cos\theta V_\theta))$$
$$T_{\theta\theta} = 2\eta(\frac{1}{r}\frac{\partial V_\theta}{\partial \theta} + \frac{V_r}{r} - \frac{1}{3}(\frac{1}{r^2}\frac{\partial r^2 V_r}{\partial r} + \frac{1}{r\cos\theta}\frac{\partial}{\partial \theta}(\cos\theta V_\theta))) \quad (14)$$
$$T_{\phi\phi} = 2\eta(\frac{V_\theta}{r}\tan\theta + \frac{V_r}{r} - \frac{1}{3}(\frac{1}{r^2}\frac{\partial r^2 V_r}{\partial r} + \frac{1}{r\cos\theta}\frac{\partial}{\partial \theta}(\cos\theta V_\theta)))$$
$$T_{\theta\phi} = \eta\frac{\cos\theta}{r}\frac{\partial}{\partial \theta}(\frac{V_\varphi}{\cos\theta})$$
$$T_{r\phi} = \eta r\frac{\partial}{\partial r}(\frac{V_\varphi}{r})$$
$$T_{r\theta} = \eta(r\frac{\partial}{\partial r}(\frac{V_\theta}{r}) + \frac{1}{r}\frac{\partial V_r}{\partial \theta}).$$

2.2 The Isotropic Radiation Transfer Equation: The Kompaneets Operator

Radiation transfer is defined as the process of transmission of electro-magnetic radiation through the space. Emitted photons from a faint source generally interact several times with the surrounding medium before they reach the observer. Depending on the density, temperature, strength of the magnetic fields and on the chemical constitution of the these media, the radiative intensity may change significantly. In particular, photons escaping from the last scattering layer, i.e., from the photosphere, generally carry with them sufficient informations about the governing physical processes operating in this layer. For example, electrons gyrating around magnetic fields lines emit cyclotron radiation, which in turn may interact with other energetic electrons. The process by which photons can gain or lose energy by scattering off thermal electrons is called Comptonization. This process is most efficient in unsaturated Comptonization regions where the Compton-Y parameter is of order unity. This parameter acquires large values in optically thick media, and small values in the corona, implying that the corona-disk interaction region and/or the innermost region of the disk are most appropriate for this process to operate efficiently. As a consequence, Comptonization in accretion flows is intrinsically two-dimensional, and therefore requires a multi-dimensional treatment. So far, Comptonization has been considered under strong assumptions that allow separation of variables and lead to the separation of the Kompaneets operator from the radiative transfer equation. Here, the radiative intensity is assumed to be time-independent, isotropic and the plasma is isothermal. In this case, the generation and Comptonization of photons can be described by a second order differential equation in the frequency space [19, 5, 18, 35, 14].

Different accretion models display different spectra. Therefore, it is essential to perform a diagnostic study to analyze their consistency with observations. This however requires solving the 7D radiation transfer equation:

$$\frac{1}{c}\frac{\partial I}{\partial t} + n \cdot \nabla I = \kappa_\nu \rho (S_\nu - I) - \sigma \rho I + \int CI \, d\acute{\Omega} d\acute{E} + \varepsilon_\nu^{\text{mod}}, \qquad (15)$$

where $I = I(t, r, \theta, \varphi, \vartheta, \phi, \nu)$ is the radiative intensity which depends on time t, the spherical coordinates (r, θ, φ), two ordinates (ϑ, ϕ) that determine the direction of the photons on the unit sphere, and on the frequency ν. κ_ν and σ are the absorption and scattering coefficients [30]. S_ν is a source function. $I_{\text{int}} \doteq \int CI \, d\acute{\Omega} d\acute{E}$ describes the scattering of photons through electrons, and C is the scattering kernel. $\varepsilon_\nu^{\text{mod}}$ is the modified synchrotron emission.

To make the problem tractable, the following approximations have been performed [33, 24, 32, 30]:

- The radiation field is axi-symmetric and isotropic, i.e., $\partial/\partial\varphi = 0$ and $J_\nu = \frac{1}{2\pi} \int I d\acute{\Omega} \approx I_\nu$.
- The source function is represented by the modified black body function, i.e.,

$$S_\nu = B_\nu^{\text{mod}} = \frac{2B_\nu}{1 + \sqrt{1 + \frac{\sigma}{\kappa_\nu}}}, \qquad (16)$$

where B_ν is the normal Planck function.
- The thermal energy of the electrons is far below its corresponding rest mass energy, i.e., $\epsilon = \frac{kT}{m_e c^2} \ll 1$, and $\frac{h\nu}{m_e c^2} \ll 1$.

Using the last approximation, I_{int} can be expanded up to second order in ϵ which reduces it to the so-called Kompaneets operator[3] [32]:

---

[3] This operator, in combination with the time-evolution of the non-relativistic photon distribution, forms the so-called Kompaneets equation.

$$I_{\text{int}} \Leftrightarrow \mathcal{K}_\nu = -\frac{\nu}{m_e c^2} \frac{\partial}{\partial \nu}(4kT - h\nu)I + \frac{kT\nu}{m_e c^2} \frac{\partial^2}{\partial \nu^2}(\nu I). \tag{17}$$

In this case, the radiative transfer equation with respect to a rest frame of reference reads:

$$\frac{1}{c}[\frac{\partial E_\nu}{\partial t} + \nabla \cdot V E_\nu] = -\lambda_\nu (\nabla \cdot V) E_\nu + \nabla \cdot [\frac{\lambda_\nu}{\chi_\nu} \nabla E_\nu]$$
$$+ \kappa_\nu \rho (S_\nu - E_\nu) + \mathcal{K}_\nu + \varepsilon_\nu^{\text{mod}}, \tag{18}$$

where $E_\nu (\doteq \frac{4\pi}{c} J_\nu)$, $\chi_\nu (\doteq \rho(\kappa_\nu + \sigma))$ and λ_ν are the frequency-dependent radiative density, mass absorption-scattering coefficient, and the flux limited diffusion coefficient [24], respectively. The latter limiter forces the radiative flux to adopt the correct form in optically thin and thick regions, i.e.,

$$\nabla \cdot [\lambda_\nu \nabla E_\nu] = \begin{cases} \nabla \cdot [\frac{1}{3\chi_\nu} \nabla E_\nu] & \text{if} \quad \tau \gg 1 \\ \nabla \cdot n E_\nu & \text{if} \quad \tau \ll 1. \end{cases} \tag{19}$$

λ_ν may provide a smooth matching between these two extreme regimes. The above two different behaviour of the operator can be combined as follows:

$$\nabla \cdot \lambda_\nu \nabla_\nu E_\nu \hookrightarrow \nabla \cdot \eta_r \nabla E_\nu, \tag{20}$$

where $\eta_r = (1-\alpha)\frac{\nabla E_\nu}{|\nabla E_\nu|} + \alpha \frac{1}{3\chi}$, $\alpha = e^{-R_{\text{FLD}}}$, and $R_{\text{FLD}} = \nabla E / \rho(\kappa_\nu S_\nu + \sigma E_\nu)$ [11].

$\varepsilon_\nu^{\text{mod}}$ in Eq. (15) corresponds to the modified synchrotron emission of photons by relativistic electrons gyrating around magnetic field lines, which reads:

$$\varepsilon_\nu^{mod} = \xi \, \varepsilon_\nu + (1-\xi) \varepsilon_\nu^{BB},$$

where $\xi (\doteq e^{-(\nu_c/\nu)^2})$ is a switch on/off operator which bridges optically thin and thick media to synchrotron radiation, and ν_c is a critical frequency (see below).

An appropriate approximation for ε_ν in optically thin medium reads [25]:

$$\varepsilon_\nu = 2.73 \times 10^{-5} \frac{\rho \nu}{K_2(1/\theta_e)} \tilde{\mathcal{I}}(\nu, B, \Theta), \tag{21}$$

where ε_ν is in ergs cm^{-3} s^{-1} Hz^{-1} units and K_2 is the Bessel function of the second kind and $\tilde{\mathcal{I}} = \frac{4.05}{\zeta^{1/6}}(1 + \frac{0.4}{\zeta^{1/4}} + \frac{0.53}{\zeta^{1/2}})e^{-1.89\zeta^{1/3}}$.

Here $\zeta = 2.38 \times 10^{-7}(\nu/B\theta_e^2)$ and $\theta_e = kT_e/m_e c^2$.

Below a certain critical frequency ν_c, the media become self-absorbing to synchrotron emission. In this case, $\varepsilon_\nu^{\text{mod}} \approx \varepsilon_\nu^{BB} = 2\pi \frac{\nu^2}{c^2} kT$. To find ν_c, we use the local non-linear Newton iteration procedure applied to the equation

$$\int_V \varepsilon_\nu dV = \int_S \varepsilon_\nu^{BB} dS. \tag{22}$$

Having obtained ν_c, the switch on/off operator ξ can then be constructed.

3 Solution Methods

3.1 Solving the Radiative MHD Equations

The set of equations in conservative form may be written in the following vector form:

$$\frac{\partial \mathbf{q}}{\partial t} + L_{r,rr}\mathbf{F} + L_{\theta,\theta\theta}\mathbf{G} = \mathbf{f}, \qquad (23)$$

where $\mathbf{F}$ and $\mathbf{G}$ are fluxes of q, and $L_{r,rr}$, $L_{\theta,\theta\theta}$ are first and second order transport operators that describe advection-diffusion of the vector variables $\mathbf{q}$ in r and θ directions. $\mathbf{f}$ corresponds to the vector of source functions.

Re-writing Eq. (23) in residual form: $R = \frac{\partial \mathbf{q}}{\partial t} + L_{r,rr}\mathbf{F} + L_{\theta,\theta\theta}\mathbf{G} - \mathbf{f} = 0$, and adopting a five star staggered grid discretization (see Fig. 4), we may apply Newton-linearization to calculate the Jacobian, $J_{m1,n1} \doteq \frac{\partial R_{m1}}{\partial q_{n1}}$, where $m1, n1$ are integers that run over the number of equations and variables, respectively. The solution can be obtained then as follows:

$$q^{i+1} = q^i - J^{-1}_{m1,n1} R^i,$$

where i is the iteration level. Inspection of the Jacobian, it can be verified that it has the following block matrix structure:

$$\frac{\delta q_{j,k}}{\delta t} + \underline{S}^r \delta q_{j-1,k} + D^r \delta q_{j,k} + \overline{S}^r \delta q_{j+1,k}$$
$$+ \underline{S}^\theta \delta q_{j,k-1} + D^\theta \delta q_{j,k} + \overline{S}^\theta \delta q_{j,k+1} = RHS^n_{j,k}, \qquad (24)$$

where $\delta q = q^{i+1} - q^i$, and which, in the linear case, reduces to time-difference of q. The subscripts "j" and "k" denote the grid-numbering in the r and θ directions, respectively, and $RHS^n = [\mathbf{f} - L_{r,rr}\mathbf{F} - L_{\theta,\theta\theta}\mathbf{G}]^n$. $\underline{S}^{r,\theta}$ and $\overline{S}^{r,\theta}$ mark the sub-diagonal and super-diagonal matrices, respectively. $D^{r,\theta}$ corresponds to the diagonal block matrices.

To outline the directional dependence of the block matrices, we re-write Eq. (24) in a more compact form:

$$+\underline{S}^r \delta q_{j-1,k} + \begin{matrix} \overline{S}^\theta \delta q_{j,k+1} \\ D_{\text{mod}} \delta q_{j,k} \\ +\underline{S}^\theta \delta q_{j,k-1}, \end{matrix} + \overline{S}^r \delta q_{j+1,k} = RHS^n_{j,k} \qquad (25)$$

where $D_{\text{mod}} = \delta q_{j,k}/\delta t + D^x + D^y$.

This equation gives rise to at least four different types of solution procedures:

(i) Classical explicit methods are very special cases in which the sub- and super-diagonal block matrices together with D^x and D^y are neglected. The only matrix to be retained here is $(1/\delta t) \times$ (the identity matrix), i.e.,

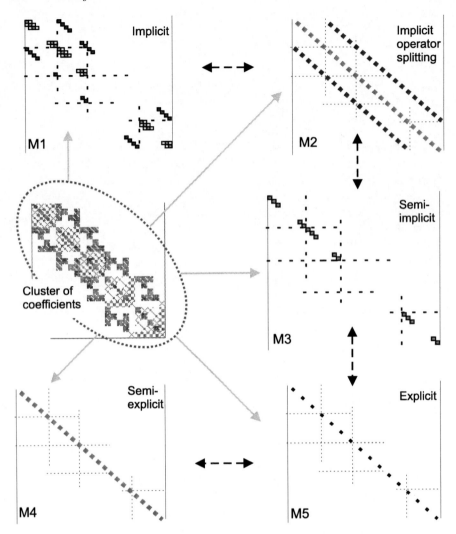

Fig. 3. A schematic description of the hierarchical solution method. A cluster of coefficients is computed in the first stage, and a matrix-generator is created that allows using various solution procedures ranging from purely explicit to fully implicit. Interchange between solution methods is possible, as modifying, adding or removing entries is directly maintainable

the first term on the LHS of Eq. 24. This yields the vector equation (see M5/Fig. 3):

$$[\frac{I}{\delta t}]\delta q_{j,k} = RHS_{j,k}^n. \qquad (26)$$

(ii) Semi-explicit methods are obtained by preserving the diagonal entries, $d_{j,k}$, of the block diagonal matrix D_{mod} (see M4/Fig. 3). This method has

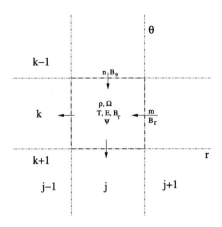

Fig. 4. Scalars, such as density, temperatures, gravitational potential as well as angular frequency ($\doteq \Omega$) and toroidal component of the magnetic field are defined at cell centers. Radial and horizontal components of the velocity and magnetic fields are defined at cell interface. This 5-star staggered grid discretization yields the sub, super and diagonal block matrices of Eq. (25)

been verified to be numerically stable even when large Courant-Friedrich-Levy (CFL) numbers are used. In particular, this method is absolutely stable if the flow is viscous-dominated.

(iii) Semi-implicit methods are recovered when neglecting the sub- and super-diagonal block matrices only, but retaining the block diagonal matrices (see M3/Fig. 3). In this case the matrix equation reads:

$$D_{\text{mod}} \delta q_{j,k} = RHS_{j,k}^n. \qquad (27)$$

We note that inverting D_{mod} is a straightforward procedure, which can be maintained analytically or numerically.

(iv) A fully implicit solution procedure requires retaining all the block matrices on the LHS of Eq. 25. This yields a global matrix that is highly sparse (M1/Fig. 3). In this case, semi-direct methods such as the "Approximate Factorization Method" [-AFM: 2] and the "Line Gauss-Seidel Relaxation Method" [-LGS: 26] are considered to be efficient preconditionings for solving the set of radiative MHD-equations within the context of defect-correction iteration method [see 12, the references therein]. Furthermore, Krylov sub-iterative methods may prove to be more efficient and robust than the above-mentioned semi-direct methods.

In the case that only stationary solutions are sought, convergence to steady state can be accelerated by adopting the so-called "Residual Smoothing Method" [see 17, and the references therein].

This method is based on associating a time step size with the local CFL-number at each grid point. While this strategy is efficient at providing quasi-

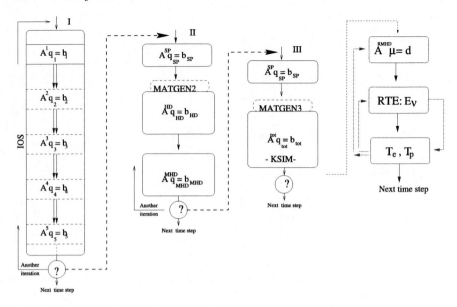

Fig. 5. A schematic description of the hierarchical algorithm for solving the radiative MHD equations. Stage I corresponds to the implicit operator splitting approach (IOS), which is most appropriate for following the early time-dependent phases of the flow. The solution obtained can then be used as initial condition for Stage II, where the hydro-equations are solved as a single coupled system, followed by the magneto component, which is again solved as a single coupled system. Here, high spatial and temporal accuracies in combination with the prolongation/restriction strategy may be used. Similarly, the solution obtained in this stage may be used as starting solutions for Stage III, where steady solutions for the fully coupled set of equations consisting of the zero moment of the radiation field and the MHD equations are sought. In this stage, pre-conditioned Krylov sub-iterative methods are considered to be robust and efficient. The very last stage, Stage IV, corresponds to the case where solutions for the internal energy equations weakly coupled with the 5D radiative transfer equation are sought

stationary solutions within a reasonable number of iterations, it is incapable at providing physically meaningful time scales for features that possess quasi-stationary behaviour. Here we suggest to use the obtained quasi-stationary solutions as initial configuration and re-start the calculations using a uniform and physically relevant time steps.

3.2 The 5D Axi-Symmetric Radiative Transfer Equation: Method of Solution

Let $\mathcal{L}E = 0$ be the equivalent operator form of radiative transfer (RT) Equation (18) in the continuous space Ω_C. $\mathcal{L}E$ consists of several terms, each of which requires a careful and different representation in the finite discretization

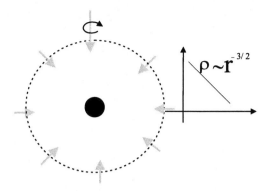

Fig. 6. Spherical symmetric free-fall of gas onto a Schwarzschild black hole. The domain of calculation is limited to the first quadrant only, and consists of 200 × 60 strongly stretched finite volume cells in the radial and horizontal directions, respectively. The advection scheme used here is of first order spatial and temporal accuracies. Symmetry boundary condition along the equator, and anti-symmetric along the axis of rotation are imposed. The flow is set to cross the outer boundary with the free-fall velocity. No specific conditions are imposed at the inner boundary as the flow is inviscid. The density of gas in free-fall obeys the $r^{-3/2}$ power law relation

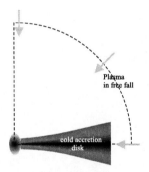

Fig. 7. Shock formation around a black hole. The initial configuration consists of a static disk surrounding a Schwarzschild black hole. Gas in free-fall is set to cross the outer boundary. The domain of calculation is sub-divided into 200 finite volume cells in the radial and horizontal directions, respectively. The advection scheme used here is of third order spatial accuracy and first order in time

space Ω_h which is defined as $[t_1, t_2, \ldots, t_N] \otimes [r_1, r_2, \ldots, r_J] \otimes [\theta_1, \theta_2, \ldots, \theta_K] \otimes [\nu_1, \nu_2, \ldots, \nu_M]$. [t], [r], [$\theta$] and [$\nu$] correspond to time, radius (spherical), latitude, and to the frequency intervals, respectively.

In most astrophysical problems, radiative effects occur on relatively short time scales compared to the hydro- or magneto-hydrodynamical ones, for which the use of unconditionally numerical schemes is essential. This requires however that all terms of Eq. (18) should be evaluated on the new time-level. The discretization used should assure that the resulting Jacobian

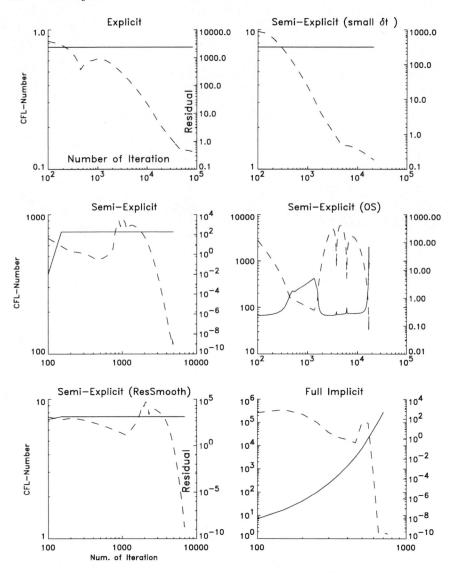

Fig. 8. The problem of free-fall of gas onto a Schwarzschild black hole. The evolution of the CFL-number (solid line) and the residual (dashed line) versus number of iterations are shown, using different solution procedures. The solution methods are: normal explicit (top/left), semi-explicit (middle/left), semi-explicit in combination with the residual smoothing strategy(bottom/left), semi-explicit using moderate CFL-numbers (top/right), semi-explicit method in which the time step size is taken to be a function of the maximum residual (middle/right), and finally the fully implicit method (bottom/right). The different forms of the semi-explicit method used here are stable and converges to the stationary solution, though at remarkably different rates

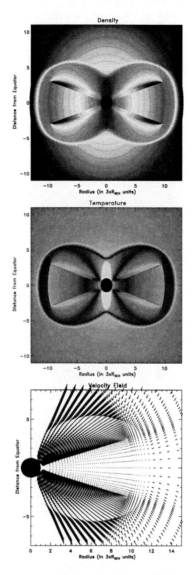

Fig. 9. Free-fall of gas onto a black hole surrounded by a static cold disk. Top: the 2D density distribution. In this figure, the color gradient is as follows: red color corresponds to large density-values, green to intermediate and blue to low values. In the gray case: light color corresponds to large values and dark to low values. In the middle figure we display the 2D distribution of the temperature. Here red color corresponds to large values, green to intermediate values and the blue to low values. In the gray case: dark color corresponds to large values and light to low values. The curved shock front, where the temperature attains maxima is obvious. Bottom: the distribution of the velocity field is shown

$A_{r\theta\nu} = \partial \mathcal{L}E/\partial E$ is diagonally dominant. Therefore, the following procedures are employed.

- The advection term $\nabla \cdot VE_\nu$ is discretized using a second order up-winding.
- The second order diffusion term $\nabla \cdot [\lambda_\nu \nabla E_\nu]$ is discretized using second order central-difference scheme on a staggered grid
- $\mathcal{K}_\nu$ contains advection and diffusion terms in the frequency space. Here up-winding discretization in the frequency space is used.

Note that the terms in these three items are functions of the velocity field, density and temperature. They depend also on the magnetic field strength indirectly through source terms (see Eq. 21), coupling thereby the RT with the other radiative MHD equations. To make the problem tractable, we may treat these extra-variables as constants temporarily, but update them by solving radiative MHD equations in an iterative manner.

Combining the contributions of all terms of Eq. (18), we obtain at each grid point the following equation:

$$\underline{S}^r E^{new}_{j-1,k,m} + \overline{S}^r E^{new}_{j+1,k,m} + \underline{S}^\theta E^{new}_{j,k-1,m}$$
$$+ \overline{S}^\theta E^{new}_{j,k+1,m} + \underline{S}^\nu E^{new}_{j,k,m-1} + \overline{S}^\nu E^{new}_{j,k,m+1}$$
$$+ (D^r + D^\theta + D^\nu) E^{new}_{j,k,m} = RHS, \tag{28}$$

where $\underline{S}^r = \partial \mathcal{L} E_\nu / \partial E_{j-1,k,m}$, $\overline{S}^r = \partial \mathcal{L} E_\nu / \partial E_{j+1,k,m}$, $D^r + D^\theta + D^\nu = \partial \mathcal{L} E_\nu / \partial E_{j,k,m}$, $\underline{S}^\theta = \partial \mathcal{L} E_\nu / \partial E_{j,k-1,m}$, $\overline{S}^\theta = \partial \mathcal{L} E_\nu / \partial E_{j,k+1,m}$, $\overline{S}^\nu = \partial \mathcal{L} E_\nu / \partial E_{j,k,m+1}$ and $\underline{S}^\nu = \partial \mathcal{L} E_\nu / \partial E_{j,k,m-1}$. The terms $\underline{S}^r, D^r$, and $\overline{S}^r$ correspond to the sub-diagonal, diagonal and super-diagonal entries of the Jacobian $A_{r\theta\nu}$ in the radial direction respectively. A similar description applies to the θ- and ν-directions.

Thus, solving the equation at all grid points, is equivalent to solve matrix equation: $A_{r\theta\nu} E^{new} = E^{old}$, or simply, $Aq = b$.

We note that the distributions of the velocity and magnetic fields, density and temperature are explicitly incorporated in the construction of the above-mentioned sub, super, and diagonal elements. During the inversion procedure of $A_{r\theta\nu}$, we may treat these extra-variables as constants temporarily.

This matrix is highly sparse, and pre-conditionings such as the Alternating Direction Implicit (ADI) and the Approximate Factorization Methods (AFM) are considered to be efficient. However, ADI is not appropriate for searching steady solution in three or more dimensions, as it is numerically unstable in high dimensions [6]. Alternatively, we have tried the AFM as a pre-conditioner. However, it turns out that the AFM converges slower than our favorite iterative method: 'Black-White-Brown' line Gauss-Seidel method [henceforth BWB-LGS, see 10, for further details]. The latter method preserves the diagonal dominance of A, and hence converges faster than AFM. It should be noted that the line Gauss-Seidel method in its classical form is

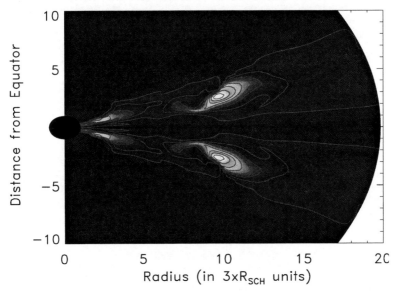

Fig. 10. The distribution of the Bernoulli number of accretion flows around a supermassive black hole. The Bernoulli number characterizes the energies of the flow in different regions. Gravitationally bound flows have negative total energy, whereas flows of positive total energy are gravitationally unbound, and potentially should expand to infinity. In this figure, the color gradient is as follows: yellow corresponds to large positive values, red to intermediate and blue to negative values. In the gray case: light color corresponds to large positive values and dark to negative values. Obviously, gravitationally unbound blobs are formed in the vicinity of the black hole, which thereafter collimate under the action of magnetic fields to form the observed highly collimated jets

not appropriate for vector and parallel machines, mainly because the vector-length is proportional to the number of unknowns in one direction. A reasonable way to extend the vector-length is to solve for all unknowns located on even-numbered grid points, and subsequently on odd-numbered grid points. The resulting vector-length in this case is proportional to the number of unknowns in the plane under consideration, and therefore enabling enhancement efficiency when using vector or parallel machines.

More specifically, in each plane we perform two sweeps: in the first sweep we consider the unknowns in the $r - \theta$ plane, i.e., we solve the system of equations:

$$\underline{S}^r \delta E^{\text{new}}_{j-1,k,m} + (D^r + D^\theta + D^\nu)\delta E^{\text{new}}_{j,k,m} + \overline{S}^r \delta E^{\text{new}}_{j+1,k,m} = RHS,$$

where $j = 1 \to J$ and k runs over odd-numbered rows. In the second sweep, we solve:

$$\underline{S}^r \delta E^{\text{new}}_{j-1,k,m} + (D^r + D^\theta + D^\nu)\delta E^{\text{new}}_{j,k,m} + \overline{S}^r \delta E^{\text{new}}_{j+1,k,m}$$
$$= RHS + \underline{S}^\theta \delta E^{\text{new}}_{j,k-1,m} + \overline{S}^\theta \delta E^{\text{new}}_{j,k+1,m},$$

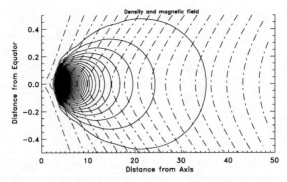

Fig. 11. The initial distribution of the density in an accretion disk ($\dot{\mathcal{M}} = 0.1\,\dot{\mathcal{M}}_{\mathrm{Edd}}$) around a Schwarzschild black hole overlied by coronal plasmas (solid lines). The dashed lines correspond to the magnetic field lines threading the disk and the corona. The distance is given in units of 2.75 R_{Sch}, where R_{Sch} is the Schwarzschild radius

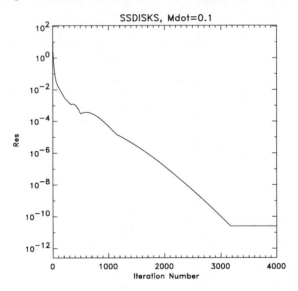

Fig. 12. The evolution of the residual in the maximum norm versus the number of iteration. The adopted density and temperature profiles correspond to standard accretion disks

where $j = 1 \rightarrow J$ and k here runs over even-numbered rows. Therefore, we actually perform 6-inversion procedures per each time step. Here the 3-dimensional problem is replaced by three one-dimensional problems that are solved iteratively to recover the solution of the original problem. The method is relatively efficient, as the overall number of arithmetic operations scales linearly with the number of grid points ($\sim 6 \times 9 \times N$). Fig. 11 and 12 show the strong convergence of the method applied to a cold accretion disk surrounding a Schwarzschild black hole.

4 Validation and Preliminary Tests

4.1 Free-Fall of Plasma onto a Schwarzschild Black Hole

A centrifugally-unsupported gas around a spinless black hole is gravitationally bound, and therefore should fall-freely onto the black hole, provided that no other external forces oppose gravity. In this case, the radial distributions of the density and velocity far from the event horizon obey the power laws: $r^{-3/2}$ and $r^{-1/2}$, respectively (see Fig. 6).

This physical problem is relevant for testing the flexibility of the hierarchical scenario at adopting various solution methods, and to test their capability to capture steady, oscillation-free and advection-dominated flows, even when a strongly stretched mesh distribution is used.

The equations to be solved in this problem are the continuity, the radial and horizontal momentum equations, and the internal energy equation. The flow is assumed to be inviscid and adiabatic ($\gamma = 5/3$). The equations have been solved using a first order accurate advection scheme both in space and time. In carrying out these calculations, the following conditions/inputs have been taken into account:

- The central object is a one solar-mass and non-rotating black hole.
- The outer boundary is 100 times larger than the inner radius, i.e., $R_{\text{out}} = 100 \times R_{\text{in}}$, where R_{in} is taken to be the radius of the last stable orbit[4] R_{LS}. To first order in V/c, the flow at this radius can be still treated as non-relativistic, though the error can be as large as 30%.
- Along the outer boundary, the density and temperature of the gas assume uniform distributions, and flow across this boundary with the free-fall velocity. Symmetry boundary conditions along the equator, and asymmetry boundary conditions along the axis of rotation have been imposed. Along the inner boundary, we have imposed non-reflecting and outflow conditions. This means that up-stream conditions are imposed, which forbid information exterior to the boundary to penetrate into the domain of calculations. In particular, the actual values of the density, temperature and momentum in the ghost zone r are erased and replaced by the corresponding values in the last zone, i.e, the zone between R_{in} and $R_{\text{in}} + \Delta R$. In the case that second order viscous operators are considered, care has been taken to assure that their first order derivatives across R_{in} are vanished.

The above set of equations are solved in the first quadrant $[1 \leq r \leq 100] \times [0 \leq \theta \leq \pi/2]$, where 200 strongly stretched finite volume cells in the radial direction and 60 in the horizontal direction are used. The advection scheme employed here is of first order spatial and temporal accuracy.

---

[4] $R_{\text{LS}} = 3 \times R_{\text{S}} = 6 \times R_{\text{g}}$, where R_{S} and R_{g} are the Schwarzschild and gravitational radii, respectively.

In Fig. 8, we show the evolutions of the CFL-number [5] and the residual as function of the number of iteration which has been obtained using various numerical approaches. The CFL-number here is set to increase in a well-prescribed manner that strongly depends on the residual. Specifically, the CFL-number may increases with decreasing the residual and vice versa.

Figure (8) shows that the convergence of the explicit and semi-explicit methods are rather slow when a relatively small time step size is used. This implies that the amplitude-limited oscillations are strongly time-dependent that may result from geometric compression. Indeed, these perturbations disappear, when relatively large time-step sizes are used (see Fig. 4, bottom/right).

In addition, the semi-explicit solver has been tested in combination with the residual smoothing strategy. As expected, this approach accelerates the convergence considerably (Fig. 4: compare the plots bottom/left with the top/right).

It is obvious from Fig. 8 that determining the size of the time step from the residual directly did not provide satisfactory convergence histories (Fig. 4, middle/right).

The results obtained here indicate that the semi-explicit method is stable and can be applied to search for stationary solutions using large time steps, or equivalently, CFL-numbers that are significantly larger than unity (Fig. 8, middle/left).

In these calculations, although the effective time reached in each of these runs is similar, the actual number of iterations performed is substantially different. Specifically, since the CPU time scales linearly with the number of computing operations performed, hence with the total computational costs (CC), the following correlation was found:

$$CC_{FI} < CC_{ERS} < CC_{Explicit} < CC_{SE\_OS},$$

where the subscripts have the following meanings: FI= fully implicit, ERS= explicit with residual smoothing and SE_OS= Semi-explicit method in which the time step is a function of the residual.

4.2 Shock Formation Around Black Holes

Similar to the forward facing step in CFD, a cold and dense disk has been placed in the innermost equatorial region: $[1 \leq r \leq 10] \times [-0.3 \leq \theta \leq 0.3]$ (see Fig. 7). We use the same parameters, initial and boundary condition as in the previous flow problem. A vanishing in- and out-flow conditions have been imposed at the boundaries of this disk. The gas surrounding the disk is taken to be inviscid, thin, hot and non-rotating. Thus, the flow configuration is similar to the forward facing step problem usually used for test calculations in CFD. The disk here serves as a barrier that forbids the gas from freely falling

---

[5]This is the Courant-Friedrich-Levy number which results from the von Neumann stability analysis of normal time explicit schemes [see 10, for further details].

onto the black hole, and instead, it forms a curved shock front around the cold disk. The purpose of this test is mainly to examine the capability of the hierarchical scenario at employing the semi-explicit method adequately and enables capturing steady solution governed by strong shocks. In solving the HD-equations, an advection scheme of third order spatial accuracy and of first order accuracy in time has been used. The domain of calculation is sub-divided into 200 strongly-stretched finite volume cells in the radial direction and 60 in horizontal direction. In Fig. 9 the configuration of the steady distributions of the density, temperature and the velocity field are shown. Similar to the calculation in the previous sub-section, the results indicate that the method employed is stable and converges to the sought steady solution even when a CFL-number of order 200 is used. However, the method converges relatively slowly compared to the implicit operator splitting approach, where steady solutions have been obtained after one thousand iterations only.

4.3 Formation and Acceleration of Proton-Dominated Jets in Active Galaxies

To study the mechanisms underlying jet formation around black holes (see Fig. 13), we have placed initially a classical accretion disk within the first 20 last stable radii, sandwiched by a hot and tenuous corona, and threaded by a large scale magnetic field. The solution procedure run as follows:

(i) The HD-equations are solved using the IOS-approach as depicted in Stage I of Fig. 5. The calculations were run to cover the viscous time scale.
(ii) Using the obtained results from the previous stage as starting conditions, Stage II of the global solution procedure is now employed to run the calculations for an additional viscous time scale. Here, the HD and the

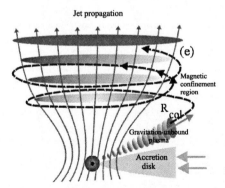

Fig. 13. Jet formation around a black hole. The initial configuration consists of a dynamical accretion disk surrounding a Schwarzschild black hole, and which is threaded by large scale magnetic fields. The domain is sub-divided into 200 × 100 finite volume cells in the radial and horizontal directions, respectively. The advection scheme used here is of second order spatial accuracy and second order in time

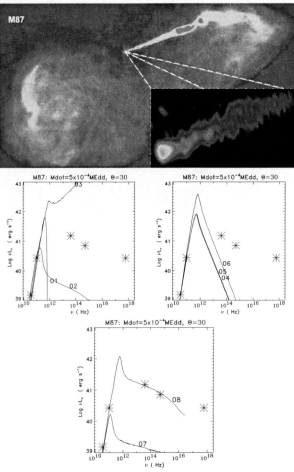

Fig. 14. A VLA radio flux image which shows the active central engine of the giant elliptical galaxy M87 (top; light color corresponds to large values and dark to low values), and a NRAO radio image of the jet apparently emanating from within 100 gravitational radii. In the following plots, the solid lines correspond to calculated profiles and the asterisks to observational data. The profiles 01 to 06 show the spectral energy distribution calculated using different magnetic field strengths, or different truncation radii, or high/low corona temperatures. In particular, the profile 07 corresponds to a model in which the toroidal magnetic field is set to vanish artificially, whereas the poloidal magnetic field is set to be in equipartition with the thermal energy of the electrons. The profile 08 is similar to 07, except that the toroidal magnetic field is allowed to develop and reach values beyond equipartition with respect to the thermal energy of the electrons in the transition layer between the disk and the overlying corona. The above spectral energy distribution has been obtained by solving the radiative transfer equation in 5-dimensions, taking into account the Kompaneets operator for consistently modeling Comptonization. 400 non-linearly distributed frequency points have been used to cover the frequency-space, and 125×40 finite volume cells to cover the spatial domain of the calculation

MHD equations are solved in a blockwise manner as described in Fig. 5. Stage III was not employed, as Alfvén-waves propagation enhances the time-dependency of the flow even more.

(iii) The final flow-configuration apparently governed by inflow and outflow plasmas. In general, outflows are gravitationally unbound, and therefore the corresponding Bernoulli number should be positive, whereas negative numbers correspond to gravitationally bound flows that should end their motion inside the black hole. Fig. 10 shows the 2D distribution of the Bernoulli number which shows the locations of the gravitationally-bound and unbound flows. The advection scheme employed here is of third order spatial accuracy and second order accurate in time.

4.4 The Spectral Energy Distribution of the in- and outflow Around The Supermassive Black Hole of the Giant Elliptical Galaxy M87

The results obtained in the previous subsection are used to construct the spectral energy distribution. Therefore, the last stage of the hierarchical scenario is now employed in combination with Stage II. Here, the solver of Stage II is activated once every several dozens iterations of the RT-solver. The accuracy of the advection scheme used here is temporally of first order and spatially of second order.

In Fig. 14 we display the results of several calculations under various conditions. The results displayed in Fig. 14 are preliminary, as the distributions of the density and temperature used here are artificial, but aimed at testing the convergence of the RT-solver.

5 The Combined Solution Procedure: The Hierarchical Scenario

In the following we describe the main steps of a possible algorithmic procedure for solving the combined set of MHD and the RT equations (see Stages III and IV of Fig. 5):

(i) Compute the RHS_i and the Jacobian $A_i = \partial Lq_i/\partial q_i$ of each physical variable $q_i (= \rho, m, n, \ldots)$, where Lq_i is the equation describing the evolution of variable q_i.
(ii) For each equation Lq_i, compute the coefficient matrices $B_i = \partial Lq_i/\partial q_j$, for which $i \neq j$. This procedure applies for advection and diffusive operators only, though not for the source terms.
(iii) Compute the coefficient matrices corresponding to the source terms only, i.e., $H_i = \partial Lq_i/\partial q_j$, for $i = 1, N$ and $j = 1, N$, and $i \neq j$.

The separation of the above-mentioned procedures is essential for enhancing the global efficiency of the hierarchical method. Specifically, the computation of each of the B_i and H_i is optional, depending on the problem in hand. For example, to solve the system of equations corresponding to the hydrodynamical and isothermal flow in 1D efficiently, the numerical algorithm should be capable of calling the relevant routines only. Thus, non-relevant routines can be switched off almost automatically, depending on the problem in hand. In particular, enlarging (reducing) dimensions, incorporating additional (excluding) variable should be algorithmically maintainable.

Taking into account that most astrophysical flows are of multi-scale by nature, we think that the hierarchical solution strategy might be a promising approach. In the following, we describe briefly the basis of this hierarchical scenario applied to set of radiative MHD and the RT equations.

(i) The hierarchical approach, or equivalently the multi-stage solution procedure, is based primarily on designing the global solver in such a manner to achieve maximum flexibility. Specifically, the numerical algorithm should be capable of solving the equations sequentially, block-sequential and/or in a fully-coupled manner. Re-ordering and using different pre-conditioning should be maintainable without changing the core of the inverter.

(ii) As far as vortex-free compressible, viscous and time-dependent flows are concerned, the implicit operator splitting approach (IOS) has been verified to be efficient and robust. IOS is most appropriate for astrophysical fluid simulations, when the sought solutions depend weakly on the initial conditions, but strongly on the boundary conditions. The IOS-method is based on solving the set of equations sequentially as described in Stage I of Fig. 5. The convergence rate of the IOS-method may depend considerably on the order in which the equations are solved, provided the number of global iterations is low.

(iii) The coupling between the equations can be enhanced gradually. From the cluster of coefficients, we may construct the Jacobians A^{HD} and A^{MHD}, which correspond to the set of HD and MHD equations (see Stage II/Fig. 5). Algorithmically, this procedure is basically a sort of re-ordering and re-organizing of the coefficients, and does not require an extensive programming. As in the previous step, the order in which the equations are solved may affect both its convergence rate and efficiency. Here, a special care should be given to assure that the inclusion of coefficients corresponding to the source terms does not enlarge the band width of A^{HD} and A^{MHD}. Test calculations have shown that careful ordering of the HD-equations may reduce the computational costs devoted for matrix inversion by 75% [12]. Furthermore, it has been verified that several equations can still be separated and solved sequentially. Namely, the Possion equation for modeling self-gravity as well as the angular momentum equation accept partial decoupling from the rest of equations, provided the flow is axi-symmetric.

(iv) Using the solutions obtained in stage II as initial conditions, we may solve the whole set of HD and MHD equations as a single set of coupled equations. The resulting Jacobian is highly sparse, for which pre-conditioned Krylov sub-iterative methods are highly appropriate.

(v) In order to solve the RT-equation consistently, the effects of advection (the LHS of Eq. 18), Synchrotron emission (see Eq. 21), collision of matter and radiation (Eq. 19) and Comptonization (Eq. 17) must be taken into account. Therefore, the distributions of the velocity and magnetic fields, density and temperature are required. These, however, can be obtained by solving the set of radiative MHD equations as described in Stage II or III of Fig. 5, and which are subsequently used as input distributions in Stage IV. We may repeat these updates in order to enhance the coupling between the RT-equation and the radiative MHD-equations. In this case, E_ν can be used to calculate the mean value of the frequency-independent radiative density:

$$E = < E_\nu > = \frac{\int \nu E_\nu d\nu}{\int \nu d\nu}.$$

E may be used then to update the RHS of Equations (2), (3), (5), (6) and (7).

By iterating over Stage II and IV, we can be sure that the resulting solution is reasonably close to sought quasi-stationary or steady solutions for the radiative MHD and radiative transfer equations. This is a consequence of:

(a) The radiative intensity in the high density regions, where the optical thickness is large, is isotropic and coincides with black-body emission. Therefore, the intensity obtained by solving the zero moment of the radiation field is sufficiently accurate in this regime.

(b) The radiative intensity obtained by solving the RT-equation in optically thin regions may differ considerably from that obtained using the gray approximation. However, radiation in such regions have negligible power and they may hardly affect the dynamics of the flow.

Consequently, the following solution method may be proposed:
- The numerical values of the variables obtained in Stage III are used as initial conditions for calculating the non-gray and time-dependent radiative intensity.
- The mean-value of the frequency-dependent intensity is computed and subsequently used as initial condition for the radiative MHD equations.
- To avoid extensive computational costs, it is suggested to solve for I_ν every 10, or 20 time-steps. However, since the radiative time-scale is extremely short compared to the hydrodynamical time scale, it is much more reasonable to solve for the time-independent intensity.

6 Summary

In this paper we have presented the hierarchical scenario for solving the set of radiative MHD equations and the 5D axi-symmetric radiative transfer equation.

The main features of this scenario are as follows:

(i) The global efficiency can be enhanced, depending on the optimal architecture of the global solver. Specifically, the algorithmic structure should be sufficiently flexible, so that scalar or set of equations in arbitrary dimensions, different accuracies and using the appropriate pre-conditionings can be solved with a reasonable efficiency.

(ii) Robustness is monitored through employing a variety of solution procedures. Depending on the particular features of the problem considered, several stages of implicitness may be used, depending on the number of coefficients used for constructing the coefficient matrix. In particular, starting with a purely explicit time-stepping scheme, the algorithm should be capable of modifying the scheme into a fully implicit method dynamically.

(iii) For implicit calculations, the hierarchical algorithm relies on using a variety of preconditioning for accelerating convergence. For example, for modeling weakly incompressible flows, it has been verified that the "Approximate Factorization Method" as pre-conditioning yields a larger convergence rate than the "Alternating Directional Implicit" or the "Line Gauss-Seidel" methods. However, the latter preconditionings provide faster convergence if the flow is compressible and advection-dominated. Therefore, depending on the problem in hand, the algorithm should be capable of employing the appropriate preconditioning at least in an explicit-adaptive manner.

(iv) The hierarchical algorithm is capable of solving the angle-averaged time-dependent radiation transfer equation, taking into account the Kompaneets operator for modeling up-scattering of soft photons by hot electrons in magnetized plasmas.

We note, however, that the assumption of isotropic radiative intensity may break down if the flow is relativistic and contains regions of significantly different optical depths. Therefore, in the near future we intend to modify the RT-solver to enable modeling the motions of ultra-relativistic plasmas in the vicinities Kerr and Schwarzschild black holes.

(v) The algorithm includes a procedure that allows solving the zero-moment MHD equations partially/loosely coupled with the radiation transfer equations. The latter coupling can be significantly enhanced through parallelization on powerful machines.

Finally, we have shown that the hierarchical algorithm presented here can be applied to study the mechanisms underlying the formation, launching and acceleration of jets in AGNs and quasars, though serious numerical and physical modifications are still required.

Acknowledgement. This research has been supported by the Deutsche Forschungsgemeinschaft (RA 306/14-1). I thank the Specialist Editor, Prof. Schoenauer, for his constructive and valuable suggestions, which significantly improved the clarity and readability of the paper. Thanks go to Dr. Spindeldreher for his carefully reading the paper.

References

[1] Balbus, S., Hawley, J., 1991, ApJ, 376
[2] Beam, R.M., Warming, R.F., 1978, AIAA, 16, 393
[3] Brandt, A., 2001, in "Multigrid", ed.: Trottenberg, U., Oosterlee, C., Schüller, A., Acad. Press, London
[4] Dongarra, J., Duff, I., Sorensen, D., van der Vorst, H.A., 1998, "Num. Linear Alg. for High-Performance Computers", SIAM, Philadelphia
[5] Felten, J.E., & Rees, M.J., 1972, "Transfer effects on X-Ray lines in optically thick sources", A&A, **21**, 139-150
[6] Fletcher, C.A.J., 1988, "Computational Techniques for Fluid Dynamics", Vol, I and II, Springer-Verlag
[7] Font, J.A., 2000, "Numerical Hydrodynamics in General Relativity", Living Rev. Relativity, **3**, 1-81
[8] Fryxell, B., Olson, K., Ricker, P., et al., 2000, "FLASH: An Adaptive Mesh Hydrodynamics Code for Modeling Astrophysical Thermonuclear Flashes", ApJS, **131**, 273-334
[9] Gammie, C.F., McKinney, J.C., Tóth, G., 2003, "HARM: A Numerical Scheme for General Relativistic Magnetohydrodynamics", ApJ, **589**, 444-457
[10] Hirsch, C., 1990, "Num. Computation of Internal and External Flows", Vol. I, and II, John Wiley & Sons, New York
[11] Hujeirat, A., Papaloizou, J.C.P., 1998, "Shock formation in accretion columns – a 2D radiative MHD approach", A&A, **340**, 593-604
[12] Hujeirat, A., Rannacher, R., 2001, "On the efficiency and robustness of implicit methods in computational astrophysics", NewAR, **45**, 425-447
[13] Hujeirat, A., Camenzind, M., Livio, M., 2002, "Ion-dominated plasma and the origin of jets in quasars", A&A, **394**, L9-L13
[14] Hujeirat, A., Camenzind, M., Burkert, A., 2002b, "Comptonization and synchrotron emission in 2D accretion flows. I. A new numerical solver for the Kompaneets equation", A&A, **386**, 757-762
[15] Hujeirat, A., Livio, M., Camenzind, M., Burkert, A., 2003, "A model for the jet-disk connection in BH accreting systems", A&A, **408**, 415-430
[16] Hujeirat, A., 2004, "A model for electromagnetic extraction of rotational energy and formation of accretion-powered jets in radio galaxies", A&A, **416**, 423-435
[17] Hujeirat, A., 2004, "A method for enhancing the stability and robustness of explicit schemes in CFD", New Astronomy Reviews, Vol. 2, Issue 3, 173-193
[18] Katz, J.A., 1976, "Nonrelativistic Compton scattering and models of quasars", ApJ, **206**, 910-916

[19] Iilarinov, A.F., & Sunyaev, R.A., 1972, "Compton scattering by thermal electrons in X-ray sources," Soviet Astr. -AJ, **16**, 45
[20] Kley, W., 1989, "Radiation hydrodynamics of the boundary layer in accretion disks. I – Numerical methods", A&A, **208**, 98-110
[21] Koide, S., Shibata, K., & Kudoh, T., 1999, "Relativistic Jet Formation from Black Hole Magnetized Accretion Disks: Method, Tests, and Applications of a General Relativistic Magnetohydrodynamic Numerical Code", ApJ, **522**, 727-752
[22] Koide, S., Shibata, K., Kudoh, T., & Meier, D.L., 2002, "Extraction of Black Hole Rotational Energy by a Magnetic Field and the Formation of Relativistic Jets", Science, **195**, 1688-1691
[23] Komissarov, S.S., 1999, "A Godunov-type scheme for relativistic magnetohydrodynamics", MNRAS, **303**, 343-366
[24] Levermore, C.D., & Pomraning, G.C., 1981, "A flux-limited diffusion theory", ApJ, **248**, 321-334
[25] Mahadevan, R., Narayan, R., & Yi, I., 1996, " Harmony of electrons: Cyclotron and Synchrotron emission by thermal electrons in magnetic fields", ApJ, **465**, 327-337
[26] MacCormack, R.W., 1985, "Current status of numerical solutions of Navier-Stokes equations", AIAA, Paper 81-0110, 1-18
[27] Martí, J.M., Müller, E., 1999, "Numerical hydrodynamics in special relativity", Living Rev. Relativity, **2**, 1-100
[28] Meier, D.L., Koide, S., & Uchida, Y., 2001, "Magnetohydrodynamic Production of Relativistic Jets", Science, **291**, 84-92
[29] Meier, D., 2003, "The theory and simulation of relativistic jet formation: towards a unified model for micro- and macroquasars", NewAR, **47**, 667-672
[30] Mihalas, D., Mihalas, B.W., 1984, "Foundations of radiation hydrodynamics", Oxford University Press, NY (MM)
[31] Ouyed, R., Pudritz, R., 1997, "Numerical simulation of astrophysical jets from Keplerian disks. II. episodic outflows", ApJ, **484**, 794-809
[32] Payne, D.G., 1980, "Time-dependent Comptonization – X-ray reverberations", ApJ, **237**, 951-963
[33] Rybiki, G.B., & Lightman, A.P., 1979, "Radiation Processes", Wiley-Interscience Publication
[34] Saad, Y., van der Vorst, H.A., 2000, "Iterative solution of linear systems in the 20-th century", J. of Comp. and Appl. Math., **123**, 1-33
[35] Shapiro, S.L., Lightman, A.P., & Eardley, D.M., 1976, "A two-temperature accretion disk model for Cygnus X-1 structure and spectrum", ApJ, **204**, 187-199
[36] Stone, J.M., Norman, M., 1992, "ZEUS-2D: A radiation magnetohydrodynamics code for astrophysical flows in two space dimensions. I – The hydrodynamic algorithms and tests", ApJS, **80**, 791-818
[37] Tóth, Keppens, R., Botchev, M.A., 1998, "Implicit and semi-implicit schemes in the Versatile Advection Code: numerical tests", A&A, **332**, 1159-1170
[38] Trottenberg, U., 2001, in Multigrid, ed.: Trottenberg, U., Oosterlee, C., Schüller, A., Acad. Press, London
[39] Uchida, Y., Nakamura, M., Hirose, S., Uemura, S., 1999, "Magnetodynamic formation of jets in accretion process of magnetized mass onto the central gravitator", Ap&SS, **264**, 195-212

[40] De Villiers, J.-P., & Hawley, J.F., 2003, "A Numerical Method for General Relativistic Magnetohydrodynamics", ApJ, **589**, 458-480

[41] Ziegler, U., 1998, "NIRVANA+: An adaptive mesh refinement code for gas dynamics and MHD", Comp. Phys. Comm., **109**, 111-123

Rapidly-Converging Methods for Solving Multilevel Transfer Problems

Eugene H. Avrett

Harvard-Smithsonian Center for Astrophysics, 60 Garden Street, Cambridge, MA 02138, USA
avrett@cfa.harvard.edu

Summary. It is well known that lambda iterations can be used to solve multilevel non-LTE transfer equations in a reasonable number of iterations when the lambda operator is preconditioned, e.g., when the diagonal part of the operator is combined with other terms analytically. This approach is currently used successfully for the solution of model atoms with many line transitions, but sometimes a very large number of iterations is needed.

Lambda iteration consists of alternate solutions of the separate transfer and rate equations. For any given line transition the transfer and rate equations can be combined so that a solution can be obtained directly for that transition with no iterations needed between the transfer and rate equations. However, iterations are needed to determine the coupling between transitions. This can be time-consuming for model atoms with a large number of transitions that are treated in this way.

Here we show that 1) a hybrid approach involving such a direct solution for a few of the strongest transitions, and lambda iterations for the rest, gives rapid convergence, often with oscillations that need to be damped, and 2) this approach should include preconditioning of the lambda operator that occurs in the radiative coupling terms.

We illustrate these results with a simple three-level hydrogen atom and a finite, plane-parallel, symmetric atmosphere resembling a solar prominence, with a temperature of 8,000 K at the center, rising to very large values at each boundary (so that hydrogen is only partly ionized at the center and fully ionized at each boundary). Lambda iterations essentially fail to give a solution for this problem, while the hydrid solution converges in 5 to 10 iterations.

1 Introduction

The fundamental problem to be solved for low-density optically thick atmospheres where local thermodynamic equilibrium cannot be assumed is the simultaneous solution of the rate and transfer equations. The rate equations are used to determine the number densities of various energy levels at any location in the atmosphere given the angle-averaged radiation intensities at that location. The transfer equation is used to determine the radiation intensity at any location and direction given the number densities along a line extending from that location in the opposite direction. For simple problems the rate and transfer equations can be combined and solved directly, but this approach becomes impractical when the number of energy levels and radiative transitions is very large, because the coupling between transitions depends on the solution and must be treated iteratively.

Rather than solving the combined rate and transfer equations, one can iterate between the two. The transfer equation can be solved for the angle- and frequency-integrated mean intensity $\overline{J}$ at each depth given starting values of the number densities throughout the atmosphere. This $\overline{J}$ then can be used in the rate equations to obtain new number densities everywhere, replacing the starting values. These updated number densities can be used to obtain a new $\overline{J}$, etc. This is called the lambda iteration method. (The lambda operator is the function used to determine the mean intensity at one depth in terms of source-function values throughout the atmosphere.) Such a direct lambda-iteration method converges too slowly to be practical, but various techniques have been developed to achieve convergence in a reasonable number of iterations.

A comprehensive review of this topic has been provided by Hubeny (2003). The term "Accelerated Lambda Iteration" (ALI) ordinarily refers to removing the diagonal elements (and sometimes principal off-diagonal elements) from the lambda operator and combining these elements analytically with other terms. "Acceleration" thus refers to the faster convergence that results. Typically the convergence is still rather slow, and monotonic, so that purely numerical acceleration techniques can be applied as well.

Here we formulate the ALI method in an equivalent way but with slightly different terminology than used in Hubeny (2003) and in earlier papers. Instead of using the term "acceleration" we refer to the special treatment of diagonal elements as "preconditioning".

We use the simple case of a two-level atom to illustrate the much faster convergence that results from preconditioning, and compare with a direct solution that requires no iteration.

For a three-level case we show how to obtain a direct solution for each transition, and how to derive the coupling terms that relate each transition to the others. We use preconditioning in these coupling terms. Only the strongest transitions need to be solved directly in this way. Preconditioned lambda iterations can be used for weaker transitions. We call this a hybrid approach.

Finally, we show an example for which the ALI method essentially fails, while such a hybrid method converges in 5 to 10 iterations.

2 The Radiative Transfer and Rate Equations

The rate equation for level m of an $\mathcal{N}$-level atom with a continuum is

$$\frac{\partial n_m}{\partial t} + \nabla \cdot (n_m V_m) = \sum_{\substack{\ell=1 \\ \neq m}}^{\mathcal{N}} n_\ell P_{\ell m} + n_\kappa P_{\kappa m} - n_m \Big(\sum_{\substack{\ell=1 \\ \neq m}}^{\mathcal{N}} P_{m\ell} + P_{m\kappa} \Big) \quad (1)$$

where n_a is the number density of level a, P_{ab} is the transition rate from a to b per atom in level a, κ refers to the next higher ionization stage, and V_m is the mean flow velocity of atoms in level m.

The bound-bound rates are

$$P_{u\ell} = A_{u\ell} + B_{u\ell}\bar{J}_{u\ell} + C_{u\ell}, \quad u > \ell \tag{2}$$

$$P_{\ell u} = B_{\ell u}\bar{J}_{u\ell} + C_{\ell u}, \quad u > \ell \tag{3}$$

where A and B are the Einstein coefficients, $C_{\ell u}$ and $C_{u\ell}$ are the collisional excitation and de-excitation rates, and $\bar{J}$ is the integrated mean intensity for the $u\ell$ transition, which must be calculated from the transfer equation.

The transfer equation for the $u\ell$ line transition is

$$\frac{dI_\nu}{ds} = -\frac{h\nu_{u\ell}}{4\pi}\varphi_\nu\left[(n_\ell B_{\ell u} - n_u B_{u\ell})I_\nu - n_u A_{u\ell}\right] - \kappa_\nu^C I_\nu + \epsilon_\nu^C \tag{4}$$

where s is geometrical depth in the direction of the intensity I_ν. Here, for simplicity, we assume complete frequency redistribution (CRD, i.e., absorption and emission have the same uncorrelated dependence on frequency ν, and have the common profile function φ_ν). Using the well-known relationships between the Einstein coefficients, we can write

$$\frac{dI_\nu}{ds} = -\kappa_\nu^L(I_\nu - S^L) - \kappa_\nu^C(I_\nu - S_\nu^C) \tag{5}$$

where the line absorption coefficient is

$$\kappa_\nu^L = \frac{h\nu_{u\ell}}{4\pi}\varphi_\nu(n_\ell B_{\ell u} - n_u B_{u\ell}) \tag{6}$$

and the line source function is

$$S_{u\ell}^L = \frac{2h\nu_{u\ell}^3/c^2}{(g_u n_\ell/g_\ell n_u) - 1}. \tag{7}$$

For simple illustrative purposes we consider the plane-parallel semi-infinite case for which $\kappa_\nu^C \ll \kappa_\nu^L$. Then the transfer equation reduces to

$$\mu\frac{dI_\nu}{d\tau_\nu} = I_\nu - S^L \tag{8}$$

where $d\tau_\nu = \kappa_\nu^L dz$, $dz = -\mu^{-1}ds$, and where μ is the cosine of the angle between the direction of I_ν (along ds) and the inward normal direction.

Solving for I_ν in terms of S^L, the mean intensity $J_\nu = \frac{1}{2}\int_{-1}^{+1}I_\nu(\mu)d\mu$ is given by

$$J_\nu(\tau_\nu) = \frac{1}{2}\int_0^\infty E_1(|t - \tau_\nu|)S^L(t)dt, \tag{9}$$

where E_1 is the first exponential integral. The mean intensity can be expressed in the discrete form

$$J_{ik} = \sum_j W^\Lambda_{ijk} S^L_j \tag{10}$$

where i and j are depth indices and k is the frequency index. Various quadrature representations can be used to determine the lambda-operator weighting coefficients W^Λ_{ijk} which depend on the monochromatic optical depths τ_{ik}.

The integrated mean intensity is

$$\overline{J} = \int \varphi_\nu J_\nu d\nu, \quad \int \varphi_\nu d\nu = 1, \tag{11}$$

so that

$$\overline{J} = \sum_j W^\Lambda_{ij} S^L_j, \quad W^\Lambda_{ij} = \int \varphi_\nu W^\Lambda_{ijk} d\nu. \tag{12}$$

The net rate coefficient $\rho_{u\ell}$ is defined by

$$n_u A_{u\ell} \rho_{u\ell} = n_u (A_{u\ell} + B_{u\ell} \overline{J}_{u\ell}) - n_\ell B_{\ell u} \overline{J}_{u\ell}, \tag{13}$$

so that

$$\rho_{u\ell} = 1 - \frac{\overline{J}_{u\ell}}{S^L_{u\ell}}. \tag{14}$$

We can write $\overline{J}_i$ (i.e., $\overline{J}$ at depth i for the $u\ell$ transition) either as

$$\overline{J}_i = \sum_j W^\Lambda_{ij} S^L_j, \quad \text{or as} \quad \overline{J}_i = S_i + \sum_j W^{\Lambda-1}_{ij} S^L_j, \tag{15}$$

where $W^{\Lambda-1}_{ij} = W^\Lambda_{ij} - U_{ij}$ and U_{ij} is the unit matrix. Then we obtain

$$\rho_i = -\frac{1}{S^L_i} \sum_j W^{\Lambda-1}_{ij} S^L_j, \tag{16}$$

to be used in the rate equations in place of $\overline{J}_{u\ell}$.

In this derivation we have assumed that the line source function is frequency-independent, but the general case with partial frequency redistribution (PRD), with the continuum included (and using either plane-parallel or spherical geometry), can be treated by methods very similar to those discussed here. (See Avrett & Loeser 1984.)

3 Preconditioning

Let $W^{\Lambda-1}_{ij} = d_i + W^r_{ij}$ where $d_i = W^{\Lambda-1}_{ii}$ and where W^r_{ij} is the same as $W^{\Lambda-1}_{ij}$ but with zero diagonal elements. Then

$$\rho_i = -d_i - \frac{1}{S^L_i} \sum_j W^r_{ij} S^L_j. \tag{17}$$

From equation (7), the S_i^L term in the denominator (for the $u\ell$ transition) can be written as

$$S_{u\ell}^L = \left(\frac{n_u}{n_\ell}\right) q_{u\ell}, \tag{18}$$

where

$$q_{u\ell} = \left(\frac{g_\ell}{g_u}\right) \frac{2h\nu_{u\ell}^3/c^2}{1 - (g_\ell n_u)/(g_u n_\ell)}. \tag{19}$$

Note that $q_{u\ell}$ has only a secondary dependence on n_u/n_ℓ except when stimulated emission is important. Equation (17) for $\rho_{u\ell}$ at depth i may then be written as

$$n_u \rho_{u\ell} = -n_u d_{u\ell} - n_\ell \chi_{u\ell}, \tag{20}$$

where, at depth i,

$$\chi_i = \frac{1}{q_i} \sum_j W_{ij}^r S_j^L. \tag{21}$$

Expressing $n_u \rho_{u\ell}$ in terms of $d_{u\ell}$ and $\chi_{u\ell}$ in the rate equation gives much better results than using $\rho_{u\ell}$ obtained directly from $S_{u\ell}^L$.

4 The Simple Two-Level Atom

The time-independent rate equation for a two-level atom without a continuum, and without mass flow, is

$$n_2(A_{21} + B_{21}\bar{J}_{21} + C_{21}) = n_1(B_{12}\bar{J}_{21} + C_{12}), \tag{22}$$

or

$$n_2(A_{21}\rho_{21} + C_{21}) = n_1 C_{12}. \tag{23}$$

Using $n_2 \rho_{21} = -n_2 d_{21} - n_1 \chi_{21}$ gives

$$n_2(-A_{21}d_{21} + C_{21}) = n_1(A_{21}\chi_{21} + C_{12}) \tag{24}$$

or

$$\frac{n_2}{n_1} = \frac{A_{21}\chi_{21} + C_{12}}{C_{21} - A_{21}d_{21}}. \tag{25}$$

Let $R = n_2/n_1$ and ignore stimulated emission. Then at depth i,

$$R_i = \frac{\sum_j W_{ij}^r R_j + \epsilon_i R_i^*}{\epsilon_i - d_i}, \tag{26}$$

where $\epsilon = C_{21}/A_{21}$ and

$$R_i^* = C_{12}/C_{21} = \left(\frac{g_2}{g_1}\right) \exp^{-h\nu_{21}/kT} = n_2^*/n_1^*. \tag{27}$$

This shows that n_2/n_1 approaches the LTE ratio n_2^*/n_1^* at large optical depths, as d_i and W_{ij}^r approach zero.

Thus we can have three forms of the equation for R_i, as a result of:

1) preconditioning, using $W_{ij}^r = W_{ij}^{A-1} - d_i$:

$$R_i = \frac{\sum_j W_{ij}^r R_j + \epsilon_i R_i^*}{\epsilon_i - d_i} \tag{28}$$

2) using W_{ij}^{A-1}:

$$R_i = \frac{\sum_j W_{ij}^{A-1} R_j + \epsilon_i R_i^*}{\epsilon_i} \tag{29}$$

3) using the usual lambda operator, W_{ij}^A:

$$R_i = \frac{\sum_j W_{ij}^A R_j + \epsilon_i R_i^*}{1 + \epsilon_i} \tag{30}$$

Example 1. Let $R_i^* = 1$ and $\epsilon_i = 0.01$ at all depths i:

With preconditioning:
$$R_i = \frac{\sum_j W_{ij}^r R_j + 0.01}{0.01 - d_i}$$

Without preconditioning:
$$R_i = \frac{\sum_j W_{ij}^A R_j + 0.01}{1 + 0.01}$$

Numerical solution:

τ	0	0.1	1	3	10	30	100
R	0.1	0.123	0.260	0.476	0.842	0.999	1.000

Number of iterations required to reach $R = 0.1 \pm 0.001$ at $\tau = 0$

	With preconditioning	Without preconditioning
initial $R = 1$	76	189
initial $R = 0$	79	334

Alternatively, as shown below, we can solve the set of linear equations for R_i directly without any iterations, and in much less time than required by the preconditioned lambda iterations.

From

$$R_i = \frac{\sum_j W_{ij}^{A-1} R_j + \epsilon_i R_i^*}{\epsilon_i} \tag{31}$$

we write

$$R_i - \frac{1}{\epsilon_i} \sum_j W_{ij}^{A-1} R_j = R_i^*. \tag{32}$$

Then, if M_{ij}^{-1} is the inverse of

$$M_{ij} = U_{ij} - \frac{1}{\epsilon_i} W_{ij}^{A-1}, \tag{33}$$

the solution is

$$R_i = \sum_j M_{ij}^{-1} R_i^*. \tag{34}$$

5 A Three-Level Atom

We illustrate the basic properties of the general multilevel case by considering a three-level atom.

The rate equations for levels 2 and 3 (again ignoring other stages of ionization for simplicity) are

$$n_2(A_{21}\rho_{21} + C_{21} + C_{23}) = n_1 C_{12} + n_3(A_{32}\rho_{32} + C_{32}) \tag{35}$$

and

$$n_3(A_{31}\rho_{31} + C_{31} + A_{32}\rho_{32} + C_{32}) = n_1 C_{13} + n_2 C_{23}. \tag{36}$$

These two equations can be solved for the two unknowns (n_2/n_1) and (n_3/n_1) if, from the transfer equations, we know the values of ρ_{21}, ρ_{31}, and ρ_{32}. Ordinary lambda iteration consists of successively solving these two rate equations and the three transfer equations.

Alternatively, we can write

$$n_u \rho_{u\ell} = -n_u d_{u\ell} - n_\ell \chi_{u\ell} \tag{37}$$

to get a better-conditioned set of equations. This is the basic preconditioning step, as in the two-level case. The preconditioned rate equations for levels 2 and 3 are

$$n_2(C_{21} - A_{21}d_{21} + C_{23} + A_{32}\chi_{32}) = n_1(C_{12} + A_{21}\chi_{21}) + n_3(C_{32} - A_{32}d_{32}) \tag{38}$$

and

$$n_3(C_{31} - A_{31}d_{31} + C_{32} - A_{32}d_{32}) = n_1(C_{13} + A_{31}\chi_{31}) + n_2(C_{23} + A_{32}\chi_{32}). \tag{39}$$

We can iterate between the χ values and the rate equations, just as in the two-level case, with slow convergence using W_{ij}^A, but with much faster convergence using W_{ij}^r.

Direct numerical solutions for each transition can be used in the multilevel case, just as in the two-level case. The above equations can be written as

$$\left(\frac{n_2}{n_1}\right)(x_{21} + y_{32}) - \left(\frac{n_3}{n_1}\right)x_{32} = y_{21}, \tag{40}$$

and

$$-\left(\frac{n_2}{n_1}\right)y_{32} + \left(\frac{n_3}{n_1}\right)(x_{31} + x_{32}) = y_{31}, \tag{41}$$

where

$$y_{u\ell} = C_{\ell u} + A_{u\ell}\chi_{u\ell}, \tag{42}$$

and

$$x_{u\ell} = C_{u\ell} - A_{u\ell}d_{u\ell}. \tag{43}$$

Then we can eliminate n_3/n_1 to obtain

$$\left(\frac{n_2}{n_1}\right)(x_{21} + \overline{x}_{21}) = y_{21} + \overline{y}_{21}. \tag{44}$$

Finally, the coefficients ϵ_{21} and B_{21}^S in the expression

$$S_{21}^L = \frac{\overline{J}_{21} + \epsilon_{21} B_{21}^S}{1 + \epsilon_{21}} \tag{45}$$

can be expressed in terms of the x and y coefficients. The results are

$$\epsilon_{21} = \epsilon_{21}^a - \beta_{21} \epsilon_{21}^b, \quad \beta_{21} = e^{-h\nu/kT}, \tag{46}$$

and

$$B_{21}^S = \alpha_{21} \beta_{21} (\epsilon_{21}^b / \epsilon_{21}), \quad \alpha_{21} = 2h\nu_{21}^3/c^2, \tag{47}$$

where

$$\epsilon_{21}^a = \frac{1}{A_{21}}(C_{21} + \overline{x}_{21}), \quad \epsilon_{21}^b = \frac{g_1}{g_2 \beta_{21} A_{21}}(C_{12} + \overline{y}_{21}). \tag{48}$$

Then, from equation (45), we can obtain n_2/n_1 at each depth i from the solution of the set of equations

$$S_i^L - \frac{1}{\epsilon_i} \sum_j W_{ij}^{\Lambda-1} S_j^L = B_i^S. \tag{49}$$

Here ϵ and B^S depend on χ_{31} and χ_{32}, but not on χ_{21}.

We can derive similar equations for S_{31}^L that depend on χ_{21} and χ_{32}, and for S_{32}^L that depend on χ_{21} and χ_{31}.

The direct approach applied to each of the transitions would consist of assuming initial values of χ_{21}, χ_{31}, and χ_{32}, and solving each set of simultaneous equations to get S_{21}^L, S_{31}^L, and S_{32}^L, thus giving new iterative values of χ_{21}, χ_{31}, and χ_{32} from equation (21). (Equation 21 represents the solution of the transfer equation.)

Preconditioned lambda iteration in this case would consist of substituting the initial χ values into equations (40) and (41) to obtain n_2/n_1 and n_3/n_1 at each depth, and then using equation (21) to obtain new χ values. Each lambda iteration needs fewer computations than the direct approach, but many more iterations are required, and sometimes the lambda iterations do not converge.

It is not necessary to apply the direct approach to all transitions, only to the strongest ones that control the large-scale behavior of the solution. Thus in Example 2 below we use such a hybrid method, solving the simultaneous equations for S_{21}^L and S_{31}^L but using preconditioned lambda iteration for the 32 transition, i.e., the χ_{32} used in equations (40) and (41) is determined directly from n_3/n_2 rather than by solving the set of simultaneous equations for S_{32}^L.

The derivation given above for a three-level atom with no continuum can be extended without difficulty to cases with an arbitrary number of bound levels together with other stages of ionization.

Example 2. Consider a 3-level hydrogen atom with a continuum (i.e., with the bound-free rates in equation 1 included), and a finite, symmetric atmosphere extending over the geometrical depth range $-700 \leq s(\text{km}) \leq +700$, with the total hydrogen density $n_H = 10^{11} \text{cm}^{-3}$, constant with depth, and the temperature varying as $T = 8000 \exp(s^2/10^5)$, so that $T = 8000$ K at $s = 0$, and $T = 10^6$ K at $s = \pm 700$. In this case $n_H \approx n_e \approx n_p$. Let n_1 represent the number density of hydrogen atoms in level 1 calculated from 1) the rate equations that include the bound-free transitions, and 2) the transfer equations for continuum radiation as well as line radiation.

Numerical solution: $n_1 = 2.4 \times 10^4$ at each surface
$n_1 = 5.6 \times 10^6$ at the center

Total line-center optical thickness: $\tau_{21} = 19$, $\tau_{31} = 3.0$, $\tau_{32} = 0.0027$

Note that in this example there are large temperature variations, but not large optical depths.

Parameters of the (21)-line solution:

	surface	center
ϵ	1.4×10^{-4}	1.4×10^{-5}
B	1.9	8.3×10^{-8}
S	4.9×10^{-7}	1.3×10^{-6}

Results for two methods of solution:

Case I

Direct solutions for (21) and (31)
Preconditioned lambda iterations
for (32).

< 0.1% convergence
after 8 iterations;
solution oscillating
with diminishing changes

computer time 10 min.

Case II

Preconditioned lambda iterations
for (21), (31), and (32)

far from convergence
after 50 iterations;
n_2(center) $= 19100$ and slowly
decreasing toward the
correct value of 129

computer time > 50 min.

6 Conclusions

1. We illustrate the well-known improvement in lambda-iteration solutions that results from preconditioning.
2. The strongest line transitions of a multilevel atom should be treated by solving the simultaneous equations corresponding to the combined rate and transfer equations, and not by lambda iteration.

3. Preconditioning should be used to determine the coupling between transitions that are needed in the simultaneous-equation solutions.

The methods described here are used in the Pandora atmosphere program (Avrett & Loeser (2003)). I am very grateful to Rudolf Loeser for his continued collaboration in this work.

References

Avrett, E.H. & Loeser, R. 1984, Line Transfer in Static and Expanding Spherical Atmospheres, in Methods in Radiative Transfer, ed. W. Kalkofen, Cambridge Univ. Press, 341-379.

Avrett, E.H. & Loeser, R. 2003, Solar and Stellar Atmospheric Modeling Using the Pandora Computer Program, in Modelling of Stellar Atmospheres, IAU Symp. No. 210, ed. N. Piskunov, W.W. Weiss, & D.F. Gray, ASP, Ann Arbor, A21.

Hubeny, I. 2003, Accelerated Lambda Iteration: An Overview, in Stellar Atmosphere Modeling, ed. I. Hubeny, D. Mihalas, & K. Werner, ASP Conf. Series, 288, 17-30.

Radiative Transfer in NLTE Model Atmospheres

Jiří Kubát

Astronomický ústav AV ČR, CZ 251 65 Ondřejov, Czech Republic
kubat@sunstel.asu.cas.cz

Summary. The method of a simultaneous solution of the spherically symmetric radiative transfer equation together with other basic equations of NLTE model stellar atmospheres (radiative equilibrium, hydrostatic equilibrium, statistical equilibrium, optical depth) is described.

1 Introduction

The main purpose of this book is to deal with radiative transfer in multiple dimensions. Thus, this contribution may seem to be a bit out of the main topic, since it deals with plane-parallel and spherically symmetric geometry, i.e. with geometrically one-dimensional atmospheres. However, there is strictly no one-dimensional radiative transfer problem. In all the cases which are being called one-dimensional, one has to cope with full spatial information of the radiation field and the one-dimensionality comes only from applied symmetries. In any case, it is necessary to perform a full spatial integration of the radiation field. In addition, sometimes it is considered that other independent variables of the radiation field serve as additional dimensions. Therefore, treating angle dependence in detail adds two dimensions and treating frequency dependence of the radiation field adds another one dimension. So, the full problem is 6-dimensional, and, the problem of spherically symmetric model atmospheres is 3-dimensional (radius, frequency, angle).

The task of calculating NLTE spherically symmetric or plane-parallel model atmospheres in hydrostatic and radiative equilibrium begins with the general input of stellar luminosity, mass, radius, and elemental abundances. Adding basic model equations of radiative equilibrium, hydrostatic equilibrium, statistical equilibrium, optical depth, and radiative transfer we obtain as a result of the model atmosphere calculation the depth dependence of temperature, density, and atomic population numbers.

There exists a number of excellent books and papers dealing with NLTE model atmospheres. An interested reader is recommended to consult the textbook of Mihalas [30] or a more recent lecture of Hubeny [15]. Recent developments in this field are summarized in the conference proceedings "Stellar Atmosphere Modeling" [20].

In this paper, we shall first discuss the particular equations governing the model atmosphere calculations and how the radiation field affects them. Then we shall describe how it is possible to add the radiative transfer equation and how to solve all these equations together most efficiently.

2 Radiative Transfer Equation

We shall deal with the radiative transfer equation in a spherically symmetric geometry. For simplicity, we assume that the coefficients of opacity, emissivity, and scattering are isotropic. The corresponding basic radiative transfer equation is (r is the radial distance, ν is the frequency, and $\mu = \cos\theta$ is the direction cosine)

$$\mu\frac{\partial I_{\nu\mu}(r)}{\partial r} + \frac{1-\mu^2}{r}\frac{\partial I_{\nu\mu}(r)}{\partial \mu} = \frac{dI_{\nu\mu}(r)}{ds} = -\chi_\nu(r)I_{\nu\mu}(r) + \eta_\nu(r) + \sigma_\nu(r)J_\nu(r), \quad (1)$$

where $I_{\nu\mu} \equiv I(\nu,\mu)$ is the specific intensity, χ_ν is the opacity, η_ν is the emissivity, and σ_ν is the scattering coefficient. For a solution of the model atmosphere problem, a moment equation using Eddington factors $f_\nu = K_\nu/J_\nu$ is very useful,

$$\frac{d^2(f_\nu q_\nu J_\nu)}{dX^2} = \frac{r^4}{q_\nu}\left(\frac{J_\nu - \eta_\nu + \sigma_\nu J_\nu}{\chi_\nu}\right) \quad (2)$$

where $dX_\nu = -q_\nu\chi_\nu dr$ and q_ν is the sphericity function [2], sometimes also referred to as a configuration function [27, 28]. These equations have to be supplemented with boundary conditions. The simplest boundary conditions are that there is no incident radiation from outside the star ($I^- = 0$) and that there is some specified intensity I^+ (for a shell) or a diffusion approximation (for a star) at the lower boundary.

3 Statistical Equilibrium

For a solution of equation (1) or (2) it is necessary to know the opacity χ_ν and the emissivity η_ν, which can't be calculated without a knowledge of the level population numbers n_i. For a more simple case of an assumption of local thermodynamic equilibrium (LTE) the latter may be determined relatively easily using the Saha and Boltzmann equations which depend only on the electron density n_e and the temperature T. On the other hand, for a physically more general (and more correct) case of the so-called NLTE (statistical equilibrium), the population numbers depend on the radiation field and they have to be determined using the set of equations of statistical equilibrium. In the following, the population numbers will be described using the departure coefficients $b = n/n^*$ [29], where n is the actual level population and n^* is the LTE level population. The meaning of n^* is not as straightforward as it may seem from first glance, in our case it is the LTE population with respect to the next higher ionization degree. The equations of statistical equilibrium read

$$b_i \sum_l (\tilde{R}_{il} + \tilde{C}_{il}) - \sum_l b_l (\tilde{R}_{li} + \tilde{C}_{li}) = 0, \qquad (3)$$

The collisional rate $\tilde{C}_{il}$ for a transition between levels i and l may be written as (cf. [23])

$$\tilde{C}_{il} = \tilde{C}_{li} = n_e n_i^* w_l q_{il}(T)$$

where $q_{il}(T)$ is the collisional excitation or ionization rate coefficient and w_l is the occupation probability of the level l. This quantity describes more correctly the lowering of the ionization energy [9]. Radiative rates for upward bound-bound transitions (excitation) are

$$\tilde{R}_{il} = n_i^* w_l \frac{4\pi}{h\nu_{il}} \int_0^\infty \sigma_{il}(\nu) J_\nu d\nu, \qquad (4a)$$

for downward bound-bound transitions (deexcitation)

$$\tilde{R}_{li} = n_l^* w_i \frac{g_i}{g_l} \frac{4\pi}{h\nu_{il}} \int_0^\infty \sigma_{il}(\nu) \left(\frac{2h\nu^3}{c^2} + J_\nu \right) d\nu, \qquad (4b)$$

for bound-free transitions (ionization)

$$\tilde{R}_{ik} = 4\pi n_i^* w_k \int_0^\infty \frac{\sigma_{ik}(\nu)}{h\nu} J_\nu d\nu, \qquad (4c)$$

and for free-bound transitions (recombination)

$$\tilde{R}_{ki} = 4\pi n_i^* w_i \int_0^\infty \frac{\sigma_{ik}(\nu)}{h\nu} \left(\frac{2h\nu^3}{c^2} + J_\nu \right) e^{-h\nu/kT} d\nu. \qquad (4d)$$

Since the equations of statistical equilibrium (3) are linearly dependent, an additional equation has to be used. For this purpose, we may use either the particle conservation equation

$$\sum_k N_k = N - n_e$$

or the charge conservation equation

$$\sum_k \sum_j q_j N_{jk} = n_e$$

for the main atom in the atmosphere (usually hydrogen). For the other atoms, we use the abundance equation

$$N_k = Y_k N_r,$$

where N_r is number density of a reference element (usually hydrogen) and Y_k is the abundance relative to the reference element.

3.1 Treatment of Non-Explicit Atomic Levels

For a correct calculation of the population number densities and the electron density it is necessary to also include the population of the levels which were not included in the explicit set of statistical equilibrium equations. The total number density of an ion j (of an atom k) is

$$N_{jk} = \sum_l n_{ljk} + \tilde{N}_{jk}$$

where $\tilde{N}_{jk}$ is the total population of all non-explicit levels. These levels are included in particle conservation, charge conservation, and abundance equations. They are assumed to be in LTE with respect to the ground level of the next higher ion and the collisional excitation to these levels is added to the corresponding collisional ionization rate.

4 Hydrostatic Equilibrium

The hydrostatic equation for the spherically symmetric case reads

$$\frac{dp_g}{dm} = \frac{G\mathcal{M}}{R^2} - \frac{4\pi}{c} \int_0^\infty \frac{1}{q_\nu} \frac{d(q_\nu f_\nu J_\nu)}{dm} d\nu, \qquad (5)$$

where the generalized column mass depth can be introduced after [32] as

$$dm = -\rho \left(\frac{R^2}{r^2}\right) dr = \rho d\left(\frac{R^2}{r}\right), \qquad (6)$$

where R is the radius of a star, defined as the radius at the Rosseland optical depth $\tau_R \sim 2/3$. The equation is supplemented with a upper boundary condition

$$\frac{p_1}{m_1} = \frac{G\mathcal{M}}{R^2} - \frac{4\pi}{c} \left(\frac{r_1}{R}\right)^2 \int_0^\infty \frac{\chi_{1\nu}}{\rho_1} \left(g_{1\nu} J_{1\nu} - H^-\right) d\nu, \qquad (7)$$

where $g_{1\nu} = \frac{\int_0^1 \mu j_{1\nu\mu} d\mu}{J_{1\nu}}$ and H^- is the incident flux.

5 Radiative Equilibrium

Radiative Equilibrium

The equation of radiative equilibrium expresses the conservation of radiative energy. Its integral form is

$$\int_0^\infty (\kappa_\nu J_\nu - \eta_\nu)\, d\nu = \int_0^\infty \kappa_\nu (J_\nu - S_\nu)\, d\nu = 0. \tag{8}$$

Unfortunately, this integral form of radiative equilibrium does not guarantee the conservation of radiation flux at large optical depths. For this case it is preferable to use the differential form

$$H_0 = \int_0^\infty \frac{\rho}{q_\nu \chi_\nu} \frac{d(q_\nu f_\nu J_\nu)}{dm}\, d\nu. \tag{9}$$

Here, $H_0 = L/(4\pi R)^2$. To ensure better stability of the solution it is a good idea to use the superposed form of the radiative equilibrium [18]

$$I_{RE} + \beta D_{RE} = 0 \tag{10}$$

where I_{RE} is the integral equation (8), D_{RE} is the differential equation (9), $\beta \ll 1$ for small depths (necessary for the stability), and $\beta \gg 1$ at great depths (necessary for the flux conservation).

Thermal Balance of Electrons

Although commonly used, the condition of radiative equilibrium is not the only one which enables the determination of the equilibrium temperature. To clarify this, we plotted a simplified energy balance diagram in the Fig. 1. The condition of radiative equilibrium describes the balance of the radiation energy. The balance of the atomic internal energy is described by equations of statistical equilibrium. The thermal energy balance is a consequence of both radiation energy and atomic internal energy balances. However, the thermal energy balance *directly* affects the temperature of the medium. Consequently, we may use the thermal energy balance as an advantageous alternative to the radiation energy balance. The overall physical picture of the energy balance does not change, but numerical properties change significantly.

In the parts of the atmosphere which are optically thin for continuum radiation and optically thick for radiation in some lines, the radiative equilibrium equation is dominated by the line (bound-bound) transitions, which *do not directly* affect the temperature. As a consequence, the numerical stability is poor and the solution often tends to divergence. However, it is easy to remove this numerical instability by simply considering the thermal energy balance instead of the radiative energy balance. Then, the large energy transfer between the radiation energy pool and the atomic energy pool does not enter the

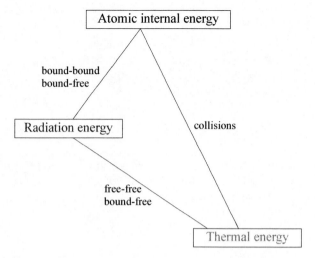

Fig. 1. Simplified energy balance in the stellar atmosphere. It describes the balance of three basic energetic pools in the stellar atmosphere and denotes the microscopic physical processes which enable the transfer of energy between individual pools

equations of thermal balance, since only bound-free, free-free, and collisional transitions enter. The condition of thermal energy balance may be expressed as

$$(Q_c^H + Q_{bf}^H + Q_{ff}^H) - (Q_c^C + Q_{bf}^C + Q_{ff}^C) = 0 \tag{11}$$

where free-free heating Q_{ff}^H (absorption), free-free cooling Q_{ff}^C (emission), bound-free heating Q_{bf}^H (absorption), bound-free cooling Q_{bf}^C (emission), collisional heating Q_c^H (deexcitation and recombination), and collisional cooling Q_c^C (excitation and ionization) are

$$Q_{ff}^H = 4\pi n_e \sum_j N_j \int_0^\infty \alpha_{ff,j}(\nu, T) J_\nu \, d\nu \tag{12a}$$

$$Q_{ff}^C = 4\pi n_e \sum_j N_j \int_0^\infty \alpha_{ff,j}(\nu, T) \left(J_\nu + \frac{2h\nu^3}{c^2}\right) e^{-h\nu/kT} d\nu \tag{12b}$$

$$Q_{bf}^H = 4\pi \sum_{l,k} n_l^* b_l w_k \int_0^\infty \alpha_{bf,\,lk}(\nu) J_\nu \left(1 - \frac{\nu_{lk}}{\nu}\right) d\nu \tag{12c}$$

$$Q_{bf}^C = 4\pi \sum_{l,k} n_l^* b_k w_l \int_0^\infty \alpha_{bf,\,lk}(\nu) \left(J_\nu + \frac{2h\nu^3}{c^2}\right) e^{-h\nu/kT} \left(1 - \frac{\nu_{lk}}{\nu}\right) d\nu \tag{12d}$$

$$Q_c^H = n_e \sum_{l,m} b_m n_l^* w_m q_{lm}(T) h\nu_{lm} \tag{12e}$$

$$Q_c^C = n_e \sum_{l,m} b_l n_l^* w_m q_{lm}(T) h\nu_{lm} \tag{12f}$$

Note that for the case of the local thermodynamic equilibrium (LTE) $b_l = b_m = 1$, which means that $Q_c^H = Q_c^C$

Using the thermal balance condition instead of the radiative equilibrium equation in the outer parts of the stellar atmosphere significantly improves the convergence properties, as can be seen from Fig. 3 in Kubát et al. [26].

6 Formal Solution of the Radiative Transfer Equation

The most important part of the model atmosphere calculation is the formal solution of the radiative transfer equation. By formal solution, we mean a solution for a *given* opacity χ_ν and emissivity η_ν. To this end we use the Feautrier method [10], which transfers the first order radiative transfer equation to the equation of the second order using symmetric and antisymmetric means of the specific intensity. Numerical details of the Feautrier method for both plane-parallel and spherically symmetric geometry may be found in a review paper of Mihalas [31]. Instead of the Feautrier method which is based on differential equations, we may use any other method for the formal solution, like, e.g., the solution of the integral equation [21, 38], or a kind of a finite element method [8, 36, 22].

Variable Eddington factor method For a treatment of the angular dependence, the so called Variable Eddington factor method [7] is used. The essence of this method is in the iterative solution for the Eddington factors $f_\nu = K_\nu / J_\nu$ instead of the direct solution of the angle dependence of the radiation field, which simplifies the handling of scattering. For spherically symmetric atmospheres, the sphericity function q_ν is also being determined iteratively using this method.

7 Solution of the Model Atmosphere Problem

To solve a model atmosphere problem we search for a *simultaneous* solution of the equations of energy equilibrium (i.e. either radiative equilibrium (8) or thermal balance (11) equation), hydrostatic equilibrium (5), statistical equilibrium (3), and of the radiative transfer equation (1)–(2).

Since the equations form a nonlinear set of integrodifferential equations, their solution is not straightforward and has to be performed efficiently. There are two very powerful methods which allow us to do this.

The first one is the Newton-Raphson method, in astrophysics often referred to as the complete linearization method [6]. In this method, all equations are solved simultaneously using an iterative procedure. Let $\boldsymbol{\psi} = (n_e, T, b_i, J_\nu)$ represents a vector consisting of all variables which are to be calculated and $F(\boldsymbol{\psi}) = 0$ represents all the equations. Assuming a known estimate of the variables, the corrections $\delta\boldsymbol{\psi}$ to the values of $\boldsymbol{\psi}$ are found using

$$\delta\psi = -\left(\frac{\partial F(\psi)}{\partial \psi}\right)^{-1} F(\psi) \tag{13}$$

The linearized system is solved by standard Gaussian elimination [30]. The complete linearization method became a basis for a number of efficient computer codes for the calculation of NLTE model stellar atmospheres [14, 33].

Another method employs the simplicity of the so-called lambda iteration, an iterative method using the formal solution of the radiative transfer equation $J_\nu = \Lambda_\nu S_\nu$, which is, unfortunately, unusable for the conditions present in stellar atmospheres, since it results in an extremely slow convergence [4].

The basic idea of the so-called *accelerated lambda iteration* is in replacing the Λ operator with an *approximate operator* Λ_ν^* [16, and references therein], which is easy to calculate and which describes as much physics as possible. Then the iteration scheme takes the form

$$J_\nu^{(n+1)} = \Lambda_\nu^* \left[S_\nu\left(J_\nu^{(n+1)}\right)\right] + (\Lambda_\nu - \Lambda_\nu^*) \left[S_\nu\left(J_\nu^{(n)}\right)\right]. \tag{14}$$

There are several possibilities of calculating the approximate lambda operator, the most efficient calculation follows directly from the formal solution using the same numerical scheme, which ensures numerical consistency of the results [37, 35]. The accelerated lambda iteration is now used in a number of computer codes which calculate a solution to the model atmosphere problem [39, 25, 40, 19, 12, 13].

However, the most effective possibility is the combination of both methods [18], i.e. of the complete linearization and accelerated lambda iteration. This may be done by eliminating the radiation field J_ν from the vector of variables ψ, which is done by using expression (14). The linearized radiation field is then expressed as $\delta J_\nu^{(n)} = \Lambda_\nu^* \delta S_\nu^{(n)} + \Delta J_\nu^{(n-1)}$ and the vector of variables is reduced to $\psi_d = (n_e, T, r, b_i)_d, d = 1, \ldots, ND$, which saves a large amount of memory needed for the calculation.

Additional memory savings may be achieved using the implicit linearization of *b*-factors [3, 1] instead of the standard explicit one. It means that instead of treating the equations of statistical equilibrium explicitly we express the changes δb using other variables, namely [24]

$$\delta b_i = \frac{\partial b_i}{\partial n_e} \delta n_e + \frac{\partial b_i}{\partial T} \delta T + \frac{\partial b_i}{\partial r} \delta r \tag{15}$$

where

$$\frac{\partial \mathbf{b}}{\partial \psi} = -\left(\mathcal{A} + \frac{\partial \mathcal{A}}{\partial \mathbf{b}} \cdot \mathbf{b} - \frac{\partial \mathcal{B}}{\partial \mathbf{b}}\right)^{-1} \cdot \left(\frac{\partial \mathcal{A}}{\partial \psi} \cdot \mathbf{b} - \frac{\partial \mathcal{B}}{\partial \psi}\right) \tag{16}$$

However, a simpler expression for the derivatives may be used (and yields faster code)

$$\frac{\partial \mathbf{b}}{\partial \psi} = -\left(\frac{\partial \mathcal{A}}{\partial \psi} \cdot \mathbf{b} - \frac{\partial \mathcal{B}}{\partial \psi}\right)$$

Further computational time may be saved using some acceleration technique. The most simple one is the Kantorovich acceleration [17], which avoids unnecessary calculations of the Jacobian in equation (13) and allows it to remain constant after accomplishing several first initial iterations. It is one of the most efficient methods of acceleration, since it is extremely simple and yields a stable accelerated code. A more sophisticated method is the Ng acceleration [34, 5] where a new estimate of accelerated variables is taken as a prediction based on values from the last preceding iterations. However, the Ng acceleration sometimes fails to converge if the predicted value overshoots the correct solution.

Output from the calculations The basic output which comes out from the computer code is the model atmosphere $(T(m), n_e(m), n_i(m), \rho(m), r(m), \tau_R(m))$ itself. This result may be subsequently used for *detailed* stellar spectrum calculations. Additional output may help to understand the physics of the results, and it may also help to identify potential inconsistencies and bugs, which are inherent to any code longer than several statements. This additional output involves checking for flux conservation, detailed output of rates in individual transitions, emergent flux H_ν, emergent specific intensities $I_{\nu\mu}$ (which may serve for correct determination of limb darkening [11]), output of the energy balance, and information about the calculations (relative changes, timing, errors, ...).

8 Summary

In this paper we briefly described a method and a code for calculation static spherically symmetric NLTE model atmospheres in hydrostatic and radiative equilibrium. In addition to classical model atmosphere calculations the code is also applicable to the so-called NLTE line formation problem (solution of the radiative transfer equation and the equations of statistical equilibrium for a given atom). The code is also able to calculate model atmospheres of shells (no diffusion approximation at the lower boundary).

Acknowledgements. The author would like to thank Daniela Korčáková and Adéla Kawka for their valuable comments to the manuscript. This research has made use of NASA's Astrophysics Data System. This work was supported by a grant GA ČR 205/01/0656. The Astronomical Institute Ondřejov is supported by projects K2043105 and AV 0Z1 003909.

References

1. Anderson, L.S.: 1987, in *Numerical Radiative Transfer*, W. Kalkofen ed., Cambridge Univ. Press, p. 163
2. Auer, L.H.: 1971, *J. Quant. Spectrosc. Radiat. Transfer* **11**, 573

3. Auer, L.H.: 1973, *Astrophys. J.* **180**, 469
4. Auer, L.H.: 1984, in *Methods in Radiative Transfer*, W. Kalkofen ed., Cambridge Univ. Press., p. 237
5. Auer, L.H.: 1991, in *Stellar Atmospheres: Beyond Classical Models*, L. Crivellari, I. Hubeny, & D.G. Hummer eds., NATO ASI Series C 341, Kluwer Academic Publishers, p. 9
6. Auer, L.H., Mihalas, D.: 1969, *Astrophys. J.* **156**, 157
7. Auer, L.H., Mihalas, D.: 1970, *Mon. Not. Roy. Astron. Soc.* **149**, 65
8. Castor, J.I., Dykema, P.G., Klein, R.I.: 1992, *Astrophys. J.* **387**, 561
9. Däppen, W., Anderson, L.S., Mihalas, D.: 1987, *Astrophys. J.* **319**, 195
10. Feautrier, P.: 1964, *C.R. Acad. Sci. Paris* **258**, 3189
11. Hadrava, P., Kubát, J.: 2003, in [20], p. 149
12. Hamann, W.-R.: 2003, in [20], p. 171
13. Höflich, P.: 2003, in [20], p. 185
14. Hubeny, I.: 1988, *Comput. Phys. Commun.* **52**, 103
15. Hubeny, I.: 1997, in *Stellar Atmospheres: Theory and Observations*, J.P. De Greve, R. Blomme, & H. Hensberge eds., Lecture Notes in Physics Vol. 497, Springer Verlag, Berlin, p. 1
16. Hubeny, I.: 2003, in [20], p. 17
17. Hubeny, I., Lanz, T.: 1992, *Astron. Astrophys.* **262**, 501
18. Hubeny, I., Lanz, T.: 1995, *Astrophys. J.* **439**, 875
19. Hubeny, I., Lanz, T.: 2003, in [20], p. 51
20. Hubeny, I., Mihalas, D., Werner, K.: 2003, *Stellar Atmosphere Modeling*, ASP Conf. Ser. Vol. 288
21. Kalkofen, W.: 1974, *Astrophys. J.* **188**, 105
22. Korčáková, D., Kubát, J.: 2003, *Astron. Astrophys.* **401**, 419
23. Kubát, J.: 1997, *Astron. Astrophys.* **326**, 277
24. Kubát, J.: 2001, *Astron. Astrophys.* **366**, 210
25. Kubát, J.: 2003, in *Modelling of Stellar Atmospheres*, IAU Symp. 210, N.E. Piskunov, W.W. Weiss & D.F. Gray eds., ASP, p. A8
26. Kubát, J., Puls, J., Pauldrach, A.W.A.: 1999, *Astron. Astrophys.* **341**, 587
27. Leung, C.M.: 1975, *Astrophys. J.* **199**, 340
28. Leung, C.M.: 1976, *J. Quant. Spectrosc. Radiat. Transfer* **16**, 559
29. Menzel, D.H.: 1937, *Astrophys. J.* **85**, 330
30. Mihalas, D.: 1978, *Stellar Atmospheres*, 2nd ed., W.H. Freeman & Comp., San Francisco
31. Mihalas, D.: 1985, *J. Comput. Phys.* **57**, 1
32. Mihalas, D., Hummer, D.G.: 1974, *Astrophys. J. Suppl. Ser.* **28**, 343
33. Mihalas, D., Heasley, J.N., Auer, L.H.: 1975, NCAR-TN/STR-104, NCAR Boulder
34. Ng, K.C.: 1974, *J. Chem. Phys.* **61**, 2680
35. Puls, J.: 1991, *Astron. Astrophys.* **248**, 581
36. Richling, S., Meinköhn, E., Kryzhevoi, N., Kanschat, G.: 2001, *Astron. Astrophys.* **380**, 776
37. Rybicki, G.B., Hummer, D.G.: 1991, *Astron. Astrophys.* **245**, 171
38. Schmid-Burgk, J.: 1975, *Astron. Astrophys.* **40**, 249
39. Werner, K.: 1986, *Astron. Astrophys.* **161**, 177
40. Werner, K., Deetjen, J.L., Dreizler, S., Nagel, T., Rauch, T., Schuh, S.L.: 2003, in [20], p. 31

The Solution of the Radiative Transfer Equation in Axial Symmetry

Daniela Korčáková and Jiří Kubát

Astronomický ústav AV ČR, CZ-251 65 Ondřejov, Czech Republic
kor@sunstel.asu.cas.cz, kubat@sunstel.asu.cas.cz

Summary. We present a new method for the solution of the radiative transfer equation in axial symmetry. The three dimensional problem is reduced to a two dimensional one using a set of independent planes. Our extended short characteristics method is applied to the radiative transfer problem in each of the planes using polar coordinates. The method is able to handle a velocity field, which is accounted for using the Doppler shift between neighboring cells.

Our method is suitable for studying the stellar wind with rotation, where the symmetry of the problem is naturally axial. Results of test calculations are shown for the case of a stellar wind of a typical B star.

1 Introduction

Problems studied in the astrophysical radiative transfer are rather diverse. There exist objects, like planetary nebulae, where the density is very low and the medium is very transparent. On the other hand, some parts of galactic nebulae can be opaque.

For the particular case of stars, their most important part is the stellar atmosphere – a region where an optically thick star fades into an optically thin interstellar medium. The values of physical quantities change by several orders of magnitude there. The transfer problem must be solved from an optical depth of 10^6 to 10^{-6}. The geometrical scale of stellar atmospheres is usually much larger than the mean free photon path and also much larger than the thermalization length. As a consequence, long distance interaction by radiation plays an extremely important role in the physical description of the medium. These facts introduce problems. Some methods fail under these conditions and become numerically unstable or they converge to the wrong solution. Due to other inherent complexities (effects of a huge number of lines, absence of a thermodynamic equilibrium, etc. – see Kubát, this volume), stellar atmospheres are usually being studied using one-dimensional radiative transfer. However, for studying the various effects we must solve the radiative transfer equation (RTE) in a more general geometry than plane-parallel or spherically symmetric ones. There are not too many methods available for solving the multidimensional problem. For optically thin regions, the Monte Carlo method [21] may be used. On the other hand, in the optically thick regions it's possible to calculate using a method which uses the diffusion approximation [10]. Both of these methods are inappropriate for stellar atmospheres.

The classical way for calculating a solution to this equation in more dimensions is the long characteristic method [2]. This method fully describes the radiation field, but the necessary computer time is rather long. Due to this reason, Kunasz & Auer [14] developed the short characteristic method, which is the best out of the available multidimensional methods. There exist several methods using short characteristics in a Cartesian grid [6, and references therein], 2D axially symmetric geometry [8], and a general method, which is also able to solve the transfer equation in spherical and cylindrical grids [19, 20]. An efficient approach is using adaptive grids [7, 18]. An excellent insight into astrophysical multidimensional radiative transfer can be achieved from a paper of Auer [1].

Another possibility is to apply the finite element method, which has been recently used for the problem of multidimensional radiative transfer by, e.g., Richling et al. [17]. However, the finite element method is not used very often due to its convergence problems. This annoyance has been overcome by Dykema et al. [5], who used a modification of the finite element method, namely the discontinuous finite element method.

For the study of a stellar wind, accretion discs, or stellar rotation we need such a method for calculating a solution of the radiative transfer equation, which is able to work in a more general geometry than the plane-parallel or spherically symmetric ones. But, it isn't necessary to include the whole three dimensional space.

In this paper, we describe a new method for calculating a solution to the radiative transfer problem in axial symmetry. The basic idea behind our method is a solution of the transfer problem in separated planes. In these planes, a combination of long and short characteristic methods is used. The results are more accurate than in the plane-parallel and spherically symmetric cases, and the calculation is faster than for the complete 3D problem. In Section 3 we test this method for the case of a stellar wind of a main sequence B star.

2 Technique

The basic idea of this method is the calculation of a solution of the radiative transfer problem not in the whole star, but in separated planes intersecting the star.

Let us consider the spherical coordinate system (r, θ, ϕ). The axis of symmetry is around $\theta = 0$. We introduce the discretization of the radial distance r and angle θ. Due to axial symmetry, the physical quantities don't vary with angle ϕ. The grid is chosen to give the best description of the system. For example, for the study of stellar winds, the grid of angles θ can be equidistant, but for accretion discs it must be finer near the equatorial plane. We choose the grid in radial distances to be very similar to a 1D problem, we want the points to be equidistant in the logarithm of the optical depth scale. A good choice is about 5 points per decade. We assume the opacity and emissivity of the stellar material to be known in the grid points. To reduce the 3D problem to 2D, we will not solve radiative transfer equation in this grid, but in a set of "longitudinal" planes, intersecting the star parallel to the plane $\phi = 0$ (see Fig. 1). In every longitudinal plane we choose the polar coordinate

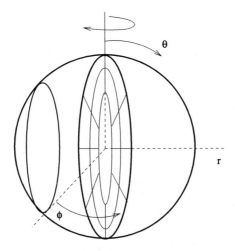

Fig. 1. Scheme of the set of longitudinal planes

system and define a grid of *concentric circles* and *radial lines*. The grid of concentric circles corresponds with a 3D grid in the planes, which intersect the region of validity of a lower boundary condition. But, we must be careful in other planes, where the grid chosen must imply the geometry of the problem as well as the velocity field. We interpolate the opacity, emissivity, and the source function to the new coordinate system. We solve the radiative transfer equation in every longitudinal plane independently.

First, we solve the radiative transfer equation using the boundary condition at the end of the stellar atmosphere (the place where the star ends and the interstellar medium begins) in the given longitudinal plane. Once the radiation field in the downward direction (towards stellar center) is known, we can continue with the solution in the opposite direction – from the center of the star towards the upper boundary. We can assume the diffusion approximation at the inner boundary of the longitudinal planes, which intersect the stellar center. In other planes (which do not intersect the stellar center), we adopt as the lower boundary condition the intensity taken from the previously calculated solution from the outer boundary to the center of the star.

To obtain the whole radiation field we take advantage of the symmetry of the problem. We know the radiation field at the grid points in all directions lying within the longitudinal planes. We choose the main plane to be the one which intersects the stellar center. We obtain the radiation field in other directions by rotating the remaining chosen planes around the axis of symmetry $\theta = 0$. For the sake of this, we must choose the grid points in the planes at the same distance from the equator (see Fig. 2).

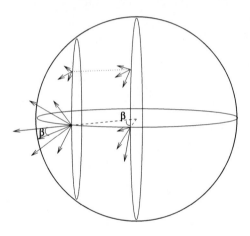

Fig. 2. Scheme for the calculation of the whole radiation field

The Solution from the Outer Boundary to the Central Regions in the Longitudinal Plane

The solution starts at the outer boundary (stellar surface), where we know the boundary condition. In each grid point we choose three rays per quadrant (see Fig. 3). Along these rays we solve the transfer equation. The angle distribution of these rays may be the same as in the plane-parallel geometry, where the angles are chosen to be the roots of Legendre polynomials in the interval $(0, 1)$ ($\mu = 0.8872983346, 0.5, 0.1127016654$) to ensure better numerical accuracy of angle integration [16, section 4.5]. We measure these angles from the normal to the grid circle at the given point. As one can see in Fig. 3, the rays end at the next grid circle. This means that it is possible that the rays intersect some grid lines. The number of rays at a given point is sufficient, because the

The Solution of the Radiative Transfer Equation in Axial Symmetry   241

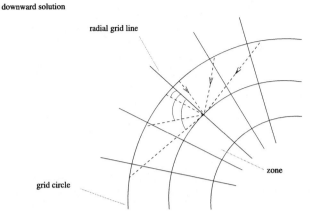

Fig. 3. The scheme for the solution of the radiative transfer in the longitudinal plane from the upper boundary to the central regions

whole radiation field will be obtained by summing the information from all longitudinal planes (see Fig. 2). So, the space description of the radiation field is sufficient.

The rays connect grid circles with the possibility of intersecting with more radial grid lines. We perform a linear interpolation of the source function and opacity to obtain their values at points B and C and of the incoming intensity for the value at point A (see Fig. 4). The optical depth difference

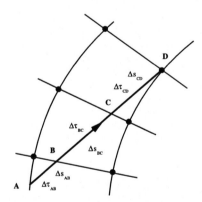

Fig. 4. The scheme for the integration along the ray

$\Delta\tau$ is calculated along the ray between the individual intersection points (AB, BC, and CD). We assume a linear dependence of the source function on the optical depth between the intersection points. We solve the equation of radiative transfer between all intersection points along the ray. For the interval AB the solution is

$$I_{(B)} = I_{(A)}e^{-\Delta\tau_{(AB)}} + \int_0^{\Delta\tau_{(AB)}} S(t)e^{[-(\Delta\tau_{(AB)}-t)]}dt. \tag{1}$$

The integration along the ray is performed in more steps due to two reasons. First, in the nonorthogonal grid the parabolic interpolation fails and the linear interpolation between points AD is a better choice. Second, if there is a velocity field presented, we can assume a constant velocity in the cells and permit a change of a velocity only at the boundary of the cells. So, it's possible to solve the *static* equation of radiative transfer in the cells and to perform the Lorentz transformation of frequency and intensity on the boundaries.

The Solution from the Central Regions to the Outer Boundary in the Longitudinal Plane

The upward solution is very similar to the previous step. From Fig. 1, one can see that there exist planes which don't intersect the inner boundary region. For these planes we adopt the intensity calculated from the previous step (solution from the upper boundary to the stellar center) as the lower boundary condition. However, the solution of the transfer equation in the central grid circle must be performed with care. We split these rays into two parts (AB and BC – see Fig. 5), in which we solve the transfer problem separately. For

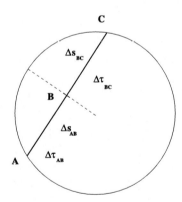

Fig. 5. The scheme for solution of the radiative transfer equation in the central region in the planes, which don't intersect the region of validity of the diffusion lower boundary condition

planes, which intersect the region of validity of the diffusion approximation (the stellar core), the situation is easier. We simply use the corresponding lower boundary condition.

Further solution is depicted in the Fig. 6. From the grid points we lead three rays, using the same angles as in the previous case. In the intersec-

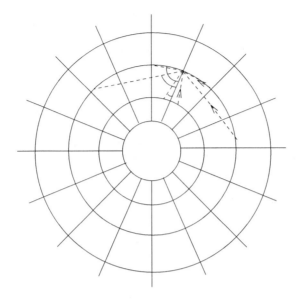

Fig. 6. The scheme for the solution from the center of the star to the upper boundary

tion points we interpolate the opacity and the source function. Between these points we calculate the optical depth and solve the transfer equation as in the previous case of downward integration (see Eq. 1).

2.1 Velocity Field

Let us consider a velocity field with a velocity gradient small enough to neglect the aberration. This assumption is valid, if $v/c \ll 0.01$ [9], [15]. We assume the velocity field to be constant in every cell. At cell boundaries we perform the Lorentz transformation of intensity and frequency and solve the *static* equation within the cell. To ensure a correct treatment of the line transfer we have to guarantee the frequency shift due to the motion is smaller than a quarter of a Doppler halfwidth (see [11]). If it is not, we must make a finer grid.

3 Test Calculations

Tests of the method are performed for a main sequence B star with an effective temperature of $17\,000\,K$, a surface gravitational acceleration of $\log g = 4.12$, and a radius of $3.26\,R_\odot$. From the hydrostatic spherically symmetric model of the atmosphere [13] we obtain the state parameters, electron density and temperature. For simplicity, we consider no incoming radiation at the outer boundary and a diffusion approximation at the inner boundary.

We adopt the beta law as a velocity field of the stellar wind (see, e. g., [3])

$$v(r) = v_\infty \left\{ 1 - \left[1 - \left(\frac{v_R}{v_\infty}\right)\right]^{\frac{1}{\beta}} \frac{R}{r} \right\}^\beta, \qquad (2)$$

with the parameter $\beta = 1$, the velocity in the photosphere $v_R = 100\,km \cdot s^{-1}$, and the terminal velocity $v_\infty = 1000\,km \cdot s^{-1}$. We adopt the Doppler profile for hydrogen lines.

The results of the calculations are shown in Fig. 7. The Hα line profile from the stellar wind (dotted line) is compared to the static profile (solid line). Since we take the input parameters from the hydrostatic model, the atmosphere is too thin for the P Cygni profile to be seen. For this reason, we chose a different geometrical depth scale ($r(d) = r(d-1) \cdot 1.0016$) with the same T_eff, n_e, and velocity at the corresponding grid points. This "extended" atmosphere line profile is plotted using the dashed line and it has a P Cygni shape.

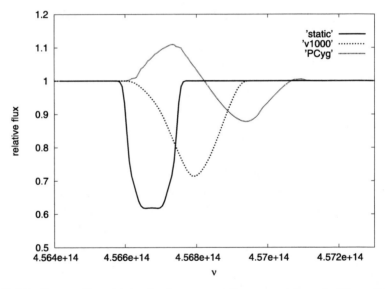

Fig. 7. The line profiles obtained using our axially symmetric code. The solid line indicates the static solution, the dotted one the calculation which includes the stellar wind, and the dashed one is the P Cygni profile obtained from the extension of the given atmosphere

3.1 The Tests of the Grid

The grid tests show more or less a linear dependence of the computing time on the geometrical depth (see Fig. 8, left panel). The dependence of the computing time on the number of frequency points is shown in the right panel of the same figure. Here, the fitting function is a polynomial of the third order.

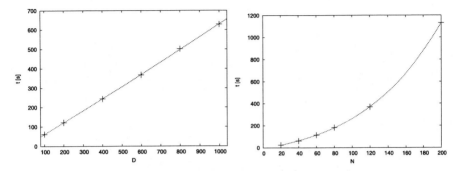

Fig. 8. The dependence of the computing time on the number of depth points (*left panel*) and on the number of frequency points (*right panel*)

4 Conclusions

A new method for a solution of the radiative transfer equation in axial symmetry was presented. The transfer problem is solved at planes intersecting the star independently. The radiative transfer equation is solved at every plane using a combination of short and long characteristics methods. This allows us to take into account the global character of the radiation field and the necessary computing time is not too long.

The velocity field (with velocity gradients) is taken into account using a Lorentz transformation of corresponding quantities at cell boundaries, which also allows the treatment of nonmonotonic velocity fields.

We performed tests for a main sequence B star with an effective temperature of $17\,000\,K$. In Fig. 7, Hα line profiles which include a velocity field are presented. Since we adopted the input parameters, electron temperature and density, from the hydrostatic model, the resulting line profiles don't show a P Cygni profile. The dependence of the computing time on the number of geometrical depth points is linear, but the dependence on the number of frequency points is a polynomial of third order.

This method is useful not only for the study of stellar winds and stellar rotation, but also for axially symmetric planetary nebulae or accretion disks. It takes advantage of the symmetry of a problem without the need of calculating a complete 3D problem.

Acknowledgements. The authors would like to thank Adela Kawka for language corrections. This research has made use of NASA's Astrophysics Data System. This work was supported by grants GA ČR 205/01/0656 and 205/04/ P224. The Astronomical Institute Ondřejov is supported by projects K2043105.

References

1. Auer, L., 2003, in *Stellar Atmosphere Modeling*, I. Hubeny, D. Mihalas, & K. Werner eds., ASP Conf. Ser. Vol. 288, p. 405
2. Cannon, C.J., 1970, ApJ, 161, 255
3. Cassinelli, J.P., Lamers, H.J.G.L.M., 1999, Introduction to Stellar Winds, Cambridge Univ. Press, Cambridge
4. Domiciano de Souza, A., Kervella, P., Jankov, S., Abe, L., Vakili, F., di Folco, E., Paresce, F., 2003, A&A, 407, L47
5. Dykema, P.G., Klein, R.I., Castor, J.I., 1996, ApJ, 457, 892
6. Fabiani Bendicho, P., 2003, in *Stellar Atmosphere Modeling*, I. Hubeny, D. Mihalas, & K. Werner eds., ASP Conf. Ser. Vol. 288, p. 419
7. Folini, D., Walder, R., Psarros, M., Desboeufs, A.C., 2003, in *Stellar Atmosphere Modeling*, I. Hubeny, D. Mihalas, & K. Werner eds., ASP Conf. Ser. Vol. 288, p. 433
8. Georgiev, L.N., Hillier, D.J., 2003, in *Stellar Atmosphere Modeling*, I. Hubeny, D. Mihalas, & K. Werner eds., ASP Conf. Ser. Vol. 288, p. 437
9. Hauschildt, P.H., Starrfield, S., Shore, S., Allard, F., Baron, E., 1995, ApJ, 447, 829
10. Kneer, F., Heasley, J.N., 1979, A&A, 79, 14
11. Korčáková, D., Kubát, J., 2003, A&A, 401, 419
12. Kroll, P., Hanuschik, R.W., 1997, in: IAU Coll. 163, 494
13. Kubát, J., 2003, in Modelling of Stellar Atmospheres, IAU Symp. 210, N.E. Piskunov, W.W. Weiss & D.F. Gray eds., ASP, A8
14. Kunasz, P., Auer, L.H., 1988, JQSRT, 39, 67
15. Mihalas, D., Kunasz, P.B., Hummer, D.G., 1976, ApJ, 206, 515
16. Press, W.H., Flannery, B.P., Teukolsky, S.A., Vetterling, W.T., 1986, *Numerical Recipes, The Art of Scientific Computing*, Cambridge Univ. Press, Cambridge
17. Richling, S., Meinköhn, E., Kryzhevoi, N., Kanschat, G., 2001, A&A, 380, 776
18. Steinacker, J., in *Stellar Atmosphere Modeling*, I. Hubeny, D. Mihalas, & K. Werner eds., ASP Conf. Ser. Vol. 288, p. 449
19. Noort, van M., Hubeny, I., Lanz, T., 2002, ApJ, 568, 1066
20. van Noort, M., Hubeny, I., Lanz, T., 2003, in *Stellar Atmosphere Modeling*, I. Hubeny, D. Mihalas, & K. Werner eds., ASP Conf. Ser. Vol. 288, p. 445
21. Wolf, S., Henning, Th., Stecklum, B., 1999, A&A, 349, 839

Multidimensional Radiation Hydrodynamics

Wolfgang Kalkofen[1,2]

[1] Harvard-Smithsonian Center for Astrophysics, 60 Garden Street, Cambridge, MA 02138, USA
[2] Kiepenheuer Institut für Sonnenphysik, Schöneckstrasse 6, 79104 Freiburg, Germany

Summary. We describe a time-dependent, multidimensional radiation-hydrodynamic problem for the H line of Ca II in the solar chromosphere and propose an empirical approach using time-independent, one-dimensional equations. The purpose of the investigation is to determine the salient features of the structure of the underlying atmosphere and to decide between two fundamentally different models of the chromosphere. We discuss the case of the oscillations of the small-scale calcium bright points in the quiet chromosphere, whose intensities fluctuate intermittently with periods near the acoustic cutoff period of 3 min. The observed profile of the H line shows that a bright point is formed by an acoustic wave traveling upward in downward-streaming gas. The shape of the oscillating gas is approximately that of a cone oriented vertically, with the apex in the photosphere and an opening angle of $\sim 90°$. The effects of multidimensional radiative transfer on the solution of the equations of radiation hydrodynamics are likely to be small compared with the effects on 3D wave propagation, and the influence on the emergent radiation field may be treated in a perturbation approach.

1 Introduction

Dynamical problems in astrophysics are described by the time-dependent, multidimensional equations of radiation hydrodynamics, in which radiative transfer is intimately coupled to the dynamics of the gas. For the solar atmosphere, such problems have never been solved in all generality in a real case since the demands on computer resources are too severe. Instead, the problems are simplified by describing them in terms of plane-parallel (1D) atmospheres. To justify this approach one would like to know its limits of applicability and to estimate the errors incurred.

In this paper I describe the dynamics of chromospheric bright points, which are observed for the H and K lines of Ca II in features that are formed in the middle chromosphere, at a height of $z \sim 1$ Mm above the photospheric level of unit optical depth at 5000 Å. The intensity of the blue emission peaks, H_{2v} and

K_{2v}, in the cores of the calcium lines fluctuates intermittently with periods near the acoustic cutoff period, P_{ac}, about 3 min in the solar photosphere. Acoustic waves at the cutoff period have the effect of exciting the atmosphere to oscillate at that period. The description of the state of the gas therefore requires the equations of hydrodynamics. Since the origin of the waves in the photosphere is in small, point-like source regions, the geometry of the wave problem is multidimensional.

Bright-point oscillations are found only at a small number of discrete locations in the quiet sun. Although the quiet chromosphere oscillates everywhere with a typical time scale of 3 min, most of these oscillations in the intensities of the H_{2v} and K_{2v} emission peaks are more or less symmetric. But the bright points show much higher intensity and distinctly asymmetrical fluctuations, with higher upward than downward excursions of the intensity, indicating a nonlinear process. The horizontal size of H_{2v} bright points is of the order of only a few times the mean free path of photons at the frequency of the emission peaks. Effects due to the presence of the boundary of the oscillating gas are therefore beginning to become noticeable. This makes also the radiative transfer problem multidimensional.

A full solution of the time-dependent, multidimensional radiation hydrodynamics problem would place prohibitive demands on current computer resources. Recourse in numerical simulations is therefore taken to one-dimensional geometry. The consequences of this assumption are more severe for the hydrodynamics than for the radiative transfer.

The purpose of this paper is to investigate the difference between the 1D and the 3D hydrodynamic solutions and to describe approximate corrections of the 1D solutions. Further work, which will not be pursued in this paper, would concern 3D solutions of the hydrodynamics for idealized media as well as a study of edge effects in the radiative transfer. In Section 2, I discuss the phenomenology of calcium bright points; in Section 3, I describe the time-dependent and the time-average temperature structures of the dynamical model; in Section 4, I use the conflict between simulated and observed H line intensities to devise a 3D correction to the 1D solution; and in Section 5, I present conclusions and suggest further work that could be performed in order to learn more about the errors incurred in the 1D solutions.

2 Calcium Bright Points

Acoustic waves traveling vertically in a stratified atmosphere are dispersive: if their period is short relative to the acoustic cutoff period, which is $P_{ac} \sim 3$ min in an isothermal gas at the temperature of the solar photosphere, then they travel like the sound waves we are familiar with. But as the period approaches the acoustic cutoff, dispersion becomes noticeable: the phase velocity of the waves increases above the sound speed and the group velocity decreases. At

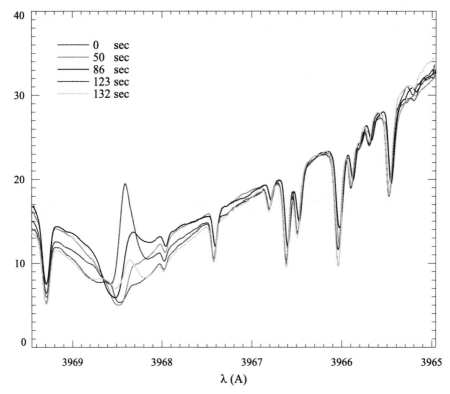

Fig. 1. Five phases in the 3 min evolution of the H line of an H_{2v} bright point (from Kariyappa 1992, 2003)

the acoustic cutoff, the phase velocity is infinite and the group velocity is zero. A plane wave at the cutoff sets the atmosphere to oscillate but does not transport any energy.

The dispersive nature of the acoustic waves is evident in the spectrum of the H line ($\lambda = 3968.49$ Å) at a bright point, observed by Kariyappa (1992, 2003), shown in Figure 1. The intensity profile of the blue emission peak at the instant of maximal brightness is shifted towards the red. As Carlsson & Stein (1994, 1995: hereafter CS) have shown with simulations of bright-point dynamics, the intensity increase is caused by an upward-propagating shock wave. The redshift of the peak seen in Fig. 1 implies that the shock is traveling in downward-flowing gas. The gas overlying the layer of formation of the blue emission peak is flowing downward as well, as is apparent by the line center, called H_3, also being redshifted. This downward motion of the gas is caused by the preceding wave, which set the atmosphere to oscillate at the acoustic cutoff period.

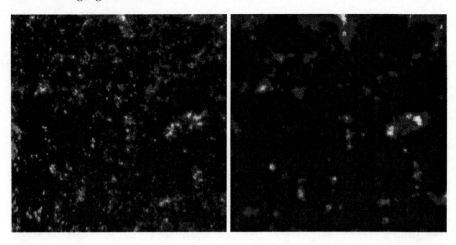

Fig. 2. K line emission with a 300 mÅ filter centered on the K_{2v} peak. Left frame: 5 s exposure, with K_{2v} bright points and network bright points. Right frame: The same area on the sun, in a 1 h exposure, with network bright points (from von Uexküll & Kneer 1995)

Plane disturbances traveling vertically in an isothermal atmosphere are described by the Klein-Gordon equation (Rae & Roberts 1982). An impulsive solution of this equation (Lamb 1909, 1932) shows the atmosphere after the pulse to oscillate at the cutoff period with decreasing amplitude. Stochastic wave excitation, as is thought to occur in the solar convection zone in the generation of the waves that heat the solar chromosphere (Sutmann, Musielak & Ulmschneider 1998) also gives rise to fluctuations at the 3 min period. This is true as well for the excitation (by an unknown mechanism) of the large-amplitude waves that cause the disturbances we observe in the middle chromosphere ($z \sim 1$ Mm) as H_{2v} and K_{2v} bright points.

In the quiet sun, the 3 min oscillations by waves associated with the general heating of the chromosphere occur everywhere in the nonmagnetic interior of supergranulation cells (Grossmann-Doerth, Kneer & v. Uexküll 1974). But as seen in Figure 2, calcium bright points occur at only 10 to 20 discrete points in a typical supergranulation cell (Brandt et al. 1992); the diameter of bright points tends to be less than 2 Mm, and their filling factor of the cell interior is 5% to 10% (von Uexküll & Kneer 1995).

The temporal evolution of the emergent intensity at the location of an H_{2v} bright point, seen in Figure 3 (from Lites, Rutten & Kalkofen 1993), shows intermittent oscillations with a period of ca. 3 min, from groups of typically 3 to 5 intensity flashes; the groups are separated from one another by somewhat longer time intervals. The intensity enhancement is almost exclusively on the blue (left) side of line center, marked in white to guide the eye. Clearly in

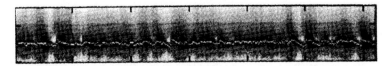

Fig. 3. Evolution of the H line intensity of an H_{2v} bright point as a function of wavelength and time. 10 min separation of tick marks on the time axis (from Lites et al. 1993)

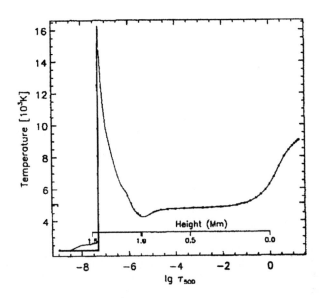

Fig. 4. Instantaneous temperature profile of the dynamical model for an H_{2v} bright point (from Carlsson & Stein 1995)

evidence is the redshift of the line center, which occurs simultaneously with the strong enhancement of the blue emission peak H_{2v}.

3 The Dynamical Model of the Chromosphere

The solution by CS of the radiation hydrodynamics equations in one-dimensional geometry results in a height- and time-dependent temperature structure, $T_{CS}(z,t)$. A snapshot, $T_{CS}(z,t_0)$, is seen in Figure 4, from CS. The background temperature decreases monotonically from 9 kK in the convection zone ($\log \tau \sim 1$) through the photosphere ($0 < z < 0.5$ Mm) and the

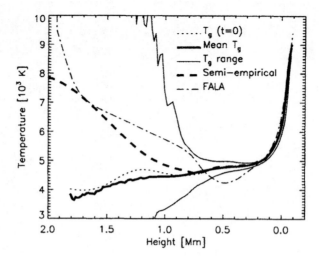

Fig. 5. Chromospheric temperatures. From top to bottom at $z = 1$ Mm: upper envelope of the time-dependent temperature in the dynamical model of CS; FALA, the coolest of the FAL models; the CS model representing the time average emission; the solar model in radiative equilibrium, which serves as the starting model of the simulations; the CS model representing the time average temperature; and the lower envelope in the dynamical model (from Carlsson & Stein 1995)

traditional height range of the chromosphere ($z > 0.5$ Mm). This temperature drop is interrupted by the upward-propagating shock, which in the snapshot has reached a height of $z \sim 1.4$ Mm. In the shock front, the temperature rises from 3 kK to 16 kK.

The low preshock temperature is caused by the absence of a background atmosphere from the model, and the high postshock temperature is a consequence of the plane-wave assumption for the acoustic waves. The excitation of the oscillations is by waves for which the power spectrum has been observed by Lites et al. (1993) in the Doppler motion of the photospheric Fe I line formed at a height of $z = 250$ km and located directly below the H_{2v} bright point. The model represents the temperature structure in the solar chromosphere at the location of the bright point that is associated with the photospheric Fe I line. It differs from the actual solar temperature structure because of simplifying assumptions and because of the reduced frequency range over which acoustic waves can be observed; the range is limited by seeing, and the effects of radiative transfer for lines subject to velocity fluctuations.

Various temperatures of the chromosphere can be seen in Fig. 5. The outer curves represent the range of fluctuations caused by the upward-propagating acoustic waves. This time-varying temperature allows two averages to be defined: (1) a direct time-average, called mean $T_g(z)$ in the model, and a so-called semiempirical temperature, $T_{\rm CS}(z)$, which reproduces the time-average of the emission of the time-dependent dynamical model. The mean $T_{\rm g}(z)$ drops monotonically in the upward direction, as a consequence of the relatively long time spent by the atmosphere at very low temperature, which drops below $T(z,t) \sim 3$ kK at a height of $z = 1.1$ Mm. The semiempirical temperature $T_{\rm CS}(z)$, on the other hand, reaches its lowest value, $T_{\rm CS} = 4.5$ kK, at a height of $z \sim 0.7$ Mm. At greater height, that temperature rises monotonically in the upward direction.

The traditional empirical temperature structure of the solar chromosphere, by Fontenla, Avrett & Loeser (1993; hereafter FAL), for which the coolest model ($T_{\rm FALA}$) is shown in the figure, has its lowest value, $T_{\rm FALA} \sim 4.2$ kK, at a height of $z = 0.5$ Mm. The dotted temperature curve is the starting model of the simulation of the dynamics. It represents a solar model in radiative equilibrium, i.e., it is without any wave heating.

Several features of the semiempirical model of Carlsson & Stein are unusual and surprising. We expected their model for the bright-point oscillations to be at least as hot as the model of the average quiet chromosphere, called FAL model C or FALC, which is hotter than FALA, the model shown in the CS paper. At the height of formation of the emission peaks $H_{\rm 2v}$ and $K_{\rm 2v}$, at $z \sim 1$ Mm, the difference in temperature between the two models, $T_{\rm FALC} - T_{\rm FALA}$, is $\delta T = 400$ K. This spread is a measure of the temperature variation of the quiet, nonmagnetic chromosphere at this formation height. But the CS model is not only cooler than the coolest empirical model. At a height of $z = 1$ Mm it is cooler than the average by $T_{\rm FALC} - T_{\rm CS} \sim 1300$ K, i.e., an amount that is large compared to any plausible temperature variation of the chromosphere. At the same time, and consistent with the low temperatures, the time-average $H_{\rm 2v}$ intensity is also low, $I_{\rm CS} = 0.2 \times I_{\rm FALC}$, i.e., it is low by a factor of 5.

Another surprising feature of the semiempirical model by CS is the magnitude of the temperature rise between the height of the base of the chromosphere, $z = 0.5$ Mm, and the height of formation of the blue emission peaks in the H and K lines, $z = 1$ Mm. In the average sun, that temperature difference is $T_{\rm FALC}(1\ {\rm Mm}) - T_{\rm FALC}(0.5\ {\rm Mm}) = 1500$ K; in the CS model, it is $T_{\rm CS}(1\ {\rm Mm}) - T_{\rm CS}(0.5\ {\rm Mm}) = 50$ K. From this comparison between the empirical FAL model, which reproduces the average emergent intensity of the chromosphere, and the dynamical model, which is intended to provide a physical explanation for it, we conclude that the dynamical model does not describe the solar chromosphere, although the mechanism of bright-point formation is explained by it. The main limitations of the time-dependent model are that it does not give a quantitative description of the emergent intensity of a bright point, and that it does not explain the origin of the acoustic waves that produce the bright points. However, the CS results do limit theories for the generation of the waves causing bright points.

4 3D Correction of the 1D Dynamical Model

We have tested the semiempirical CS model by comparing it with the average sun as described by the empirical FAL models, and found the CS model to be too cool and the chromospheric temperature rise too small to be a plausible representation of the sun. Another test is the comparison of the time-dependent, emergent intensity of the model with that of the sun. A convenient feature is the H_{2V} intensity at maximal brightness. Our expectation is that the instantaneous value of the simulated intensity is lower than that of the observed intensity, because the time-average temperature of the model is lower than that of the sun and, more importantly, because the frequency range of the observed velocity spectrum is reduced from that of the actual solar velocity spectrum by the modulation transfer function, which accounts for the property of the acoustic waves to cause only line broadening, but no Dopper signal, when the wavelength of the waves is too short relative to the width of the radiation contribution function of a spectral line that is used for measuring the velocity spectrum. The power in the acoustic waves serving as the boundary condition in the simulation is therefore lower than in the solar atmosphere. As a consequence, we would expect the simulated intensity to be lower than the observed intensity.

The comparison shows the surprising result (cf. Carlsson & Stein 1997, Fig. 18) that the simulated H_{2V} intensity at maximal brightness is higher than observed, indicating a problem with the assumed geometry of wave propagation. We conclude that the dynamical model has two defects as a quantitative representation of the chromosphere: The time-average intensity is too low, and the instantaneous H_{2V} intensity at maximal brightness is too high.

It is evident that the CS model requires two separate and independent corrections in order to be a physical description of the sun. We now investigate the excess intensity of the H_{2V} emission peak at maximal brightness.

The reason for the excess brightness of the instantaneous H_{2V} intensity at its maximum is the assumption that the acoustic waves propagate as plane waves, which prevents horizontal spreading of the upward propagating wave and, therefore, dilution of the energy flux. This property of the waves in the sun is shown by the increase of the area oscillating with large amplitude at the 3 min period; its filling factor grows from 5% − 10% at a height of 1 Mm (von Uexküll & Kneer 1995) to 50% at the top of the chromosphere at $z = 2$ Mm (Carlsson, Judge & Wilhelm 1997). It is also shown by the systematic increase of the diameter of the bright points from their sources in the photosphere, where their size is estimated to be of the order of the width of intergranular lanes, about 100 km, to the top of the chromosphere, where they have a diameter of 5–6 Mm (Carlsson et al. 1997). This growth implies that the channels in which acoustic waves travel upward have the form of vertical cones with an opening angle of more than 90° (Kalkofen 2003).

Assuming that a cone has an opening angle of 90° and that its apex is at $z = 0$ we can estimate the increase of the cross-sectional area of the disturbed region between the layer of formation of the Fe I line in the photosphere, at $z = 250$ km, where the observed Doppler motion provides the driving of the waves in the simulations, and the height of formation of the H_{2v} and K_{2v} emission peaks, at $z = 1$ Mm. We find the cross-sectional area to increase by a factor of at least 10 and estimate therefore that the simulated solar intensity must be lowered by a factor of about 10 in order to account for the expansion of the wave. Thus, correcting the time-average intensity for the effects of 3D wave propagation, we find that $I_{CS,3D} = 0.02 \times I_{FALC}$. Converting this intensity into an equivalent temperature, the corrected CS temperature is $T_{CS,3D}(1 \text{ Mm}) = 4640$ K. This is lower than the CS temperature at the base of the chromosphere, $T_{CS}(0.5 \text{ Mm}) = 4800$ K, by $\delta T = 160$ K. Thus, the CS atmosphere, when corrected for the effects of spherical wave propagation, does not have a chromospheric temperature rise between the base of the chromosphere and the height of $z = 1$ Mm. Since the area correction of the 1D wave increases approximately as the square of the height (Bodo et al. 2000) we conclude that the CS atmosphere, when corrected for the effects of 3D wave propagation, has no chromosphere.

We draw several important conclusions from our analysis of the dynamical model of CS:

1. The time-dependent, dynamical model correctly accounts for the mechanism of calcium bright-point formation and reproduces signature features, such as the simultaneous occurrence of maximal brightening of the H_{2v} and K_{2v} emission peaks and redshift of the line center.

2. The time-average (semiempirical) model of CS, when corrected for the effects of 3D wave propagation, lacks a chromospheric temperature rise. It therefore cannot be a model of the temperature structure of the solar chromosphere.

3. The propagation of acoustic waves in a stratified atmosphere is in the form of 3D waves, which travel upward in horizontally expanding channels. This is shown here only for the acoustic waves of chromospheric bright points, but it is indicated also for the acoustic waves in chromospheric heating. It is likely to be true generally for the propagation of acoustic waves in a stratified atmosphere.

5 Summary

We describe a time-dependent, three-dimensional (3D) problem in radiation hydrodynamics from solar physics. Numerical solutions have been obtained by Carlsson & Stein (1994, 1995) only in the 1D approximation. We find the correction for 3D wave propagation on the hydrodynamics to be large and estimate the magnitude of the effect. In the radiative transfer, the 3D effects are likely to be small, but their investigation, including the study of effects of the boundary between the region disturbed by the acoustic waves and the undisturbed background atmosphere is left as a challenge. Solar observations of high spatial and temporal resolution would provide constraints on the numerical simulations.

The case discussed in detail is that of 3 min oscillations at calcium bright points in the quiet (i.e. nonmagnetic) solar chromosphere. We show that the aim to heat the chromosphere with only the energy dissipated in bright-point oscillations cannot be realized since their energy flux even at the location of bright points is too small by two orders of magnitude.

The main conclusions of the hydrodynamic modeling are that dissipation of the acoustic waves which are responsible for the bright-point dynamics does not account for the hot chromosphere, and that the waves propagate as three-dimensional waves, emanating from point-like sources regions. It is surmised that the 3D wave propagation is a general property of acoustic-wave propagation in stratified atmospheres.

Acknowledgements. I thank my colleagues at the Kiepenheuer Institut and its director, Oskar von der Lühe, for their hospitality and in particular Reiner Hammer for stimulating discussions; I also thank R. Kariyappa for providing the profiles of the H line. I am grateful to the University of Freiburg for a Mercator guest professorship funded by the DFG.

References

Bodo, G., Kalkofen, W., Massaglia, S. & Rossi, P. 2000, AA, 354, 296
Brandt, P.N., Rutten, R.J., Shine, R.A. & Trujillo Bueno, J. 1992, ASP Conf. Ser. 26, 161
Carlsson, M., Judge, P.G. & Wilhelm, K. 1997, ApJ, 486, L63
Carlsson, M. & Stein, R.F. 1994, in Proc. Mini-Workshop, Chromospheric Dynamics, ed. M. Carlsson, Oslo University, 47 (CS)
Carlsson, M. & Stein, R.F. 1995, ApJ, 440, L29 (CS)
Carlsson, M. & Stein, R.F. 1997, ApJ, 481, 500
Fontenla, J.M., Avrett, E.H. & Loeser, R. 1993, ApJ, 406, 319 (FAL)
Grossmann-Doerth, U., Kneer, F. & v. Uexküll, M. 1974, Solar Phys., 37, 85
Kalkofen, W., 2003, PASP Conf. Ser. 286, 385
Kariyappa, R. 1992, Thesis, Bangalore Univ., Bangalore, India
Kariyappa, R. 2003, private communication
Lamb, H. 1909, Proc. London Math. Soc., ser. 2, 7, 122
Lamb, H. 1932, Hydrodynamics, Cambridge Univ. Press
Lites, B.W., Rutten, R.J. & Kalkofen, W. 1993, ApJ, 414, 345

Rae, I.C. & Roberts, B. 1982, ApJ, 256, 761
Sutmann, G., Musielak, Z.E. & Ulmschneider, P. 1998, AA, 340, 556
von Uexküll, M. & Kneer, F. 1995, AA, 294, 252

Probing the Initial Conditions for Star Formation with Monte Carlo Radiative Transfer Simulations

Dimitris Stamatellos and Anthony P. Whitworth

School of Physics & Astronomy, Cardiff University, 5 The Parade, Cardiff CF24 3YB, Wales, UK
D.Stamatellos@astro.cf.ac.uk
A.Whitworth@astro.cf.ac.uk

Summary. Monte Carlo radiative transfer is a method that can be used for the treatment of physical systems with arbitrary geometries. We review the basics of the method, and present a recent idea that allows the use of this method in radiative equilibrium without iteration. We also discuss special techniques that can be used to improve its performance, and extend the method to treat density fields that result from hydrodynamic simulations. We finally present applications of Monte Carlo radiative transfer in the study of prestellar cores, i.e. condensations in molecular clouds that are expected to collapse and form stars.

1 Introduction

The Monte Carlo radiative transfer method has been applied successfully to a variety of radiative transfer problems such as protostellar disks ([WLB02]), protostellar envelopes ([WWV03]), prestellar cores ([SW03]), dusty disk galaxies ([BD02]), and supernovae ([Luc99b]). The main advantage of the method is its 3D nature, which makes it attractive for use in a variety of problems.

2 Monte Carlo Radiative Transfer

The Monte Carlo method uses a large number of monochromatic luminosity packets (hereafter referred to as L-packets) to represent the radiation being transported through the computational domain. This radiation originates from discrete sources within the domain (i.e. stars), from diffuse emission within the domain (i.e. radiative cooling), and from the background radiation field incident on the boundary of the domain. The method is based on the fundamental principle of Monte Carlo methods according to which a quantity $\xi \in [\xi_1, \xi_2]$ can be sampled from a probability distribution function (PDF) p_ξ

using uniformly distributed random numbers $\mathcal{R} \in [0,1]$, by picking ξ such that $\int_{\xi_1}^{\xi} p_{\xi'} d\xi' / \int_{\xi_1}^{\xi_2} p_{\xi'} d\xi' = \mathcal{R}$. This formula is used here to chose the frequency of the emitted luminosity packets, their direction, their mean free path and the kind of interactions (absorption or scattering) they have.

L-packet Emission: The frequency of an *L*-packet emitted by a source is chosen using the source radiation field I_ν as the frequency PDF. Each of the *L*-packets is injected stochastically into the medium, either from a specific point (for a point star, see e.g. [YMW84]) or from the boundaries of the system (for background radiation, see e.g. [SWA04]). Each *L*-packet is also assigned a random optical depth, using $\tau_\nu = -\ln \mathcal{R}_\tau$, $\mathcal{R}_\tau \in [0,1]$, and this determines how far the packet propagates into the medium before it interacts with it.

Radiative Transfer Cell Construction: The computational domain in which the *L*-packets propagate is divided into a number of cells (RT cells). In regions where the density or the temperature gradients are large, more cells are needed. Both conditions can be fulfilled by constructing cells with dimensions less than, or on the order of, the local directional density and temperature scale-heights (see [SW03]).

L-packet Propagation: If τ_{total} is an *L*-packet's total optical depth then in order to calculate the distance it propagates into the system before it interacts with it, we need to calculate the line integral along the path of the packet,

$$\Delta S = \int_0^{\tau_{\text{total}}} \frac{d\tau}{\kappa_\lambda \rho} . \qquad (1)$$

In the general case it is not possible to calculate the preceding integral analytically. Our approach is to approximate this integral with a sum: $\Delta S = \sum_i (\delta \tau_i / \kappa_\lambda \rho_i) = \sum_i \delta S_i$. The element step δS_i is chosen according to the following condition:

$$\delta S_i = \text{MIN}\{\eta_\rho h_\rho, \eta\, l, (\tau_i + \epsilon)\, l, \eta_r |\mathbf{r}|\} , \qquad (2)$$

where $l = (\kappa_\lambda \rho_i)^{-1}$, and η_ρ, η, η_r are constants that determine the accuracy we demand (typical values are between 0.1 and 1). The first term ($\eta_\rho h_\rho$) ensures that the density does not change much in one element step (h_ρ is the density scale height in the direction that the *L*-packet propagates), the second term ($\eta\, l$) ensures that the element step is less than the mean free path of the packet and the third term [$(\tau_i + \epsilon)\, l$] takes effect on the last step. The last term ($\eta_r |\mathbf{r}|$) ensures that the distance the *L*-packet travels in one element step is less than the distance from the luminosity source. This term comes into effect when a gap exists around the source. The smaller the factors η_ρ, η, η_r are chosen, the better the accuracy in propagating *L*-packets.

L-packet Interactions (Absorption, Reemission and Scattering): When an L-packet reaches an interaction point within the medium (at the end of its optical depth), it is either scattered or absorbed, depending on the albedo. If the packet is absorbed its energy is added to the medium and raises the local temperature. To ensure radiative equilibrium the L-packet is immediately reemitted ([Luc99a], [BW01]). The new temperature of the cell that absorbed the packet is determined by equating the absorbed luminosity

$$L_i^{\text{abs}} = k_i\,\delta L\,, \tag{3}$$

to the emitted luminosity

$$L_i^{\text{em}} = 4\pi\,m_i \int_0^\infty \kappa_\nu B_\nu(T_i)\,\mathrm{d}\nu\,, \tag{4}$$

where the index i refers to the cell where the absorption takes place, k_i is the number of packets that have been absorbed by the cell, T_i the temperature of the cell, m_i the mass of the cell, and δL the luminosity of each packet. The frequency of the reemitted L-packet is chosen using a PDF so as to correct for the packets that were reemitted previously from the cell with an incorrect frequency distribution. The adjustment PDF is constructed from the difference in the emissivity of the cell before and after the emission of the L-packet ([BW01])

$$p(\nu)\mathrm{d}\nu = \frac{\kappa_\nu[B_\nu(T_i+\Delta T_i)-B_\nu(T_i)]\mathrm{d}\nu}{\int_0^\infty \kappa_\nu[B_\nu(T_i+\Delta T_i)-B_\nu(T_i)]\mathrm{d}\nu}\,, \tag{5}$$

provided that the cell is in radiative equilibrium ([BS04]). Using this procedure the correct temperature distribution and spectrum of the system are obtained at the end of the simulation, when all the packets have been propagated through the medium and escaped, *without iteration*. If an L-packet is scattered then it is assigned a new direction, using the scattering phase function due to [HG41].

Code Implementation and Efficiency: Our radiative transfer code PHAETHON is described in detail in [Sta03]. A simple code efficiency analysis, indicates that the *L-packet propagation* routine takes about 25–50% of the computational time, depending on the specific problem. Thus, this is the routine that should be targeted by any efforts to diminish the running time of the code. Time efficiency is very important since a large number of L-packets is needed for good statistical results. To reduce the running time of the code, whilst maintaining good results, the specific nature of the system we examine should be taken into account (for example, in the case of a uniform density medium we can propagate each L-packet in a single step), any kind of symmetry should be exploited (e.g. for spherically symmetric systems), and look-up tables should be used to solve for the cell temperature after absorbing a packet and then to calculate the reemission frequency of the new packet.

Code Parallelisation: The nature of this radiative transfer method means that each L-packet is treated independently. To calculate the temperature of a cell we just need to know the number of luminosity packets absorbed in that cell; the order does not matter. Thus, each processor of a parallel machine could be used to treat a single L-packet until it escapes from the computational domain, meaning that it would be quite easy to parallelise the code using an automated parallelisation protocol such as OPENMP. The only problem would arise if an L-packet were absorbed by a cell while another L-packet had just been absorbed by the same cell and calculations of the cell temperature and L-packet reemission were ongoing. The problem that would arise is that during the calculation the value of the cell luminosity variable changes. The probability of this happening would be relatively small, and even in this case the absorption of one more L-packet would change the luminosity absorbed by the cell only by a small fraction and, consequently, the temperature would also change only by a small fraction. Thus, in any case the frequency of the re-emitted L-packet would be calculated with high precision.

Code Tests: [IGM97] used three different, well established radiative transfer codes using different numerical schemes to solve a set of benchmark spherical geometry problems. In all cases, these methods gave differences smaller that 0.1% and, as [IGM97] noted, the solution should be considered exact. PHAETHON reproduced those results and also the results of [BW01] for a disk-like structure embedded in an envelope. [SW03] introduced the thermodynamic equilibrium test, in which a physical system is exposed to a uniform, isotropic blackbody radiation field of a given temperature T. Thermodynamic equilibrium dictates that every part of the system will adopt the same temperature T. This also means that the intensity of the radiation coming from the system is the same as that of the illuminating blackbody field. This test can be applied to any physical structure, it is very robust, and it is suggested that any continuum radiative transfer code should be tested in this way. PHAETHON also performed well in this test.

Discussion: Advantages, Disadvantages and Special Techniques

The Monte Carlo radiative transfer method conserves energy exactly, accounts for the diffuse radiation field, can be implemented for any geometrical structure, and is very efficient, making it attractive for use in a variety of problems. However, it can be implemented, without iteration, only when the opacity is independent of temperature, so the method is useful for treating radiation transport against opacity due to dust grains which are large enough to be in thermal equilibrium. The method solves for the dust temperature and also calculates continuum spectral energy distributions (SEDs), intensity and polarisation maps at different wavelengths and at different viewing angles.

The main disadvantage of the method is that a large number of L-packets needs to be simulated, so that the sources' radiation field and the interactions with the dust are adequately sampled. The number of L-packets needed depends on the specific problem. Typical numbers for obtaining temperatures and SEDs with low statistical noise are 10^6–10^7 packets, and for good isophotal maps $> 10^8$–10^9 packets. Another disadvantage of the method is that it does not behave well in low optical depth regions (because not enough packets are absorbed) or in regions with high optical depths (because packets interact too many times and thus spend too much time in such regions).

The method can be made more efficient by optimising the subroutine that propagates the luminosity packets in the computational domain, and/or by using techniques that extract more information from a lower number of packets. In the *weighted L-packets* technique packets are allowed to have different luminosities, i.e. different weights (e.g. [Wit77]). All packets start off with the same weight (=1), but during the simulation their weight is adjusted using the enforced interaction technique (packets are enforced to interact with the medium and their weight is reduced according to their probability of interaction) or the peel-off technique (packets are enforced to escape towards the observer after an interaction event and their weight is reduced according to the probability they have to be observed) (e.g. [YMW84]). In the *mean intensity* technique ([Luc99a]) the performance of the method is improved in low-optical depth regions by estimating the energy that is absorbed by each cell from the mean intensity of the radiation field, which, in turn is calculated from the time that the photons spend within each cell. Other techniques include the "net-exchange" Monte Carlo method (e.g. [LDH02], Soufiani this volume), the use of the formal integral of intensity to calculate SEDs and isophotal maps after the Monte Carlo simulation (e.g. [Luc99a]), and a switch to the diffusion approximation in regions of high-optical depth (e.g. [Gen01], [BWW02]).

3 Monte Carlo Radiative Transfer and SPH

Smoothed Particle Hydrodynamics (SPH) is a Lagrangian method that uses a large number of particles to represent a fluid (see review by [Mon92]). The particles interact with each other through gravity and hydrodynamic forces (pressure and viscosity forces). The physical properties of the fluid (e.g. density) at a given point are calculated as weighted averages over the local neighbourhood. The inclusion of radiative transfer in hydrodynamic simulations in an efficient way is computationally very challenging. In this first approach to this problem, we focus on performing radiative transfer calculations on single snapshots of SPH simulations ([Sta03], [SW04]).

The radiative transfer simulations applied to SPH density fields produce dust temperature distributions, SEDs and intensity maps at different wavelengths. Hence, they can be used for comparing the results of hydrodynamic simulations with observations. However, the radiative transfer is not treated consistently within SPH simulations, in the sense that it is not combined with an energy equation. Hence, we do not capture the full effect of radiation transfer on the evolution of the system. This may be important in some systems (e.g. a fragmenting protostellar disk).

Construction of RT Cells: The radiative transfer cells absorb, scatter and reemit L-packets. The cell construction depends on the specific system under study, and, thus, in a hydrodynamic simulation the RT cells need to be constructed at every step of the simulation. Therefore, a robust algorithm for the cell construction is needed. To construct RT cells in an SPH simulation snapshot we take advantage of the fact that SPH uses an octal tree structure to group particles together when calculating gravity forces. The SPH tree is a hierarchical division of the computational domain into virtual cells and subcells ([BH86]). When the SPH tree is being built, we give identity numbers to each of the virtual cells created at every level of the tree. We also record information about the size of each cell and the number of SPH particles it contains. These cells serve as *potential RT cells*. The mass and the density of each potential RT cell are determined from the mass of the SPH particles in the cell and the size of the cell. We then choose which of these cells will be the actual *active RT cells*. This is done by choosing a maximum number, $N_{\max}$, of SPH particles that should be contained in each RT cell. The above condition leads to a unique subgroup of SPH cells that constitute the active RT cells. This method has the advantage that the size of the RT cells is adjusted automatically, so that all cells have comparable masses. We choose $N_{\max} \sim 100$, i.e. somewhat larger than the mean number of SPH neighbours, so that the size of the RT cells is on the order of the smoothing length h, i.e. the density resolution of the SPH simulation. This method constructs an adaptive radiative transfer grid (for other adaptive grids see [WHS99], [KH01], [SBH02]), that uses the SPH tree structure already built within SPH, and hence it can be implemented within an SPH simulation without greatly increasing the computational time.

L-Packet Propagation: The propagation of L-packets in the computational domain is performed in small steps, δS_i, that gradually decrease the optical depth τ of the L-packet until $\tau = 0$, whereupon the L-packet reaches an interaction point. To choose the propagation step δS_i, we use the same condition as in Section 1, with the only difference being the addition of the term $\eta_c \, S_c$ in place of the directional density scale-height term $\eta_\rho \, h_\rho$. This term ensures that the L-packet does not travel a distance that it is too large in comparison to the size of the local cell, which in effect is the directional density scale-height. As an L-packet propagates through the computational domain, we use the SPH tree to find which cell the L-packet is in. The search starts from the root cell and continues to the lower level cells, until an active RT cell is reached. A more efficient way would be to perform the search upwards, starting from the cell where the L-packet was before the last step.

Tests & Problems: The method performs well against the thermodynamic equilibrium test. However tests against the [IGM97] benchmark problems show that the method does not perform well when regions of high temperature gradients are present in the physical system (e.g. in regions around stars). The reason for this problem is that the RT cells are too big to be able to capture the high temperature gradients. We solve this problem by using an additional 1D or 2D spherical grid around each star in the simulation. After this addition the method performs well. Another solution to the problem is to further subdivide RT cells in regions of high temperature gradients to smaller RT subcells (for details see ([Sta03], [SW04]).

4 Radiative Transfer Models of Prestellar Cores

Star formation is believed to happen when dense regions inside molecular clouds collapse. This occurs when self-gravity dominates over the forces that support the cloud (e.g. thermal pressure, magnetic fields, turbulence). The details of this process are not very well understood but recent observations, theoretical models and numerical simulations indicate that star formation is a dynamical, violent process in which magnetic fields, turbulence, the presence of ionised gas around hot young stars, ejection of mass from newly-born stars and shock waves induced by supernova explosions, may play an important role (see review by [Lar03]).

Our study focuses on the study of prestellar cores in molecular clouds that collapse and form stars. These cores appear to be on the verge of collapse or already collapsing. They are believed to represent the initial stage of star formation. Prestellar cores have extent $\gtrsim 0.2\text{--}4 \times 10^4$ AU and masses $0.05\text{--}10$ M$_\odot$ (see review by [AWB00]). They do not have an internal energy source and they are heated by the ambient radiation field. They have approximately uniform density in their central regions, and the density then falls off in the envelope. If the density in the envelope is fitted with a power law, $n(r) \propto r^{-\eta}$, then $\eta \sim 2 - 4$. Here $\eta \sim 2$ is characteristic of more extended prestellar cores in dispersed star formation regions (e.g. L1544, L63 and L43), whereas $\eta \sim 4$ is characteristic of more compact cores in protoclusters (e.g. ρ Oph and NGC2068/2071). These features are conveniently represented by a Plummer-like density profile ([Plu15]), that is modified to include azimuthally symmetric departures from spherical symmetry. We study flattened cores, using density profiles of the form

$$n(r,\theta) = n_0 \frac{1 + A\left(\frac{r}{r_0}\right)^2 \sin^p(\theta)}{\left[1 + \left(\frac{r}{r_0}\right)^2\right]^{(\eta+2)/2}}, \quad (6)$$

where n_0 is the density at the centre of the core, and r_0 is the extent of the region in which the density is approximately uniform (Fig. 1, left). The parameter A determines the equatorial-to-polar optical depth ratio e, i.e. the maximum optical depth from the centre to the surface of the core ($\theta = 90°$), divided by the minimum optical depth from the centre to the surface of the core ($\theta = 0°$ and $\theta = 180°$). The parameter p determines how rapidly the optical depth from the centre to the surface rises with increasing θ, i.e. going

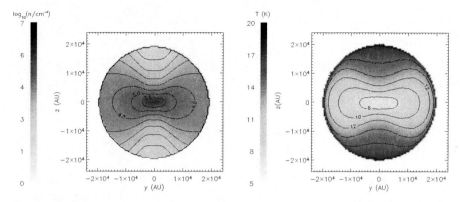

Fig. 1. Left: Density on the $x = 0$ plane for a flattened asymmetric core with equatorial-to-polar optical depth ratio $e = 2.5$ and $p = 4$. We plot isopycnic contours every $10^{0.5} \text{cm}^{-3}$. The central contour corresponds to $n = 10^{5.5} \text{cm}^{-3}$. **Right:** Temperature on the $x = 0$ plane for the same model, calculated with a Monte Carlo RT simulation. We plot isothermal contours from 8 to 18 K, every 2 K

from the north pole at $\theta = 0°$ to the equator at $\theta = 90°$. We assume that the core has a spherical boundary at radius $R_{\text{core}} = 2 \times 10^4$ AU, and use $n_0 = 10^6 \text{ cm}^{-3}$, $r_0 = 2 \times 10^3$ AU, $\eta = 2$.

Simulation Setup

Cell Construction: The code used for this problem is optimised for the study of cores having azimuthal symmetry. The core itself is divided into a number of cells by spherical and conical surfaces. The spherical surfaces are evenly spaced in radius, and there are typically 50–100 of them. The conical surfaces are evenly spaced in polar angle, and there are typically 10–20 of them. Hence the core is divided into 500–2000 cells. The specific number of cells used is chosen so that the density and temperature differences between adjacent cells are small.

L-Packet Frequency: We assume that the radiation field incident on the core is the [Bla94] interstellar radiation field (hereafter BISRF), which consists of radiation from giant stars and dwarfs, thermal emission from dust grains, mid-infrared emission from transiently heated small grains, and the cosmic background radiation.

L-Packet Emission: The L-packets are injected from the outside of the core with injection point and injection direction chosen to mimic an isotropic radiation field incident on the core (see [SWA04]).

Dust Properties: The dust composition in prestellar cores is uncertain, but in such cold and dense conditions, dust particles are expected to coagulate and accrete ice. We use the [OH94] opacities for a standard interstellar grain mixture (53% silicate and 47% graphite) that has coagulated and accreted thin ice mantles over a period of 10^5 yr at a density of 10^6 cm^{-3}.

Results: Dust Temperature, SED and Intensity Maps

The results of RT simulations are presented in detail in [SWA04]. In non-embedded cores, i.e. cores that are directly heated by the BISRF, the dust temperature drops from around 17 K at the edge of the core to 7 K at the centre of the core (Fig. 1, right). We also see that the dust temperature inside cores with disk-like asymmetry is θ dependent: the equator of the core is colder than the poles, as expected.

The core being very cold emits most of its radiation in the FIR region (Fig. 2, left). The SED of the core we examine, is the same at any viewing angle, because the core is optically thin to the radiation it emits (FIR and longer wavelengths).

The isophotal maps of the core depend on the observer's viewing angle and on the wavelength of observation. Our code calculates images at any wavelength, and therefore provides a useful tool for comparison with observations. We distinguish two broad wavelength regions which yield complementary information: (i) the submm and mm region (we choose 850 μm as a representative wavelength), and (ii) the region near the peak of the core emission (150–250 μm; we choose 200 μm as a representative wavelength).

In the first wavelength region (e.g. 850 μm) the core emission is mainly regulated by the column density (Fig. 3, left). Thus, the intensity is larger at

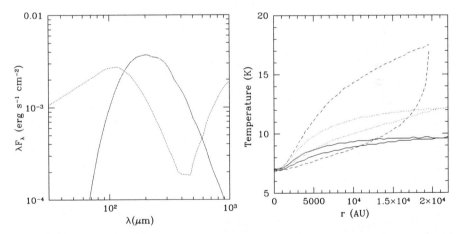

Fig. 2. Left: SED for the core in Fig. 1. The SED is independent of the observer's viewing angle. The dotted line on the graph corresponds to the incident radiation. **Right**: The effect of the parent cloud on cores. Temperature profiles of a non-embedded core (dashed lines), and of the same core at the centre of an ambient cloud with visual extinction $A_V = 4$ (dotted lines), and $A_V = 13$ (solid lines). The upper curve of each set of lines corresponds to the direction towards the pole of the core ($\theta = 0°$), and the bottom curve to the direction towards the core equator ($\theta = 90°$). The core is colder when it resides inside a thicker parent cloud

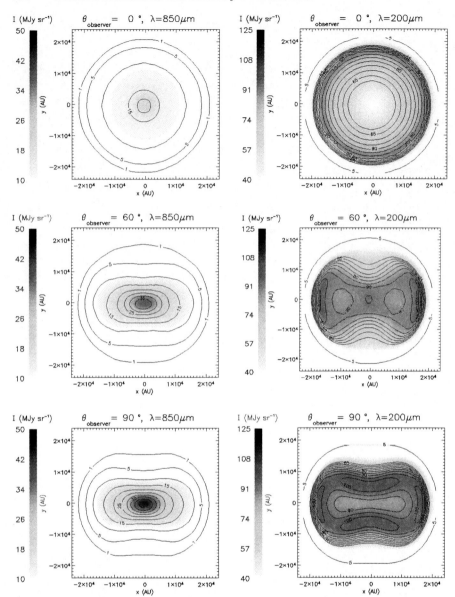

Fig. 3. Isophotal maps at viewing angles 0°, 60° and 90°, for a flattened prestellar core with equatorial-to-polar optical depth ratio $e = 2.5$ and $p = 4$. **Left:** 850 μm maps. We plot an isophotal contour at 1 MJy sr^{-1} and then from 5 to 50 MJy sr^{-1}, every 5 MJy sr^{-1}. The core appears elongated when viewed at an angle other than $\theta = 0°$. **Right:** 200 μm maps. We plot an isophotal contour at 5 MJy sr^{-1} and then from 60 to 110 MJy sr^{-1}, every 5 MJy sr^{-1}. There are characteristic symmetric features due to core temperature and orientation with respect to the observer. (We note the axes (x, y) refer to the plane of sky as seen by the observer)

the centre, where the column density is larger. For the same reason the core appears flattened when is viewed from any direction other than pole-on.

At 200 μm (Fig. 3, right) the core appears circular when viewed pole-on and flattened when viewed edge-on. The outer parts of the core can be more or less luminous than the central parts, depending on the core temperature and the observer's viewing angle. At any viewing angle the appearance of the core is determined by a combination of temperature and column density effects. This interplay between core temperature and column density along the line of sight results in characteristic features on the images of the cores, such as the two intensity maxima, at symmetric positions relative to the centre of the core, on the images at 90°(Fig. 3, right). Thus isophotal maps at 200 μm contain detailed information, and sensitive, high resolution observations at 200 μm, could be helpful in constraining the core density and temperature structure, and the orientation of the core with respect to the observer.

Embedded Prestellar Cores: Cores are generally embedded in molecular clouds, with visual optical depths ranging from 2 to 40. The ambient cloud absorbs the energetic UV and optical photons and re-emits them in the FIR and submm (because the ambient cloud is generally cold, $T_{cloud} \sim 20\text{--}100$ K). Therefore, the radiation incident on a core that is embedded in a cloud is reduced in the UV and optical, and enhanced in the FIR and submm ([MMP83]). We have performed radiative transfer calculations for a core embedded at the centre of an ambient cloud with $A_V = 4$ (Fig. 2, right, dotted lines) that show that the embedded core is colder ($T < 12$ K) and that the temperature gradients inside the core are smaller than in the non-embedded core. If the core is embedded in a thicker ambient cloud with $A_V = 13$, then the core is even colder (Fig. 2, right, full lines).

5 Conclusion

Monte Carlo radiative transfer is a powerful method that allows the treatment of a variety of physical systems with arbitrary geometry. The efficiency of the method has been considerably improved by new techniques that avoid iteration, propagate luminosity packets faster, and extract more information from a given number of packets. However, even better efficiency is required if this method is to be used to treat the time-dependent radiative transfer within hydrodynamic simulations.

Acknowledgement. We gratefully acknowledge support from the EC Research Training Network "The Formation and Evolution of Young Stellar Clusters" (HPRN-CT-2000-00155).

References

[AWB00] André, P., Ward-Thompson, D., & Barsony, M., Protostars and Planets, IV, 59 (2000)
[BD02] Baes, M. & Dejonghe, H., MNRAS, **335**, 441 (2002)
[BH86] Barnes, J. & Hut, P., Nature, **324**, 446 (1986)
[Bla94] Black, J.H., ASP Conf. Ser. 58: The First Symposium on the Infrared Cirrus and Diffuse Interstellar Clouds, 355 (1994)
[BS04] Baes, M., Stamatellos, D., Davies, J., Whitworth, A.P., et al., submitted (2004)
[BW01] Bjorkman, J.E. & Wood, K., ApJ, **554**, 615 (2001)
[BWW02] Bjorkman, J.E., Whitney, B.A., & Wood, K., Bulletin of the American Astronomical Society, **34**, 1185 (2002)
[Gen01] Gentile, N.A., 2001, Journal of Computational Physics, **172**, 543 (2001)
[HG41] Henyey, L.C. & Greenstein, J.L., ApJ, **93**, 70 (1941)
[IGM97] Ivezić, Z., Groenewegen, M.A.T., Men'shchikov, A., & Szczerba, R., MNRAS, **291**, 121 (1997)
[KH01] Kurosawa, R. & Hillier, D.J., A&A, **379**, 336 (2001)
[Lar03] Larson, R.B., ArXiv Astrophysics e-prints, 6595 (2003)
[LDH02] de Lataillade, A., Dufresne, J.L., El Hafi, M., Eymet, V., & Fournier, R., Journal of Quantitative Spectroscopy and Radiative Transfer, **74**, 563 (2002)
[Luc99a] Lucy, L.B., A&A, **344**, 282 (1999)
[Luc99b] Lucy, L.B., A&A, **345**, 211 (1999)
[MMP83] Mathis, J.S., Mezger, P.G., & Panagia, N., A&A, **128**, 212 (1983)
[Mon92] Monaghan, J.J., ARA&A, **30**, 543 (1992)
[OH94] Ossenkopf, V. & Henning, T., A&A, **291**, 943 (1994)
[Plu15] Plummer, H.C., MNRAS, **76**, 107 (1915)
[SBH02] Steinacker, J., Bacmann, A., & Henning, T., Journal of Quantitative Spectroscopy and Radiative Transfer, **75**, 765 (2002)
[Sta03] Stamatellos, D.: Monte Carlo Radiative Transfer in Prestellar Cores & Protostellar Disks. Ph.D. Thesis, Cardiff University, Wales, UK (2003) (www.astro.cf.ac.uk/pub/Dimitrios.Stamatellos/phd.thesis)
[SW03] Stamatellos, D. & Whitworth, A.P., A&A, **407**, 941 (2003)
[SW04] Stamatellos, D. & Whitworth, A.P., in preparation (2004)
[SWA04] Stamatellos, D. & Whitworth, A.P., André, P., & Ward-Thompson, D., A&A, submitted (2004)
[WHS99] Wolf, S., Henning, T., & Stecklum, B., A&A, **349**, 839 (1999)
[Wit77] Witt, A.N., ApJS, **35**, 1 (1977)
[WLB02] Wood, K., Lada, C.J., Bjorkman, J.E., Kenyon, S.J., Whitney, B., & Wolff, M.J., ApJ, **567**, 1183 (2002)
[WWV03] Whitney, B.A., Wood, K., Bjorkman, J.E., & Wolff, M.J., ApJ, **591**, 1049 (2003)
[YMW84] Yusef-Zadeh, F., Morris, M., & White, R.L., ApJ, **278**, 186 (1984)

Radiative Transfer Through the Intergalactic Medium

Avery Meiksin

Institute for Astronomy
University of Edinburgh
Royal Observatory
Edinburgh EH9 3HJ, United Kingdom
aam@roe.ac.uk

1 Introduction

The baryons produced in the Big Bang are believed to have recombined around a redshift of $z \sim 1000$. This has been corroborated by measurements of fluctuations in the Cosmic Background Radiation (CBR), believed to have arisen during the recombination epoch, when the radiation produced in the Big Bang was last scattered before reaching us. Recently, signatures in the CBR detected by the Wilkinson Microwave Anisotropy Probe (*WMAP*) indicate that the baryons were subsequently reionized through some as yet un-identified process [S03], but widely speculated to be a consequence of photoionization by Quasi-Stellar Objects (QSOs) and young galaxies.

Further evidence for the reionized gas is provided by measurements of H I and He II Lyα absorption (actually scattering out of the line-of-sight) in the spectra of high redshift ($z > 2$) QSOs. The spectra reveal that the IGM has fragmented into many systems per unit redshift, with H I column densities ranging over 10^{11}–10^{21} cm^{-2}, and Doppler parameters for the non-radiatively damped systems (H I column densities smaller than 10^{19} cm^{-2}) of typically $20 \, \mathrm{km \, s^{-1}}$. The Doppler parameters indicate the temperatures expected for photoionized gas.

Numerical simulations of the formation of cosmological structures in the Universe have successfully reproduced many of the measured characteristics of the Intergalactic Medium (IGM) [C94, ZAN95, HKWM, ZANM97], provided the intergalactic photoionization rate is comparable to that expected from the combined effects of QSOs and galaxies. The simulations have generally assumed instantaneous reionization throughout the simulation volume at a given time, neglecting any possible role played by radiative transfer. Detailed statistical comparisons between the simulation predictions and the statistics of the measured properties of the absorption features show this may be an inadequate approximation. For example, the widths of the H I systems optically thin in Lyα are predicted to be substantially narrower than measured [MBM01]. Comparisons between the H I and He IILyα optical depths near redshifts of $z = 3$ also reveal substantial variations in their ratio, indicating a high amount of local variability in the shape of the ambient UV photoionizing radiation field, as may arise from local radiation sources, or the possibility that the reionization of He II in the IGM is largely incomplete at these redshifts [R97].

In order to address these issues, as well as the larger question of the reionization of the IGM itself, it is necessary to incorporate the transfer of ionizing radiation into the numerical simulations. Doing so, however, is very computationally demanding, requiring highly efficient numerical schemes. The description of one such scheme, relying on a probabilistic approach to radiative transfer, is described here.

2 Numerical Reionization

2.1 Direct Integration Requirements

The need for an efficient means of computing advancing I-fronts in the IGM may be illustrated by considering the direct integration of the H I ionization equation for gas at a distance r from a source of specific luminosity L_ν

$$\frac{dn_{\rm HII}(r)}{dt} = n_{\rm HI}(r)\Gamma_{\rm HI}(r) - n_e(r)n_{\rm HII}(r)\alpha_{\rm HI}(T), \quad (1)$$

where $n_{\rm HI}(r)$ and $n_{\rm HII}(r)$ are the neutral and ionized number densities of hydrogen at r, respectively, $n_e(r)$ is the electron number density at r, $\alpha_{\rm III}(T)$ is the radiative recombination rate to H I at temperature T, and $\Gamma_{\rm HI}$ is the photoionization rate per unit atom, given by

$$\Gamma_{\rm HI}(r) = \frac{1}{4\pi r^2} \int_0^\infty d\nu \frac{L_\nu}{h\nu} \exp(-\tau_\nu) \sigma_\nu. \quad (2)$$

Here σ_ν is the photoelectric cross-section of H I at frequency ν, and τ_ν is the optical depth to photoelectric absorption, given by

$$\tau_\nu = \int_0^r dr' n_{\rm HI}(r') \sigma_\nu. \quad (3)$$

An accurate integration (to 10% accuracy in the position of the ionization front), requires grid spacings that correspond to a photoelectric optical depth through the HI of no greater than about 1/4 per grid zone. For a neutral IGM at $z = 8$, this corresponds to a length scale of 0.16 kpc. The simulation box size requirement is roughly 20,000 times larger. This incompatibility of length scales places an inordinate demand on computer memory resources even with judiciously placed grid zones made fine only near the I-fronts. The characteristic timescale is on the order of the photoionization rate for optically thin gas, which is typically 10^{12} sec (and even shorter very near the sources), while the simulations must be run for at least a few billion years ($\sim 10^{17}$ sec), so that the range in timescales is also impractical. Together, the direct integration approach makes the problem unfeasible with current computational resources.

2.2 A Probabilistic Approach to Radiative Transfer

An alternative probabilistic approach was suggested for the photoionization of a single species (hydrogen) by [ANM99]. Their scheme is extended here to include hydrogen and both ionization states of helium (see [BMW04]). The advantage of this scheme over the direct integration outlined above is that energy is conserved independently of numerical resolution, which better ensures that I-fronts propagate at the correct speeds in our simulations. In practice, this means that larger step sizes may be taken on the simulation space grid without the associated loss of accuracy which would occur when solving the radiative transfer equation through direct numerical integration. The probabilities for the absorption of an ionizing photon by H I, He I and He II, respectively, are

$$P_{\text{abs}}^{\text{HI}} = p_{\text{HI}} q_{\text{HeI}} q_{\text{HeII}} \left[1 - \exp\left(-\tau_\nu^{\text{tot}}\right)\right]/D, \tag{4}$$

$$P_{\text{abs}}^{\text{HeI}} = q_{\text{HI}} p_{\text{HeI}} q_{\text{HeII}} \left[1 - \exp\left(-\tau_\nu^{\text{tot}}\right)\right]/D, \tag{5}$$

$$P_{\text{abs}}^{\text{HeII}} = q_{\text{HI}} q_{\text{HeI}} p_{\text{HeII}} \left[1 - \exp\left(-\tau_\nu^{\text{tot}}\right)\right]/D. \tag{6}$$

Here, auxiliary absorption and transmission probabilities have been defined, given by $p_i = 1 - \exp(-\tau_\nu^i)$, $q_i = \exp(-\tau_\nu^i)$, i denoting the species being referred to, $\tau_\nu^{\text{tot}} = \tau_\nu^{\text{HI}} + \tau_\nu^{\text{HeI}} + \tau_\nu^{\text{HeII}}$, where τ_ν^i is the optical depth for a given species, and $D = p_{\text{HI}} q_{\text{HeI}} q_{\text{HeII}} + q_{\text{HI}} p_{\text{HeI}} q_{\text{HeII}} + q_{\text{HI}} q_{\text{HeI}} p_{\text{HeII}}$. These probabilities are used to calculate the ionization rate for a given species per unit volume, $n_i \Gamma_i$, as follows. If in a time δt, $\delta t \dot{N}_\nu^{l-1}$ photons enter grid zone l from zone $l-1$, then the number of photons that will be absorbed in zone l by species i is $\delta t \dot{N}_\nu^{l-1} P_{\text{abs}}^i$, where P_{abs}^i is the absorption probability for species i within zone l. The ionization rate of species i in zone l of volume V^l is then

$$n_i^l \Gamma_i^l = \frac{1}{V^l} \sum_g \dot{N}_{\nu_g}^{l-1} P_{\text{abs}}^i(\nu_g), \tag{7}$$

where the frequencies have been divided into 100 discrete groups g evenly spaced between ν_L^{HI} and $10\nu_L^{\text{HI}}$. (While this spacing was found to be adequate, no attempt has been made to optimize it.) These are used in the following set of coupled equations to solve for the positions of the three I-fronts over time, for which the effects of cosmological expansion have now been included:

$$\frac{dn_{\rm HII}}{dt} = n_{\rm HI}\Gamma_{\rm HI} - n_e n_{\rm HII}\alpha_{\rm HI}(T) - 3\frac{\dot a}{a}n_{\rm HII}, \tag{8}$$

$$\frac{dn_{\rm HeII}}{dt} = n_{\rm HeI}\Gamma_{\rm HeI} + n_e n_{\rm HeIII}\alpha_{\rm HeII}(T) - n_{\rm HeII}\Gamma_{\rm HeII}$$
$$- n_e n_{\rm HeII}\alpha_{\rm HeI}(T) - 3\frac{\dot a}{a}n_{\rm HeII}, \tag{9}$$

$$\frac{dn_{\rm HeIII}}{dt} = n_{\rm HeII}\Gamma_{\rm HeII} - n_e n_{\rm HeIII}\alpha_{\rm HeII}(T) - 3\frac{\dot a}{a}n_{\rm HeIII}, \tag{10}$$

where n_i denotes number density, Γ_i (s^{-1}) is the photoionization rate per atom, $\alpha_i(T)$ is the total radiative recombination coefficient, and a is the cosmological expansion factor.

The time evolution of the gas temperature T is given by:

$$\frac{dT}{dt} = \frac{2(G-L)}{3kn} + \frac{T}{n}\frac{dn_e}{dt} - 3\frac{\dot a}{a}\left(\frac{2}{3}T + \frac{n_e}{n}T\right) \tag{11}$$

where $n = n_{\rm HI}+n_{\rm HeI}+n_{\rm HII}+n_{\rm HeII}+n_{\rm HeIII}+n_e$, $n_e = n_{\rm HII}+n_{\rm HeII}+2n_{\rm HeIII}$, k is the Boltzmann constant, and G and L (J m^{-3} s^{-1}) are the atomic heating and cooling rates, respectively. The last term is the adiabatic cooling term resulting from cosmological expansion, which will dominate the thermal effects of gas motions at the densities considered. While computing the photoionization, the gas over-density (not the density) is assumed to stay frozen, which is a good approximation on the scales relevant to the Lyα forest [ZMAN98].

The heating rate G^l in cell l is due to the photoionization of H I, He I and He II according to $G^l = G^l_{\rm HI} + G^l_{\rm HeI} + G^l_{\rm HeII}$, where for each species i, G^l_i is evaluated in a similar manner to the ionization rate per unit volume as given in Eq. (7),

$$n^l_i G^l_i = \frac{1}{V^l}\sum_g (h\nu_g - \chi_i)\dot N^{l-1}_{\nu_g} P^i_{\rm abs}(\nu_g), \tag{12}$$

where χ_i is the ionization potential of species i. The atomic cooling rate L includes radiative recombination cooling to H I, He I and He II, electron excitation of H I, and Compton cooling off the Cosmic Microwave Background photons. The radiative recombination and cooling rates are taken from [M94].

2.3 Numerical Integration Methods

Eqs. (8)–(11) are solved using explicit forward Euler integration. Although an implicit numerical scheme such as backward Euler integration results in a more stable solution at low numerical resolution [AZAN97], to obtain the required numerical accuracy as well as to maintain stability, both methods require similar numerical resolution. For the sake of speed and simplicity, the forward method is preferred. The timestep is restricted to being no more than several times greater that the hydrogen ionization timescale, $\Gamma^{-1}_{\rm HI}$, for numerical accuracy.

The photon-conserving algorithm was tested on the photoionization of gas with uniform density around a point source. The resulting solutions for the ionization and temperature profiles were accurate to better than 10 per cent for optical depths at the Lyman edge of up to $\Delta\tau_\nu \simeq 20$ per cell on the space grid. This results in a reduction by about a factor of 80 in the computational resources compared with direct numerical integration of the radiative transfer equation.

3 Results

The method is applied to the problem of a QSO source ionizing the IGM at the redshift $z = 6$. The QSO frequency specific luminosity is taken to be $L_\nu^Q = 10^{23}(\nu/\nu_{\rm HI})^{-1.5}$ W Hz^{-1}, where $\nu_{\rm HI}$ is the frequency at the H I Lyman edge. This spectrum is typical of that of QSOs. The assumed mass fraction of helium in the IGM is $Y \simeq 0.235$. The source is placed at a comoving distance of $10h^{-1}$ Mpc from the left edge of the density run from a cosmological numerical simulation ([MW04]). It is assumed the gas surrounding the QSO up to that point has already been ionized. Placing the source at this distance avoids having to impose the restriction that the I-fronts propagate no faster than the speed of light. The displaced position of the source is indicated in the figure below by starting the (comoving) spatial axes at $R = 10h^{-1}$ Mpc.

Two cases are compared: computing the ionization and temperature profiles by solving for the radiative transfer using the probabilistic method above, and a second case neglecting radiative transfer, assuming instead an instantaneous uniform ionization rate across the whole line-of-sight, as is usually done in simulations of the Lyα forest. The goal is to compare the final gas temperatures to determine how large an effect including radiative transfer may have.

Fig. 1 shows a comparison between the IGM temperatures at $z = 5$ computed with and without radiative transfer. The inclusion of radiative transfer results in a significant boost to the temperature of the ionized IGM. The positions of the H II and He II I-fronts at $z = 5$ in the radiative transfer simulation are $24h^{-1}$ comoving Mpc from the source, while the He III front lags behind at about $19h^{-1}$ comoving Mpc; their speeds of propagation are limited by the reduction in the ionization rate due to the increasing optical depth through the simulation volume. Consequently, the heating of the IGM, which is dominated by the photoionization of neutral hydrogen at the H II I-front, is also constrained to lie behind the H II I-front. All gas beyond $24h^{-1}$ Mpc is still much cooler. In contrast, the temperature computed by the non-radiative transfer simulation is much lower and the heating extends across the entire line-of-sight.

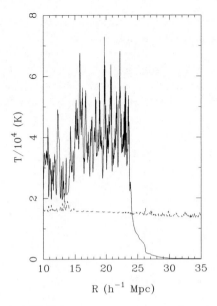

Fig. 1. Comparison of the IGM temperature at $z = 5$ against comoving distance from the ionizing source. The solid line is computed using radiative transfer, while the dashed line result assumes a uniform ionization field through the entire line of sight

References

[ANM99] Abel, T., Norman, M.L., Madau, P.: Photon-conserving Radiative Transfer around Point Sources in Multidimensional Numerical Cosmology. Astrophys. J., **523**, 66–71 (1999)

[AZAN97] Anninos, P., Zhang, Y., Abel, T., Norman, M.: Cosmological hydrodynamics with multi-species chemistry and nonequilibrium ionization and cooling. New Astron., **2**, 209–224 (1997)

[BMW04] Bolton, J., Meiksin, A., White, M.: Radiative transfer through the intergalactic medium. Mon. Not. Roy. Astron. Soc., **348**, 43–48 (2004)

[C94] Cen, R., Miralda-Escudé, J., Ostriker, J.P., Rauch, M.: Gravitational collapse of small-scale structure as the origin of the Lyman-alpha forest. Astrophys. J., **437**, L9–12 (1994)

[HKWM] Hernquist, L., Katz, N., Weinberg, D., Miralda-Escudé, J.: The Lyman-Alpha Forest in the Cold Dark Matter Model. Astrophys. J., **457**, L51–55 (1996)

[M94] Meiksin, A.: The structure and evolution of Lyman-alpha forest clouds in the minihalo model. Astrophys. J., **431**, 109–122 (1994)

[MBM01] Meiksin, A., Bryan, G.L., Machacek, M.E.: Hydrodynamical simulations of the LyA forest: data comparisons. Mon. Not. Roy. Astron. Soc., **327**, 296–322 (2001)

[MW04] Meiksin, A., White, M.: The effects of UV background correlations on LyA forest flux statistics. Mon. Not. Roy. Astron. Soc.(2004) (in press)

[R97] Reimers, D., Köhler, S., Wisotzki, L., Groote, D., Rodriguez-Pascual, P., Wamsteker, W.: Patchy intergalactic He II absorption in HE 2347-4342. II. The possible discovery of the epoch of He-reionization. Astron. & Astrophys., **327**, 890–900 (1997)

[S03] Spergel, D.N., et al.: First-Year Wilkinson Microwave Anisotropy Probe (WMAP) Observations: Determination of Cosmological Parameters. Astrophys. J. Suppl. Ser., **148**, 175–194 (2003)

[ZAN95] Zhang, Y., Anninos, P., Norman, M.L.: A Multispecies Model for Hydrogen and Helium Absorbers in Lyman-Alpha Forest Clouds. Astrophys. J., **453**, L57–60 (1995)

[ZANM97] Zhang, Y., Anninos, P., Norman, M.L., Meiksin, A.: Spectral Analysis of the Ly alpha Forest in a Cold Dark Matter Cosmology. Astrophys. J., **485**, 496–516 (1997)

[ZMAN98] Zhang, Y., Meiksin, A., Anninos, P., Norman, M.L.: Physical Properties of the Ly alpha Forest in a Cold Dark Matter Cosmology. Astrophys. J., **495**, 63–79 (1998)

Radiative Transfer with Finite Elements: Application to the Lyα Emission of High-Redshift Galaxies

Sabine Richling

Institut d'Astrophysique de Paris, 98 bis, Boulevard Arago, 75014 Paris, France
richling@iap.fr

Summary. The Lyα line is a prominent emission line of many high-redshift galaxies. These young galaxies are supposed to be very irregular. Starbursts or relativistic jets emerging from central compact objects are able to compress the gaseous material to huge high-density shells, which expand, merge and eventually fragment. The finite element method is an appropriate tool to study resonance lines in such an environment, where strong density gradients occur. Recent results are discussed for the shell model and its application to individual high-redshift galaxies.

1 Introduction

What are high-redshift galaxies? The diversity of present-day galaxies evolved from the density fluctuations left over from the big bang. After about 100 million years the first stars and quasars reionized the universe. The first protogalaxies appeared after about 1 billion years, before they merged and finally evolved into modern galaxies [1]. The term 'high-redshift galaxies' is usually attributed to very young galaxies which are still in the process of formation and which have a redshift greater than $z \sim 2$–4. At a redshift of $z \sim 4$ the age of the universe was about one billion years which is approximately 10% of its present age [2].

Plenty of information on high-redshift galaxies was provided by deep surveys like the Hubble Deep Field [3] or more recently by the FORS Deep Field [4]. FORS is an instrument at the ESO Very Large Telescope and was able to detect nearly 9000 objects in a single 6.8' × 6.8' field of view. Spectroscopic follow-up observations [5] reveal that many of these objects are galaxies in the redshift range between 2 and 5. Furthermore, it was found that many high-redshift galaxies are strong Lyα emitters, i.e. the Lyα line is the dominating feature in the spectrum over a wide range of wavelengths [5, 6, 7].

High-resolution spectra often show that the Lyα line profiles of radio galaxies [8] as well as starburst galaxies [9] are double-peaked or even multiple-peaked. The different height of the peaks indicates that global velocity fields are involved. Often profile fitting methods based on multiple emission and absorption components with plain Gaussian or Voigt profiles are used to derive physical parameters for these galaxies. But such methods do not consider the effects of scattering and frequency redistribution, which are very important for the formation of a resonance line like Lyα. The call for an accurate radiative transfer modeling is reinforced by the fact that the spatial distribution of the Lyα emission in high-redshift galaxies is often very extended and shows a clumpy morphology [10].

This paper is organized as follows. Some basic aspects of radiative transfer in resonance lines and the features of the numerical method are summarized in Sect. 2. In Sect. 3, results of radiative transfer calculations for the shell model are presented. The application of this model to high-redshift galaxies is discussed in Sect. 4. And a short summary can be found in Sect. 5.

2 Radiative Transfer in Resonance Lines

2.1 The Transfer Equation

In the comoving frame, the non-relativistic radiative transfer equation for the specific intensity I in a spectral line can be written as:

$$\mathbf{n} \cdot \nabla I(\mathbf{x}, \mathbf{n}, \nu) - \nu \mathbf{n} \cdot \nabla (\mathbf{n} \cdot \mathbf{v}(\mathbf{x})/c) \frac{\partial}{\partial \nu} I(\mathbf{x}, \mathbf{n}, \nu) =$$
$$-\chi(\mathbf{x})\phi(\nu)I(\mathbf{x}, \mathbf{n}, \nu) + \kappa(\mathbf{x})\phi(\nu)B(T(\mathbf{x}), \nu) \quad (1)$$
$$+ \frac{\sigma(\mathbf{x})\phi(\nu)}{4\pi} \int_0^\infty \int_{4\pi} \phi(\hat{\nu})I(\mathbf{x}, \hat{\mathbf{n}}, \hat{\nu}) \, d\hat{\omega} \, d\hat{\nu}.$$

In three dimensions the specific intensity I is a function of six variables. It depends on the space variable $\mathbf{x}$, the direction $\mathbf{n}$, and the frequency ν. Eq. 1 considers frequency shifts due to the Doppler effect which is described by the second term, where $\mathbf{v}(\mathbf{x})$ is the velocity field and c the speed of light. The terms on the right hand side describe the extinction, the emission and the scattering of photons, respectively. In the case of a resonance line, the scattering coefficient $\sigma(\mathbf{x})$ is very large and the coupling of the system of differential equations due to the integration over the unit sphere and the whole frequency range in the scattering term becomes very strong.

The remaining symbols are the extinction coefficient $\chi(\mathbf{x}) = \sigma(\mathbf{x}) + \kappa(\mathbf{x})$, the absorption coefficient $\kappa(\mathbf{x})$ and the Planck function $B(T(\mathbf{x}), \nu)$ which in turn depends on the spatial temperature distribution $T(\mathbf{x})$. In the case of large turbulent velocities, $v_{\text{turb}} > 100 \, \text{km s}^{-1}$, and moderate optical depths, $\tau < 10^6$, it is sufficient to use a Doppler profile for the profile function $\phi(\nu)$ [11, 12].

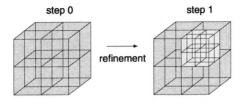

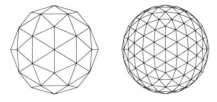

Fig. 1. Discretization of the computational domain in a) physical space and b) angular space

2.2 The Finite Element Code

For the solution of the transfer equation Eq. 1 which considers isotropic scattering in space and complete frequency redistribution, we use a parallelized 3D finite element code. This code uses an adaptive grid in space (Fig. 1a) which is refined by the means of an error indicator. The discretization of the unit sphere is fixed and based on a refined icosahedron (Fig. 1b). For the solution of the resulting linear system, a Krylov space method is employed. The basic monochromatic version of this code was developed by G. Kanschat [13] and applications of this version to astrophysical test problems can be found in [14]. The polychromatic version which is used here is described in detail in [15, 16].

3 The Expanding Shell Model

3.1 Description of the Model

A reasonable model for high-redshift galaxies is the so-called shell model: A low-density region is surrounded by an expanding high-density shell of neutral hydrogen. High-density shells are expected to form e.g. in the case of starburst galaxies, where expanding shells of individual supernova events overlap and eventually merge to a single super-bubble [17]. High-density shells can also be the result of the interaction of a relativistic jet originating from a quasar in the center of a young massive galaxy with its environment [18].

The Lyα emission regions are located somewhere inside the shell. Photons traveling through the neutral shell towards a distant observer are scattered in space and frequency. Depending on the optical depth of the shell and the details of the velocity field, the initial emission profile can be strongly modified. Here, we investigate a shell model with two small emission regions near the center of the shell.

3.2 Results

Fig. 2 displays intensity maps from three calculations of the expanding shell model with different optical depth. In the case of low optical depth, $\tau = 10$, only the two emission regions are visible. At larger optical depth, the bright emission regions are surrounded by a diffuse halo of scattered photons whose intensity increases with τ. The spatial coordinates are in normalized units, so that the results can be appropriately scaled to specific applications. Here, the inner and outer radius of the shell are 0.8 and 0.9, respectively.

Fig. 3 illustrates, how adaptive grid refinement proceeds in the case of a 2D calculation with a somewhat smaller shell. The first panel shows the initial grid. It consists of quadratic cells. Each dot marks a corner of a cell. Note, that even the initial grid is not uniform but is prerefined at the location of the shell and of the emission regions near the center.

Our solution method of Eq. 1 as described in [15] requires that we iterate several times through the frequency grid to achieve the correct result. The number of iterations n_i strongly depends on the optical depth of the shell. Table 1 gives typical values for n_i. Once obtained the result on the initial

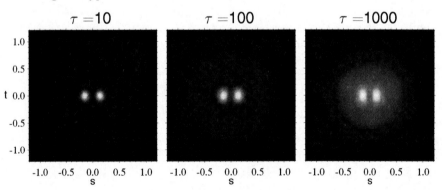

Fig. 2. Frequency integrated Lyα intensity maps for different optical depth τ of the shell

Table 1. Number of necessary iterations n_i through the frequency grid for the coarse initial grid

τ	1	10	100	1000	5000
n_i	50	120	250	1500	6000

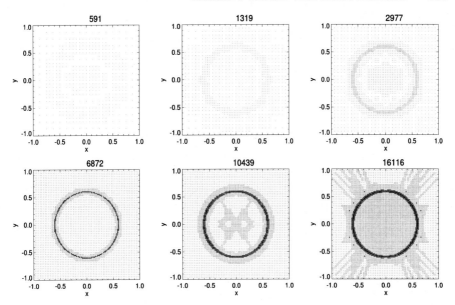

Fig. 3. Grid refinement for a shell model. The dots mark the locations of the corners of the cells. Each panel is labelled with the total number of cell corners

grid in this way, it is sufficient to iterate 2–5 times on each refined grid in order to update the solution. Nevertheless, in the case of high optical depth the iteration process on the initial grid consumes most of the computing time. This limits the applicability of the present implementation of the code to model configurations with $\tau < 10^4$.

The remaining panels in Fig. 3 show the refined grids. They are labelled with the total number of cell corners. In each refinement step, the resolution of the high-density shell increases. This example nicely demonstrates that the adaptive algorithm used in the code is able to locate and resolve strong density gradients.

The line profiles corresponding to the intensity maps in Fig. 2 are displayed in Fig. 4. The profiles are double-peaked for $\tau > 1$. The width and depth of the central absorption feature are increasing with τ. The flux in the red peak is stronger than the flux in the blue peak as expected for an expanding configuration [15]. The degree of asymmetry between the red and blue wing of the line profile depends on the details of the global velocity field.

The results discussed so far are obtained assuming that the turbulent velocity in the neutral shell v_{turb} is equal to the Doppler velocity of the emission line profile v_{emis}. This assumption leads to the extremely wide absorption features of the profiles in Fig. 4. How the shape of the line profile changes for $v_{\text{turb}} < v_{\text{emis}}$, is demonstrated in Fig. 5, where we compare line profiles for a shell model with $\tau = 10$ for different ratios $v_{\text{turb}}/v_{\text{emis}}$. If v_{turb} decreases

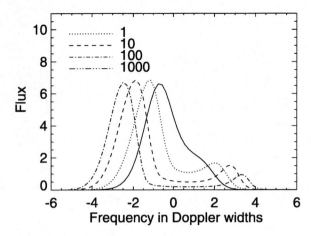

Fig. 4. Line profiles for the expanding shell model. The line styles refer to different optical depth τ as indicated

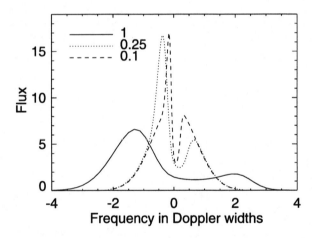

Fig. 5. Line profiles for a shell model with $\tau = 10$. The line styles refer to different ratios between the turbulent velocity in the shell and the Doppler velocity of the emission line profile, $v_{\mathrm{turb}}/v_{\mathrm{emis}}$, as indicated

with respect to v_{emis}, the absorption feature becomes narrower. Note that the wings of the emission profile are hardly effected by the redistribution of photons due to scattering in the case of small turbulent velocities in the shell.

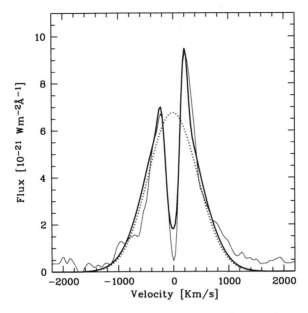

Fig. 6. Lyα line profile of FDF-4691 (thin line) and a calculated line profile based on the shell model (thick line). The dotted line is the line profile of the emission regions [19]

4 Application to High-Redshift Galaxies

The Lyα emitting galaxy FDF-4691 at a redshift of $z \sim 3.3$ is a good example for the application of the shell model. The medium-resolution spectrum of this young starburst galaxy (Fig. 6) shows that the Lyα line profile is double-peaked and has a very small and deep absorption feature [19]. The theoretical profile shown in Fig. 6 is calculated for a shell model with a broad emission profile and a slowly expanding shell with an optical depth of $\tau = 5000$ and a turbulent velocity of $v_{\text{turb}} \sim 0.1 v_{\text{emis}}$. The calculated profile is convolved with the instrumental profile and is able to fit the red and blue wing of the observed line profile. A better fit of the deep absorption feature in the line center would be possible with shells of optical depths beyond the capability of the code.

Additional calculations which consider a frequency-independent dust absorption term $\kappa_d(x)$ in the total extinction coefficient $\chi(\mathbf{x}, \nu) = \chi(\mathbf{x})\phi(\nu) + \kappa_d(\mathbf{x})$ show that the line profile of a $\tau = 5000$ shell with normal interstellar dust content is hardly effected by dust absorption. This is in agreement with the analysis of other parts of the spectrum [19]. The shape of the line profile only changes for exceptional high dust content with $\log \beta > 10^{-5}$ (Fig. 7), where $\beta = \tau_d/\tau$ is the ratio between the continuum optical depth $\tau_d = \int \kappa(\mathbf{x}) dr$ and the line center optical depth $\tau = \int \chi(\mathbf{x})\phi(0) dr$. The total flux is only considerably reduced by dust absorption for $\log \beta > 10^{-4}$ (Fig. 8).

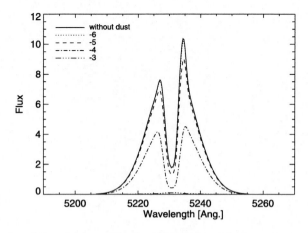

Fig. 7. Calculated Lyα line profiles for FDF-4691 with dust absorption. The line styles refer to different values of $\log \beta$ as described in the text

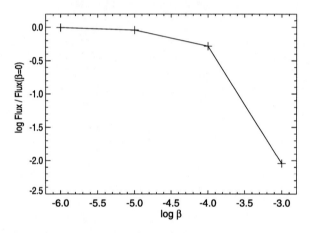

Fig. 8. Flux reduction due to dust absorption as a function of β for the line profiles shown in Fig. 7

5 Conclusions

The finite element method is an appropriate tool to study the Lyα emission of high-redshift galaxies. A comparison of the observed line profiles with the results from rather simple models like the shell model which is defined by relatively few free parameters helps to constrain the physical properties of these galaxies. But some observations definitely require the modeling of configurations with optical depth far greater than 10^4. In future, optical depths of this order of magnitude will possibly be feasible with an improved version of the finite element code which makes use of the ideas presented in the contribution by E. Meinköhn in this volume.

References

1. Larson, R.B., Bromm, V. 2001, Scientific American 285, No. 6, p. 64
2. Hamann, F., Ferland, G. 1999, Annual Review of Astronomy and Astrophysics 37, pp. 487–531
3. Williams, R., Adorf, H.-M., Blacker, B., et al. 1996, Astronomical Journal 112, p. 1335
4. Heidt, J., Appenzeller, I., Gabasch, A., et al. 2003, Astronomy and Astrophysics 398, pp. 49–61
5. Noll, S., Mehlert, D., Appenzeller, I., et al. 2004, Astronomy and Astrophysics, in press
6. Pentericci, L., Kurk, J.D., Röttgering, H.J.A., et al. 2000, Astronomy and Astrophysics 361, pp. L25-L28
7. Dey, A., Spinrad, H., Stern, D., Graham, J.R., Chaffee, F.H. 1998, Astrophysical Journal Letters 498, p. L93
8. van Ojik, R., Röttgering, H.J.A., Miley, G.K., & Hunstead, R.W. 1997, Astronomy and Astrophysics 317, p. 358
9. Dawson, S., Spinrad, H., Stern, Dey, A., van Breugel, W., de Vries, W., Reuland, M. 2002, Astrophysical Journal 570, p. 92
10. Kurk, J.D., Röttgering, H.J.A., Miley, G.K., Pentericci, L. 2002, Revista Mexicana de Astronomía y Astrofísica (Serie de Conferencias) 13, p. 191
11. Gayley, K.G. 1998, Astrophysical Journal 497, p. 458
12. Richling, S. 2003, Mon. Not. R. Astron. Soc 344, p. 553
13. Kanschat, G. 1996, Ph.D. Thesis, Univ. of Heidelberg, (http://archiv.ub.uni-heidelberg.de/volltextserver/volltexte/2006/6331/)
14. Richling, S., Meinköhn, E., Kryzhevoi, N., & Kanschat, G. 2001, Astronomy and Astrophysics 380, p. 776
15. Meinköhn, E., Richling, S. 2002, Astronomy and Astrophysics 392, p. 827
16. Meinköhn, E. 2003, Ph.D. Thesis, Univ. of Heidelberg
17. Mori, M., Ferrara, A., Madau, P. 2002, Astrophysical Journal 571, p. 40
18. Krause, M. 2002, Astronomy and Astrophysics 386, p. L1
19. Tapken, C., Appenzeller, I., Mehlert, D., Noll, S., Richling, S. 2004, Astronomy and Astrophysics, in press

Radiative Transfer Problem in Dusty Galaxies: Ray-Tracing Approach

Dmitrij Semionov and Vladas Vansevičius

Institute of Physics, Savanorių 231, LT-03154 Vilnius, Lithuania `dima@itpa.lt`

Summary. A new code for evaluation of light absorption and scattering by interstellar dust grains is presented. The radiative transfer problem is solved using ray-tracing algorithm in a self-consistent and highly efficient way. The code demonstrates performance and accuracy similar or better than that of previously published results, achieved using Monte-Carlo methods, with accuracy better than ∼ 3% for most cases. The intended application of the code is spectrophotometric modelling of disk galaxies, however, it can be easily adapted to other cases that require a detailed spatial evaluation of scattering, such as circumstellar disks and shells using both point and distributed light sources.

1 Problem Statement

The purpose for the developing radiative transfer problem solving code, described in this article, was to model spatial and spectral energy distribution (SED) observed in external galaxies. The nature of this problem requires 'self-consistency' of a solution – the resulting SED of a model must depend only on the SED of stellar sources and assumed properties of dust without any preconditioning on light and attenuation distribution within galaxy [TVA03].

While galaxies in general are a complex objects with three-dimensional (3D) distribution of radiation and mater, in most cases they are dominated by axial symmetry (2D), allowing to significantly simplify the model geometry. However, the model should account for presence of macroscopic structure within galaxies, possibly including elements having other symmetry, such as bars and spiral arms (2D+).

Most present day astrophysical radiative transfer codes employ either a Monte-Carlo (MC, e.g. [CF01]) or a ray-tracing (RTR, e.g. [RS99], [RM02]) methods. Some of implementations of these methods were compared by [BD01] for 1D and by [DT02] for 2D cases. The RTR approach allows the optimization of solution for a given system geometry, which was the main reason to use it as a basis for the Galactic Fog Engine (hereafter 'GFE'), a program for self-consistent solution of radiative transfer problem in dusty media with primarily axisymmetric geometry.

2 Algorithm

2.1 Model Geometry

GFE iteratively solves a discrete bidirectional radiative transfer problem, producing intensity maps of the model under arbitrary inclination at a given wavelength set. This paper concentrates primarily on ultraviolet-to-optical wavelength range assuming exclusively coherent scattering, therefore in most equations the wavelength dependence will be omitted.

The initial SED contributes to: escaped energy, that reaches an external observer; energy, absorbed by grains, that is eventually emitted as thermal radiation; and scattered energy distribution. Those three parts make up an energy balance equation, used to control iterative solution. The loop is repeated, substituting scattered energy as initial distribution for the next iteration, accumulating resulting escaped and absorbed energy, until certain convergence criteria are met, either a fixed number of iterations, or remaining scattered energy being below specified threshold. The remaining scattered energy may then be used to correct the final energy balance. After convergence is reached, dust temperature is calculated using distribution of absorbed energy. If it is necessary to calculate self-scattering of thermal radiation on the grains, resulting dust emission SED can be input into scattering evaluation loop and the process repeated until the final radiative energy distribution is obtained, and then used to produce SED as seen by an external observer.

Calculations are performed within a cylinder with a radius r_m and height above midplane z_m, which is subdivided into a set of layers of concentric, internally homogeneous rings ('bins') of arbitrary radial and vertical thickness. Since the linear extent of each individual light source (star) is negligible compared to the size of system, it is possible to solve the radiative transfer problem considering every volume element of the model having both attenuating (light scattering and absorption by interstellar dust) and emitting (light emission by the stars and thermal radiation of the dust particles) properties per unit volume. Each ring, therefore, has two properties: its emissivity $j' = j(r, z, \alpha, \delta)$, combined from internal light sources and energy scattered within its volume, and total absorption $k' = \kappa(r, z)$ per unit volume. Here r and z are ring indexes and α and δ are angles defining the direction of radiation propagation.

2.2 Radiative Transfer Equation

For any given direction we can describe light path as a series of intervals traversing rings until crossing the outer boundary of the model. If the viewing

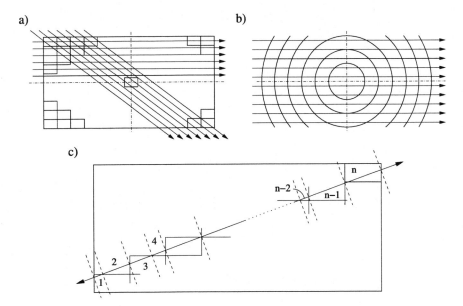

Fig. 1. The model geometry. Panel a) shows the diametral, panel b) – central plane cross-section of the model. The distribution and density of ray-tracing paths (shown as arrows) are computed to produce even and sufficient sampling of the model volume. Panel c) illustrates the discrete radiative transfer in cross-section parallel to the model Z axis. Limits of plane-parallel layers for one-directional treatment are shown as dotted lines while boundaries of the individual rings are outlined with solid lines

solid angle, containing that direction (ray), can be held small, the radiative transfer along this ray can be solved as in a plane-parallel homogeneous layer case (fig. 1c).

GFE uses static ray-casting geometry, precalculated at the start of the program execution. The program establishes sets of rays ensuring a required degree of sampling of the model volume (fig. 1a and 1b). For each ray a list of all crossed ring boundaries is compiled and converted into arrays of scattered and absorbed intensity, contributed by each crossed ring.

In a form suitable for computer implementation an incident intensity on a given point for a light path separated into n intervals (numbered outwards from that point) is written as

$$I_{\text{inc}} = \sum_{i=1}^{n} \left(\prod_{j=1}^{i-1} e^{-k'_j l_j} \right) \frac{j'_i}{k'_i} \left(1 - e^{-k'_i l_i}\right). \tag{1}$$

Similarly, the intensity of radiation scattered with albedo ω from a given direction ($\alpha : \delta$) into all other directions within a certain interval denoted by index "1" is

$$I_{1,\alpha,\delta} = \omega j_1' l_1 - \omega \left(1 - e^{-k_1' l_1}\right) \left[\frac{j_1'}{k_1'} - \sum_{i=2}^{n} \left(\prod_{j=2}^{i-1} e^{-k_j' l_j}\right) \frac{j_i'}{k_i'} \left(1 - e^{-k_i' l_i}\right)\right]. \tag{2}$$

When considering azimuthally inhomogeneous model configuration (3D case), each ring is subdivided into required number of azimuthal segments. The number and the direction of ray cast through the system has to be modified accordingly to include new sets of rays in azimuthal direction. However, the ray-tracing part of the algorithm is unchanged. The computational time scales as $N_{\text{bin}}^{3/2} \times \log N_{\text{bin}}$ for 2D and $N_{\text{bin}}^{4/3} \times \log N_{\text{bin}}$ for 3D cases.

2.3 Scattering Phase Function

Since angular distribution of radiation at each point in the model is non-isotropic, it must be described using a numerical phase function (matrix), providing the radiation intensity towards a set of predefined reference directions ('RDs') described by angular coordinates ($\alpha_0 : \delta_0$). There exists a number of ways to distribute RDs on a sphere, however, those methods that produce a set of RDs arranged in iso-latitude rows are the most efficient in this particular model geometry, allowing both efficient storage and retrieval of scattered intensity and fast rotation of the phase matrix around model Z-axis. The memory requirements and the overall algorithm's performance have also to be taken into consideration.

In this work the following methods of RD distribution were compared: HEALPix[1] [GHW99], HTM[2] [KST01], a trivial iso-latitude triangulation (hereafter 'TT'), fig. 2a) and a square matrix with elements ('texels') corresponding to evenly spaced ($\alpha_0 : \delta_0$) coordinates (hereafter 'Texel'). For triangulation schemes and HEALPix the radiation intensity towards a given point was interpolated between 3 nearest RDs using either 'flat' (fig. 2b) or 'spherical' (fig. 2c) weights.

3 Computational Precision

3.1 Scattering Phase Function Interpolation

To compare used sphere subdivision and interpolation algorithms a following test model (hereafter a 'standard model') was employed: a cylinder with height to radius ratio $r_m/z_m = 0.2$, divided into $N_{\text{bin}} = 441$ (21 × 21) equally spaced rings, filled with radiating and absorbing particles whose density follows a double exponential law

$$\rho(r,z) = \rho_0 e^{-r/r_0} e^{-z/z_0} \tag{3}$$

---

[1] http://www.eso.org/science/healpix/
[2] http://www.sdss.jhu.edu/htm/

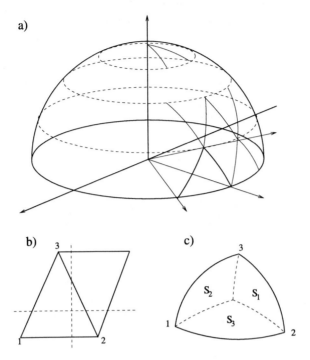

Fig. 2. The reference point structure used for interpolation of the scattering phase function. Panel a) shows the trivial iso-latitude triangulation for one hemisphere with reference directions arranged symmetrically against diametral and horizontal planes. Panels b) and c) represent two implemented interpolation schemes, 'flat' and 'spherical'. In case of spherical interpolation, input from each triangle vertex is weighted by the area defined by shortest distances from the given direction to the vertices (S_1 for 1-st vertex and so on)

with $r_0 = 0.2r_m$ and $z_0 = 0.2z_m$. The Henyey & Greenstein [HG41] scattering phase function

$$\Phi(\theta) = \frac{1 - g^2}{(1 + g^2 - 2g\cos\theta)^{3/2}} \quad (4)$$

was used with parameter $g = 0.75$, model central optical depth perpendicularly to the central plane $\tau_{ct} = 25$.

The primary quality criteria of a given algorithm is its ability to represent the angular intensity distribution of anisotropic scattering. If the phase function representation is exact, the distribution of values $(\Phi'(\theta) - \Phi(\theta))/\Phi(\theta)$ (where $\Phi'(\theta)$ is a resulting numerical phase matrix) would be a δ-function. However, since employed methods introduce different types of numerical errors, the actual distribution form depends strongly on the phase function sampling and the interpolation algorithm. An examples of resulting error distributions (as relative numerical phase matrix deviation from its analytical form) for the algorithms tested are shown on fig. 3.

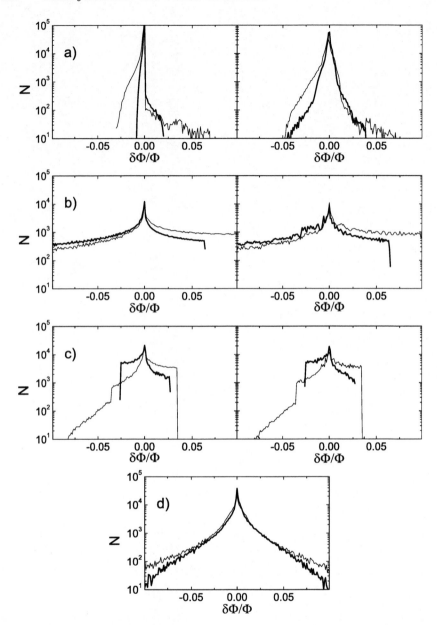

Fig. 3. The distribution of relative numerical phase matrix deviation from its analytical form for different sphere subdivision algorithms. Panels a) – d) correspond to TT, HTM, HEALPix and Texel methods. For the first three methods the results obtained using both 'flat' (left panels) and 'spherical' (right panels) weighted interpolation are presented. Thin line shows the results obtained for approximately 3100, thick line – for approximately 12000 reference directions

As can be seen, methods providing uniform sphere coverage produce more preferable error distributions. With the increasing number of RDs the representation of the scattering phase function improves, reducing maximal possible deviation from the true value, particularly for TT (fig. 3a) and HEALPix (fig. 3c) methods, with TT algorithm showing slightly better error distribution form. 'Spherical' interpolation scheme (fig. 3, right column in panels a – c) produces symmetrical error distributions, while 'flat' interpolation (left column) in some cases introduces additional numerical errors. When compared with other methods, attempt to reproduce scattering phase function using simple matrix with no interpolation between its elements (Texel scheme, fig. 3d) produces results of average quality, its error distribution quickly reaching a 'saturated' form with increasing number of RDs. The described error distribution is somewhat dependent on the orientation of the scattering phase function relative to the set of RDs, this dependence being minimal for the methods with identical size of interpolation elements (HEALPix). All methods that use interpolation display similar performance for a given number of RDs (table 1).

Table 1. The normalized computing time for models using different sphere subdivision and scattering phase function interpolation algorithms

Interpolation method	HEALPix	HTM	TT	Texel
'flat'	2.5	2.8	2.5	1.0
'spherical'	8	10	8	–

3.2 Volume Sampling and Subdivision

The problem encountered applying numerical methods is error accumulation. In case of iterative ray-tracing it arises from sampling and interpolation errors. The source of a sampling errors is mostly incomplete/inadequate spatial sampling of the system, while interpolation errors are related to the scattering phase function approximation, as described in the previous section. Both error types independently affect every ray traced through the system, thus the accumulated error increases with the increasing number of bins and rays. This makes oversampling undesirable not only due to increasing computational time, but also for a reason of minimizing numerical errors.

As a measure of method's quality a defect in energy balance E_{err} as a percentage of total energy radiated within system E_{tot}

$$E_{\mathrm{err}} = \frac{E_{\mathrm{tot}} - E_{\mathrm{abs}} - E_{\mathrm{sca}} - E_{\mathrm{esc}}}{E_{\mathrm{tot}}} \qquad (5)$$

is used. Here E_{esc}, E_{abs} and E_{sca} are the parts of a total radiated energy that escaped the system, was absorbed and remained to be scattered within the system, respectively.

To determine the sampling and gridding influence on the model precision the following two tests were performed. Firstly, the radiation field in the standard model using TT algorithm with scattering phase function represented by a set of 182 RDs was computed with different number of rays N_{ray}, cast through each ring. The results, presented on fig. 4a, show a significant error accumulation effect. Then, keeping a number of rays per ring constant the number of rings N_{bin} in model was changed (fig. 4b). As can be seen, as the model sampling improves the approximation errors decrease to a certain minimum, limited by internal errors of a chosen scattering phase function interpolation method. This geometric configuration can be considered optimal, as with further increase in a number of bins and rays the quality of the solution begins to deteriorate.

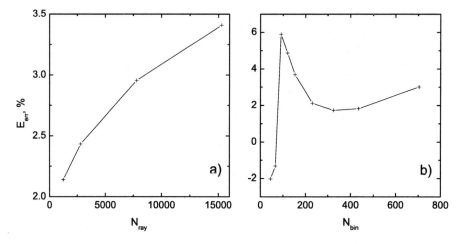

Fig. 4. The influence of subdivision and sampling of the model on the energy balance. Panel a) displays the energy error for a given number of rays cast per model bin; panel b) shows the same quantity for a models consisting of different number of bins

3.3 Dust Optical Properties

Other important aspect of a numerical radiative transfer solution is its sensitivity to variations in scattering parameters: albedo ω and asymmetry parameter g. Model precision and stability for different ω and g values place a constraint on the wavelength range where a given method can be applied. The scattering in the optical spectral region was analyzed using the standard model with $N_{\text{bin}} = 441$ and $\tau_{\text{ct}} = 10$.

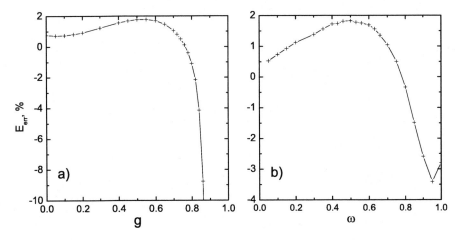

Fig. 5. The dependence of the energy losses within model on the grain scattering parameters: scattering phase function asymmetry g (panel a), with albedo assumed to be $\omega = 0.5$ for all g values), and albedo ω for $g = 0.6$ (panel b)

The dependence of overall model precision on scattering asymmetry parameter g is shown on fig. 5a. The total energy error after 9 iterations show some variation with the increasing g up to the limit imposed by the angular scattering phase function gridding (182 RDs) used in the calculations, after which the energy losses in the scattering phase matrix render results invalid.

The effects of the grain albedo on the model accuracy and stability were investigated using similar method. All computations were performed for 7 iterations assuming $g = 0.6$. The results are shown in fig. 5b. Within the range of ω values, applicable to astrophysical dust grains, those errors stay in acceptable limits, and do not influence the stability of the solution.

4 Summary

The code described in this paper has undergone an extensive testing and shows the flexibility and performance satisfying the requirements for the models of the global radiation transfer in dusty galaxies [SV02]. It has been successfully applied to model both integral and position dependent SEDs of several galaxies, some of the first results presented in [SSV03].

The main limiting factor affecting the applicability of the described code is the scattering asymmetry parameter g_λ. In order to correctly treat the scattering with g_λ approaching 1, the number of required reference directions rises sharply, compromising the performance and precision of the method. Other model properties, such as optical depth τ_λ and the relative amount of scattered radiation (dependent on albedo ω_λ) seem to have relatively little effect on the quality of the solution.

The application of this code is not restricted to the systems with dispersed sources and absorbers, the algorithm being easily extended to include treatment of interaction between radiation field and surfaces of macroscopic objects.

Acknowledgement. Some of the results in this paper have been derived using HEALPix [GHW99] package.

References

[BD01] Baes, M., Dejonghe, H.: Radiative transfer in disc galaxies I – A comparison of four methods to solve the transfer equation in plane-parallel geometry. Mon. Not. R. Astron. Soc., **326**, 722–732 (2001)

[CF01] Ciardi, B., Ferrara, A., Marri, S., Raimondo, G.: Cosmological reionization around the first stars: Monte Carlo radiative transfer. Mon. Not. R. Astron. Soc., **324**, 381–388 (2001)

[DT02] Dullemond, C., Turolla, R.: An efficient algorithm for two-dimensional radiative transfer in axisymmetric circumstellar envelopes and disks. A&A, **360**, 1187–1202 (2000)

[GHW99] Górski, K., Hivon, E., Wandelt, B.: Analysis Issues for Large CMB Data Sets. In: Banday, A., Sheth, R., Da Costa, L. (eds.) Proceedings of the MPA/ESO Cosmology Conference "Evolution of Large-Scale Structure". PrintPartners Ipskamp, NL, 37–42 (1999)

[HG41] Henyey, L., Greenstein, J.: Diffuse radiation in the Galaxy. ApJ, **93**, 70–83 (1941)

[KST01] Kunszt, P., Szalay, A., Thakar, A.: The Hierarchical Triangular Mesh. In: Banday, A., Zaroubi, S., Bartelmann, M. (eds.) Mining the Sky: Proc. of the MPA/ESO/MPE workshop, Garching. Springer-Verlag, Berlin Heidelberg, 631–637 (2001)

[RS99] Razoumov, A., Scott, D.: Three-dimensional numerical cosmological radiative transfer in an inhomogeneous medium. Mon. Not. R. Astron. Soc., **309**, 287–298 (1999)

[RM02] Razoumov, A., Michael, M., Abel, T., Scott, D.: Cosmological Hydrogen Reionization with Three-dimensional Radiative Transfer. ApJ, **572**, 695–704 (2002)

[SV02] Semionov, D., Vansevičius, V.: Radiative transfer problem in dusty galaxies: ray-tracing vs. Monte-Carlo. Balt. Astron. **11**, 537–545 (2002)

[SSV03] Semionov, D., Stonkutė, R., Vansevičius, V.: Modelling the radial color profile of M31. Balt. Astron. **12**, 633–636 (2003)

[TVA03] Takagi, T., Vansevičius, V., Arimoto, N.: Spectral Energy Distributions of Dusty Galaxies. Publications of the Astronomical Society of Japan, **55**, 385–407 (2003)

Shape Reconstruction for an Inverse Radiative Transfer Problem Arising in Medical Imaging

Oliver Dorn

Departamento de Matemáticas, Universidad Carlos III de Madrid, Spain
oliver.dorn@uc3m.es

1 Introduction

Nowadays, a standard method in medical imaging is X-ray tomography, even though it is well-known that the use of high-energy X-rays can be harmful to the human tissue. Due to the potential danger of this classical imaging method, research has been going on during the last twenty years to replace it, at least in some applications, by a less harmful but still inexpensive imaging technology. One very promising candidate for such a novel imaging technology is Diffuse Optical Tomography (DOT), which uses harmless near-infrared light for the imaging task. This light in the range of 700 to 1000 nm can easily be created by a standard laser. However, due to the low energy of this light, photons are heavily scattered in tissue, which makes the corresponding mathematical inverse problem extremely ill-posed and difficult to solve. As a correct forward model the radiative transfer equation is generally agreed on, such that the reconstruction task essentially constitutes an inverse problem for this equation, although most approaches so far only use the diffusion approximation instead. This approximation to the radiative transfer equation is easier to handle mathematically, but its validity in the physical imaging problem is not clear. In many cases, for example the imaging of the human head, it is most likely a very poor approximation due to the presence of non-scattering 'clear regions', filled with cerebrospinal fluid. Therefore, we make an attempt in this paper to treat the DOT inverse problem in the framework of radiative transfer theory, having specifically this latter application in mind. Our numerical experiments will be in two spatial dimensions, but the derivation of the algorithm will be valid in 3D as well.

The basic idea of the approach presented in this paper is to treat the inverse problem as a shape reconstruction problem. The absorption parameter of the tissue with respect to near-infrared light is decomposed into an approximately known background distribution (which might for example be derived from standard anatomical values for the area of interest, or, in long-time monitoring applications, by an initializing independent imaging method), and into an unknown 'contrast function'. This contrast function represents the quantity of interest. It might be produced by some hidden tumors in the breast, by a hematoma in the brain, or by some accumulation of an injected contrast liquid whose propagation in the body needs to be monitored. Our basic assumption in this paper will be that this contrast function has a large value compared to the (approximately) known background distribution, and is locally concentrated in the sense that it can be described by a 'shape' with more or less clear boundaries. The topology of this shape is unknown and might be quite complicated. For simplicity, the contrast value itself is assumed here to be constant and also (approximately) known inside the shape, although the algorithm can be extended to recover this value as well. Our goal will be to reconstruct the spatial distribution (i.e. the shape) of this contrast function from the near-infrared data.

We mention that more traditional DOT-algorithms typically have difficulties in imaging such high contrast situations. Due to the severe ill-posedness of the inverse problem at hand, strong regularization is required when using those methods, which has the effect that high contrast objects are reconstructed as severely 'smeared-out' versions of the correct shapes with much lower contrast values. This is the main motivation for us to introduce instead a shape-based reconstruction scheme for these applications. In the following, we will describe the mathematical and physical setup of our method.

2 The Physical Experiment in DOT

The propagation of photons in tissue is modeled by the time-dependent radiative transfer equation (or 'linear transport equation' [CZ67])

$$\frac{\partial u}{\partial t} + \theta \cdot \nabla u(x,\theta,t) + (a(x) + b(x))u(x,\theta,t)$$
$$- b(x) \int_{S^{n-1}} \eta(\theta \cdot \theta') u(x,\theta',t) d\theta' = q(x,\theta,t) \quad \text{in} \quad \Omega \times S^{n-1} \times [0,T] \quad (1)$$

with initial condition

$$u(x,\theta,0) = 0 \quad \text{in} \quad \Omega \times S^{n-1} \quad (2)$$

and boundary condition

$$u(x,\theta,t) = 0 \quad \text{on} \quad \Gamma_-. \quad (3)$$

Here,

$$\Gamma_\pm := \{(x,\theta,t) \in \partial\Omega \times S^{n-1} \times [0,T], \quad \pm\nu(x)\cdot\theta > 0\}.$$

Ω is a convex, compact domain in $\mathbb{R}^n$, $n=2,3$, with smooth boundary $\partial\Omega$. In our numerical experiments, we will only consider the case $n=2$, but the algorithm extends in a straightforward way to $n = 3$. $\nu(x)$ denotes the outward unit normal to $\partial\Omega$ at the point $x \in \partial\Omega$, and $u(x,\theta,t)$ describes the density of particles (photons) which travel in Ω at time t through the point x in the direction θ. The velocity c of the particles is assumed to be normalized to $c = 1$ cm s^{-1} and has been neglected in the formulation of (1).

$a(x)$ is the absorption cross-section (in short 'absorption'), $b(x)$ is the scattering cross-section, and $\mu(x) := a(x) + b(x)$ is the total cross-section or attenuation. These parameters are assumed to be real, strictly positive functions of the position x. The quantity μ^{-1} is the mean free path of the photons. Typical values in DOT are $a \approx 0.1$–1.0 cm^{-1}, $b \approx 100.0$–200.0 cm^{-1}, $\mu^{-1} \approx 0.005$–0.01 cm [Ar99, OF97]. The scattering function $\eta(\theta \cdot \theta')$ describes the probability for a particle entering a scattering process with the direction of propagation θ' to leave this process with the direction θ. It is normalized to

$$\int_{S^{n-1}} \eta(\theta \cdot \theta')d\theta = 1, \qquad (4)$$

which expresses particle conservation in pure scattering events. The dot-product in the argument indicates that η depends only on the cosine of the scattering angle $\cos\vartheta = \theta \cdot \theta'$ and, in particular, is independent of the location of the scattering event x (an assumption which simplifies the following calculations, but which can be relaxed in principle). In our numerical experiments, we will use (a 2D-adapted version of) the following Henyey-Greenstein scattering function

$$\eta(\theta \cdot \theta') = \frac{1 - g^2}{2(1 + g^2 - 2g\cos\vartheta)^{3/2}}, \qquad (5)$$

with $-1 < g < 1$. The parameter g in (5) is the mean cosine of the scattering function. Values of g close to one indicate that the scattering is primarily forward directed, whereas values close to zero indicate that scattering is almost isotropic. In our numerical experiments, we will choose g to be 0.9, which is a typical value for DOT.

The initial condition (2) indicates that there are no photons moving inside of Ω at the starting time of our experiment. The boundary condition (3) indicates that during the experiment no photons enter the domain Ω from the outside. All photons inside of Ω originate from the source q which, however, can be situated at the boundary $\partial\Omega$.

We consider the problem (1)–(3) for p different sources q_j, $j = 1, \ldots, p$, positioned at $\partial\Omega$. Typical sources in applications are delta-like pulses transmitted at time $t = 0$ at the position $s_j \in \partial\Omega$ into the direction θ_j, which can be described by the distributional expressions

$$\tilde{q}_j(x,\theta,t) = \delta_x(s_j)\delta_t(0)\delta_\theta(\theta_j), \qquad j = 1, \ldots, p, \qquad (6)$$

with $\nu(s_j) \cdot \theta_j < 0$. We assume that a given source q_j gives rise to the physical fields $\tilde{u}_j(x,\theta,t)$ which are solutions of (1)–(3). Our measurements consist of the outgoing flux across the boundary $\partial\Omega$ which has with (3) the form

$$\tilde{G}_j(x,t) = \int_{\nu(x)\cdot\theta>0} \nu(x) \cdot \theta \tilde{u}_j(x,\theta,t) d\theta \quad \text{on} \quad \partial\Omega \times [0,T], \tag{7}$$

for $j = 1,\ldots,p$. We will assume in the following that we know the scattering functions η and b, and we want to reconstruct the coefficient $a(x)$ (more precisely its contrast function as defined below) from the data.

As mentioned in the introduction, a specific difficulty arises in our application of DOT due to the presence of so-called 'clear regions' in the imaging domain. These are regions where the scattering coefficient $b(x)$ is very small or zero, such that the dominant behavior of the traveling photons inside these regions is transport rather than diffusion. This makes purely diffusion-based imaging techniques useless in these applications. We refer for more details to [OF97, RD00].

3 A Level-Set Based Shape Reconstruction Strategy

As described above, we will formulate our inverse problem as a shape reconstruction problem. In order to solve this problem, we will need a powerful tool for describing the unknown shapes numerically, and for propagating these shapes during the iterative reconstruction process. We have chosen to use the *level set technique* for this purpose, since it is able to model easily topological changes during the evolution of the shapes. The level set method has been developed originally by Osher and Sethian for the modelling of flame propagation [OS88], and has been used since then very successfully in a broad range of applications. For a more detailed introduction into the level set technique and its applications, we refer to the two recent monographs [OF03, Se99].

In our situation, the level set scheme can be described roughly as follows. Assume that we are given a domain $D \subset\subset \Omega$. We call a (sufficiently smooth) function $\phi : \Omega \to \mathbb{R}$ a *level set representation of D* if $\phi(x) \leq 0$ for all $x \in \overline{D}$ and $\phi(x) > 0$ for all $x \in \Omega\setminus\overline{D}$. The boundary $\Gamma = \partial D$ of the domain D is defined as the zero level set of ϕ. We will use the notation $\partial D[\phi]$ and $D[\phi]$ in the following to indicate this relationship between a domain and the level set function which is used for its representation.

If a shape evolves in time, $D = D(t)$, its describing level set function evolves as well in time $\phi = \phi(t)$ (where we have suppressed the x-dependence of ϕ in the notation). The basic idea of the level set technique is to reverse this relationship. In order to evolve a shape numerically in time, find a corresponding evolution law for the describing level set function ϕ such that $D[\phi(t)]$ always coincides with $D(t)$. In other words, the zero level set will always coincide with the boundary of the evolving shape.

We refer for more details concerning the level set technique to [OF03, Se99], where many examples are discussed. For the application of solving an inverse problem, the physical time in the evolution is replaced by an artificial evolution time measuring the progress of our reconstruction. This idea has been first introduced by Santosa [Sa96], and has been applied to DOT first in [Do02]. References to applications in various other imaging situations can as well be found in [Do02].

In our particular case, the evolution of the shapes will be driven by discrete updates of the level set function ϕ. These are calculated in a way such that, at the end of this evolution, and in the absence of noise, the measured data will be satisfied by the shape corresponding to the final level set function. In the following we will present a short outline of this method, as well as some recent numerical results. More details and further references can be found in [Do02].

4 Solving the Shape Reconstruction Problem

4.1 The Forward Operators

Given a constant $\hat{a}$ (representing the absorption value inside the unknown shapes) and a bounded function $a_\mathrm{b} : \Omega \to \mathbb{R}$ (representing the background distribution). Then, with each level set function ϕ a uniquely determined 'contrast function' a_s and a 'contrast operator' Λ acting on ϕ are associated by putting

$$a_s(x) = \Lambda(\phi)(x) = \begin{cases} \hat{a} - a_\mathrm{b}(x), & \phi(x) \leq 0 \\ 0, & \phi(x) > 0. \end{cases} \tag{8}$$

Let us assume now that we have collected some data $\tilde{G}_j$ which correspond to the 'true' absorption distribution $\tilde{a} = a_\mathrm{b} + \tilde{a}_\mathrm{s}$, where $\tilde{a}_\mathrm{s}$ is the 'true' contrast function. We consider for general a_s the following *measurement operators*

$$G_j(a_\mathrm{s})(x, t) := \int_{S^{n-1}} \nu(x) \cdot \theta\, u_j(x, \theta, t)\, d\theta, \tag{9}$$

$j = 1, \ldots, p$, where u_j solves (1)–(3) with the source $q_j \in Y$ and the parameter $a = a_\mathrm{b} + a_\mathrm{s}$. For the 'true' parameters $\tilde{a}_\mathrm{s}$ we ask (9) to coincide with the data $\tilde{G}_j$

$$G_j(\tilde{a}_\mathrm{s})(x, t) = \tilde{G}_j(x, t) \quad \text{for } j = 1, \ldots, p \tag{10}$$

on $\partial\Omega \times [0, T]$. Our goal is to determine $\tilde{a}_\mathrm{s}$ such that (10) is valid. For given data $\tilde{G}_j$, $j = 1, \ldots, p$, we define furthermore the *residual operators*

$$R_j(a_\mathrm{s})(x, t) = G_j(a_\mathrm{s})(x, t) - \tilde{G}_j(x, t). \tag{11}$$

From (10) it follows that for the 'true' obstacle the residuals vanish,

$$R_j(\tilde{a}_\mathrm{s}) = 0 \quad \text{for } j = 1, \ldots, p, \tag{12}$$

if the data are noise-free.

The *forward operators* T_j which map a given level set function $\phi \in \Phi$ into the corresponding mismatch in the data are defined by

$$T_j(\phi) = R_j(\Lambda(\phi)) \tag{13}$$

for $j = 1, \ldots, p$. The goal is to find a level set function $\tilde{\phi} \in \Phi$ such that

$$T_j(\tilde{\phi}) = 0 \quad \text{for } j = 1, \ldots, p. \tag{14}$$

We mention that all three operators Λ, R_j and T_j are nonlinear.

4.2 A Nonlinear Kaczmarz-Type Approach

Our algorithm works in a 'single-step fashion' as follows. Instead of using the data (10) for all sources at once, we only use the data for one source at a time while updating the linearized residual operator after each determination of the corresponding incremental correction $\delta\phi$.

To be more specific, let us assume that we are given a level set function $\phi^{(n)}(x)$ representing the contrast $a_s^{(n)}(x)$. Using a data set $\tilde{G}_j$ corresponding to the fixed source position q_j, we want to find an update $\delta\phi^{(n)}$ to $\phi^{(n)}$ such that the residuals in the data corresponding to this source vanish

$$T_j(\phi^{(n+1)}) = T_j(\phi^{(n)} + \delta\phi^{(n)}) = 0. \tag{15}$$

Applying a Newton-type approach, we get from (15) a correction $\delta\phi^{(n)}$ for $\phi^{(n)}$ by solving

$$T_j'[\phi^{(n)}]\delta\phi^{(n)} = -T_j(\phi^{(n)}) = -(M_j u_j - \tilde{G}_j) \tag{16}$$

where u_j satisfies (1)–(3) with

$$a(x) = a_b(x) + \Lambda(\phi^{(n)})(x) = \begin{cases} \hat{a}, & x \in D[\phi^{(n)}] \\ a_b(x), & x \in \Omega \setminus D[\phi^{(n)}]. \end{cases} \tag{17}$$

The operator $T_j'[\phi^{(n)}]$ in (16) is the linearized forward operator. Since we have only few data given for one source, equation (16) usually will have many solutions (in the absence of noise), such that we have to pick one according to some criterion. We choose to take that solution which minimizes the L^2-norm of $\delta\phi^{(n)}$

$$\text{Min } \|\delta\phi^{(n)}\|_2 \quad \text{subject to} \quad T_j'(\phi^{(n)})\delta\phi^{(n)} = -(M_j u_j - \tilde{G}_j). \tag{18}$$

This solution can be formulated explicitly. It is

$$\delta\phi_{\text{MN}}^{(n)} = -T_j'[\phi^{(n)}]^* \left(T_j'[\phi^{(n)}]T_j'[\phi^{(n)}]^*\right)^{-1} (M_j u_j - \tilde{G}_j), \tag{19}$$

where $T_j'[\phi^{(n)}]^*$ denotes the adjoint operator to $T_j'[\phi^{(n)}]$.

After correcting ϕ by $\phi \to \phi + \delta\phi_j$, where $\delta\phi_j$ is given by (19), we use the updated residual equation (16) to compute the next correction $\delta\phi_{j'}$. Doing this for one equation after the other, until each of the sources q_j has been considered exactly once, will yield one complete sweep of the algorithm. This procedure is similar to the Kaczmarz method for solving linear systems, or the algebraic reconstruction technique (ART) in x-ray tomography. For related approaches in a variety of imaging problems we refer to [NW01].

4.3 The Adjoint Linearized Operators

In order to calculate the minimal norm solution (19), we will need practically useful expressions for the adjoint linearized operators $T'_j[\phi]^*$. In the following, we will give such an expression which we will use in our numerical experiments. A more formal derivation can be found in [Do02].

Let $z \in U^*$ be a solution of the (adjoint) transport equation

$$-\frac{\partial z}{\partial t} - \theta \cdot \nabla z(x, \theta, t) + (a(x) + b(x)) z(x, \theta, t)$$
$$- b(x) \int_{S^{n-1}} \eta(\theta \cdot \theta') z(x, \theta', t) d\theta' = 0 \quad \text{in} \quad \Omega \times S^{n-1} \times [0, T] \quad (20)$$

with 'initial' condition

$$z(x, \theta, T) = 0 \quad \text{on} \quad \Omega \times S^{n-1} \quad (21)$$

and boundary condition

$$z|_{\Gamma_+} = \tilde{z}(x, t) := \zeta(x, t). \quad (22)$$

Define

$$I_j(a_s) := \int_{[0,T]} \int_{S^{n-1}} u_j(x, \theta, t) z_j(x, \theta, t) \, d\theta dt, \quad (23)$$

with u_j being a solution of (1)–(3) with source q_j and with $a = a_b + a_s$. Here, $\zeta(x,t)$ is applied uniformly into all directions θ with $\nu \cdot \theta > 0$. Then we have the following expression for the adjoint of the linearized forward operator $T'_j[\phi]$.

Theorem 1. *The adjoint operator $T'_j[\phi]^*$ acts on a given vector ζ in the following way*

$$T'_j[\phi]^* \zeta = \frac{[\hat{a} - a_b(x)]}{|\nabla \phi(x)|} C_\rho(\Gamma) \chi_{B_\rho(\Gamma)}(x) I_j(\Lambda(\phi)) \quad (24)$$

where $I_j(\Lambda(\phi))$ is given by (23) and where u_j solves (1)–(3) and z_j solves (20)–(22) with a_s replaced by $\Lambda(\phi)$.

Here, the expression $C_\rho(\Gamma)\chi_{B_\rho(\Gamma)}(x)$ serves as an approximation to the Dirac delta distribution concentrated on the boundary Γ of the most recent shape. More precisely, $\chi_{B_\rho(\Gamma)}(x)$ is the characteristic function of a small ρ-neighborhood of the boundary Γ, and $C_\rho(\Gamma)$ is the corresponding normalization factor.

4.4 The Algorithm

Table 1. The level set shape reconstruction algorithm

Initialization: $\quad n = 0; \quad \phi^{(0)} = \phi^{(0)}(D^{(0)})$.

Reconstruction loop:

> FOR $i = 1 : I_m$ $\quad$ perform I_m sweeps
> $\quad$ FOR $j = 1 : p$ $\quad$ march over source positions in each sweep
> $\quad\quad \zeta_j = \hat{C}_j^{-1}(M_j u_j - \tilde{G}_j); \quad u_j$ solves (1)–(3) with $a^{(n)}$, q_j
> $\quad\quad \delta\phi^{(n)} = -(\hat{a} - a_b(x))I_j(\Lambda(\phi^{(n)}))\chi_{B_\rho(\Gamma)}; \quad z_j$ solves (20)–(22) with $a^{(n)}$, ζ_j
> $\quad\quad \phi^{(n+1)} = C_{LS}^{(n)}(\phi^{(n)} + \eta \frac{C_\rho(\Gamma)}{|\nabla\phi(x)|}\delta\phi^{(n)}); \quad$ update level set function
> $\quad\quad a^{(n+1)} = a_b + \Lambda(\phi^{(n+1)}); \quad n = n + 1; \quad$ reinitialization $n \to n+1$
> $\quad$ END
> END

Extract final shape: $\quad$ N=n; $\quad D^{(N)} = D[\phi^{(N)}]$.

The nonlinear Kaczmarz-type method for shape reconstruction using level sets is described in brief algorithmic form in Table 1. Here, η is a (typically small) relaxation parameter for the update of the level set function and is determined empirically. The operator $\hat{C}_j^{-1}$ is an approximation to the difficult to calculate operator $(T'_j[\phi^{(n)}]T'_j[\phi^{(n)}]^*)^{-1}$ in (19) and is for the moment taken to be simply a multiple of the identity, which has proven to yield satisfactory results in our numerical experiments. The constant ρ is chosen here between one and two grid cells. The scaling factor $C_{LS}^{(n)}$ is determined after each update to keep the global minimum of the level set function at a constant value.

5 A Numerical Experiment

Figure 1 shows on the left the setup of our numerical experiment. We model the human head for simplicity as a two-dimensional square domain of size 5×5 cm². It is composed of a background material with three different clear regions embedded. The clear regions have parameters $a = 0.01$ cm^{-1} and $b = 0.01$ cm^{-1}. One of these three imbedded clear regions has a band-like structure parallel to the boundary, symbolically representing a layer of cerebro-spinal fluid beneath the scull. The other two clear regions are small pockets of liquid located in the interior of the head. The background material is assumed to have absorption parameter values randomly distributed between $a = 0.07$ cm^{-1} and $a = 0.13$ cm^{-1} (each pixel picks a value from a uniform distribution) and a homogeneous scattering parameter of $b = 100.0$ cm^{-1}. However, for the reconstruction we assume that we know only the scattering parameter of the background medium and an average value ($a = 0.1$ cm^{-1}) of the absorption

Shape Reconstruction for an Inverse Radiative Transfer Problem   307

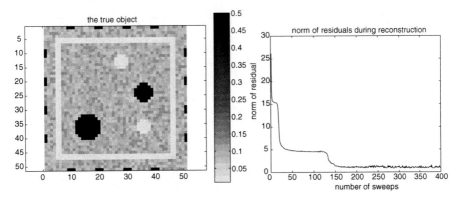

Fig. 1. Left figure: the true object. Right figure: evolution of the norm of the residuals during the shape reconstruction process. The x-axis shows the number of sweeps

parameter. We do not attempt to reconstruct the random fluctuations in the absorption parameter, but rather treat them as modelling noise.

Along the boundary, we have indicated the 16 source positions by dark bars having the extensions of these sources. At these positions, successively ultrashort laser pulses are emitted perpendicular to the boundary into the head. The measurements are taken on half of the boundary, at the opposite side of the current source position. The two dark regions in the interior of the head symbolize the objects of interest. They might represent regions of blood accumulation (e.g. hematoma) or similar objects. We want to recover these unknown shapes from the measured data corresponding to the 16 source positions.

For our numerical experiments, we use a simple finite difference discretization of the radiative transfer equations (1)–(3) and (20)–(22). The whole domain is discretized into 50×50 pixels, and we monitor 100 discrete time-steps of the propagating photons, which captures most part of the time-dependent data. 12 different directions are used for discretizing the angular variable. As scattering law, we use the Henyey-Greenstein function (5) with $g = 0.9$. We run our finite differences code on the above described model (which we call 'true model') in order to simulate our data. Then we use these simulated data as input for our shape reconstruction method. In our shape reconstruction algorithm we assume that we know the (averaged) homogeneous background distribution of $a = 0.1$ cm^{-1} with the clear regions included. We do not know the random fluctuations of the background absorption distribution and we do not know the number and shapes of the hidden objects. As mentioned above, the random fluctuations serve as modelling noise, whereas the unknown shapes are required to be reconstructed from the data.

We start our shape reconstruction process with some initial guess which is shown at the top left image of Figure 2. As initial level set function we use

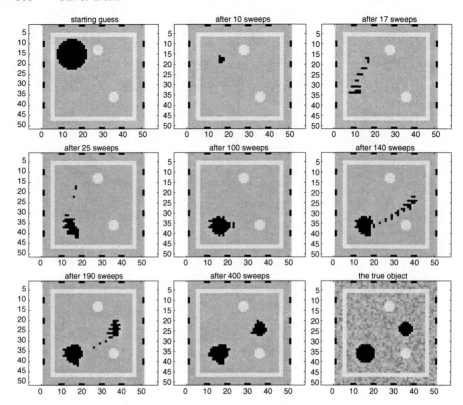

Fig. 2. Shape evolution during the level set based shape reconstruction process. Top row from left to right: Starting guess, after 10 sweeps, after 17 sweeps. Middle row from left to right: after 25, 100, 140 sweeps. Bottom row from left to right: after 190 sweeps, final reconstruction after 400 sweeps, the true object

a so-called 'signed distance function' (see [OF03, Se99]) representing a disc-shaped region as shown in the image. The other images of Figure 2 show the evolution of the shape during the reconstruction process. The final shape is shown in the second image of the bottom row of Figure 2. For comparison, we display in addition the 'true object' at the right of the bottom row of Figure 2. In Figure 1 on the right the evolution of the norm of the residuals during the reconstruction versus the sweep number is shown. It can be seen that the residual norm levels off after about 200 sweeps at some small nonzero number, which we interpret as the 'noise-level' associated with our random background fluctuations of the absorption parameter. In fact, our numerical experiments have shown that this value increases monotonically with increasing magnitude of these background fluctuations. Concluding, we observe in Figure 2 that both unknown objects have been reconstructed very well by the level set based shape reconstruction algorithm. In our future work we plan to extend the presented algorithm to capture more aspects of the underlying inverse problem (e.g. the additional reconstruction of contrast parameters and of the structure of the clear regions) and to apply it to data which have been created by an independently developed Monte Carlo simulation for this situation. These data will also incorporate different kinds of measurement and modelling noise, such that we can investigate the performance of our presented method under these more realistic constraints. The results so far make us very confident that the method will perform well also in these situations.

References

[Ar99] Arridge, S.R.: Optical tomography in medical imaging. Inverse Problems, **15** (2), R41–R93 (1999).
[CZ67] Case, K.M., Zweifel, P.F.: Linear Transport Theory. Plenum Press, New York (1967)
[Do98] Dorn, O.: A transport-backtransport method for optical tomography. Inverse Problems, **14**, 1107–1130 (1998)
[Do02] Dorn, O.: Shape reconstruction in scattering media with voids using a transport model and level sets. Canad. Appl. Math. Quart., **10** (2), 239–275 (2002).
[NW01] Natterer, F., Wübbeling, F.: Mathematical Methods in Image Reconstruction. Monographs on Mathematical Modeling and Computation 5, SIAM 2001.
[OF97] Okada, E., Firbank, M., Schweiger, M., Arridge, S.R., Cope, M., Delpy, D.T.: Theoretical and experimental investigation of near-infrared light propagation in a model of the adult head. Appl. Opt., **36** (1), 21–31 (1997)
[OS88] Osher, S., Sethian, J.: Fronts propagation with curvature dependent speed: Algorithms based on Hamilton-Jacobi formulations. J. Comput. Phys., **56**, 12–49 (1988)
[OF03] Osher, S., Fedkiw, R.: Level Set Methods and Dynamic Implicit Surfaces. Springer, New York (2003)
[RD00] Riley, J., Dehghani, H., Schweiger, M., Arridge, S.R., Ripoll, J., Nieto-Vesperinas, M.: 3D Optical Tomography in the Presence of Void Regions. Opt. Exp., **7** (13), 462 ff. (2000)
[Sa96] Santosa, F.: A Level-Set Approach for Inverse Problems Involving Obstacles. ESAIM: Control, Optimization and Calculus of Variations, **1**, 17–33 (1996)
[Se99] Sethian, J.A.: Level Set Methods and Fast Marching Methods (2nd ed), Cambridge University Press (1999)